Engineering Materials

This series provides topical information on innovative, structural and functional materials and composites with applications in optical, electrical, mechanical, civil, aeronautical, medical, bio- and nano-engineering. The individual volumes are complete, comprehensive monographs covering the structure, properties, manufacturing process and applications of these materials. This multidisciplinary series is devoted to professionals, students and all those interested in the latest developments in the Materials Science field, that look for a carefully selected collection of high quality review articles on their respective field of expertise.

Indexed at Compendex (2021) and Scopus (2022)

Peeyush Phogat · Shreya Sharma · Ranjana Jha ·
Sukhvir Singh

Electrochemical Devices

Principles to Applications

Springer

Peeyush Phogat
Department of Physics
Netaji Subhas University of Technology
New Delhi, Delhi, India

Shreya Sharma
Department of Physics
Netaji Subhas University of Technology
New Delhi, Delhi, India

Ranjana Jha
Department of Physics
Netaji Subhas University of Technology
New Delhi, Delhi, India

Sukhvir Singh
Department of Physics
Netaji Subhas University of Technology
New Delhi, Delhi, India

ISSN 1612-1317 ISSN 1868-1212 (electronic)
Engineering Materials
ISBN 978-981-96-0526-2 ISBN 978-981-96-0527-9 (eBook)
https://doi.org/10.1007/978-981-96-0527-9

This Springer imprint is published by the registered company Springer Nature Singapore Pte Ltd.
The registered company address is: 152 Beach Road, #21-01/04 Gateway East, Singapore 189721, Singapore

Preface

Welcome to the journey of exploring Next-Generation Electrochemical Devices. In recent decades, the field of electrochemical devices has witnessed remarkable advancements and innovations, driving transformative changes across various industries and scientific disciplines. From renewable energy generation to biomedical diagnostics, from environmental monitoring to telecommunications, electrochemical devices play a pivotal role in enabling sustainable development, enhancing quality of life, and addressing global challenges. This book endeavours to serve as a comprehensive guide for both beginners and seasoned researchers, offering insights into the diverse array of electrochemical devices and their intricate dependencies. It aims to provide a comprehensive overview of electrochemical devices, spanning from fundamental principles to cutting-edge applications. By bringing together insights from materials science, chemistry, physics, engineering, and beyond, we strive to offer a holistic understanding of the underlying mechanisms, design strategies, and practical considerations associated with these devices.

Our journey begins with an exploration of the fundamental principles of electrochemistry, laying the groundwork for understanding electrochemical reactions, charge transfer processes, and device operation mechanisms. Building upon this foundation, we delve into various types of electrochemical devices, including solar cells, photodetectors, sensors, batteries, and more. Each chapter explores the state-of-the-art technologies, materials, fabrication techniques, and performance metrics associated with these devices, while also discussing key challenges and future prospects. Throughout this book, we emphasize the interdisciplinary nature of electrochemical device research, highlighting the synergistic interactions between different scientific disciplines and the importance of collaboration in driving innovation. We also recognize the ethical, environmental, and societal implications of electrochemical technologies, urging readers to consider the broader impacts of their research and applications.

One of the central themes of this book is the interplay between electrochemistry and microstructural properties within the electrochemical devices. The microstructural architecture of electrodes, electrolytes, and interfaces plays a fundamental role in governing electrochemical performance, influencing parameters such as charge transfer kinetics, ion diffusion rates, and surface reactivity. Through meticulous characterization techniques such as electron microscopy, X-ray diffraction, and spectroscopic analyses, researchers can elucidate the correlations between microstructural features (e.g. grain boundaries, defects, porosity) and electrochemical phenomena. Understanding these relationships enables tailored design strategies to optimize device performance, enhance durability, and mitigate degradation mechanisms. Moreover, advancements in materials synthesis and processing techniques allow for precise control over microstructural properties, offering avenues for tailoring electrochemical device architectures at the nanoscale. By elucidating the intricate connections between electrochemical and microstructural attributes, this book aims to provide insights that inform the rational design and engineering of next-generation electrochemical technologies.

Each chapter is meticulously crafted to provide a blend of theoretical insights, practical considerations, and future research directions, thereby equipping readers with a holistic understanding of electrochemical phenomena and their applications. Whether you are a student, researcher, or industry professional, *Electrochemical Devices: Principles to Applications* aims to be a valuable resource for expanding your knowledge and driving innovation in this dynamic field. As authors, we have endeavoured to distil our collective expertise and insights into this comprehensive resource, aiming to serve as a valuable reference for researchers, students, engineers, and practitioners working in the field of electrochemical devices. We hope this book serves as a catalyst for further exploration, collaboration, and breakthroughs in electrochemistry, ultimately contributing to the advancement of science and technology for the betterment of society.

Together, let us embark on a voyage of discovery, innovation, and impact as we explore the fascinating world of electrochemical devices.

Happy exploring!

New Delhi, India

Peeyush Phogat
Shreya Sharma
Ranjana Jha
Sukhvir Singh

Acknowledgements

First and foremost, we thank the Almighty for providing us the strength, wisdom, and perseverance to complete this book, *Electrochemical Devices: Principles to Applications*. Without His grace, this endeavour would not have been possible.

We extend our heartfelt gratitude to the Vice-Chancellor of Netaji Subhas University of Technology, New Delhi, for his visionary leadership and continuous encouragement. His support has provided the academic environment essential for our research and writing.

To our parents and loved ones, we owe immense gratitude for their unwavering love, patience, and belief in us throughout this journey. Their constant encouragement has been a source of strength.

We are also deeply thankful to the students and faculty members of Netaji Subhas University of Technology, New Delhi, who have provided relentless support and inspiration. Their enthusiasm and dedication have been instrumental in shaping the ideas presented in this work.

A special thanks goes to Prof. Ranjana Jha, whose guidance, unrestricted love, and continuous support have been invaluable throughout this process. Her mentorship has been a cornerstone in the development of this book.

Finally, we express our sincere appreciation to Dr. Sukhvir Singh, whose determination and unwavering trust in his students have been an enduring source of motivation. His belief in us has made this work possible.

Thank you all for being part of this journey with us.

Contents

Chapter 1
Fundamentals of Electrochemistry

The chapter "Fundamentals of Electrochemistry" provides a comprehensive overview of the core principles governing electrochemical processes. It begins with an introduction to electrochemical processes, laying the groundwork for understanding the underlying mechanisms. Subsequently, the chapter delves into electrochemical thermodynamics and kinetics, elucidating the driving forces and rate-determining steps of electrochemical reactions. Charge transfer mechanisms are then explored, focusing on the pathways through which electrons and ions migrate within electrochemical systems. Electrochemical cell configurations are discussed, highlighting the various setups employed in experimental studies and practical applications. Furthermore, the chapter examines the principles of light absorption and charge transfer, underscoring the role of light in initiating electrochemical reactions. Finally, electrochemical reactions induced by light are investigated, showcasing the diverse range of photo electrochemical processes and their applications in energy conversion and sensing technologies. Overall, this chapter serves as a foundational resource for understanding the fundamental concepts and mechanisms that underpin electrochemistry.

1.1 Introduction to Electrochemical Processes

Electrochemistry is the branch of chemistry that deals with the study of chemical reactions that involve the transfer of electrons between species, typically mediated by an external electric circuit. This field encompasses a wide range of phenomena, from simple redox reactions to complex processes governing energy storage, corrosion, and biological electron transport. At its core, electrochemistry revolves around the fundamental principles of oxidation and reduction, collectively known as redox reactions. In these reactions, one species loses electrons (oxidation) while another gains electrons (reduction), resulting in changes in their oxidation states. The flow

P. Phogat et al., *Electrochemical Devices*, Engineering Materials,
https://doi.org/10.1007/978-981-96-0527-9_1

of electrons from the oxidized species to the reduced species generates an electric current, which can be harnessed for various practical applications. One of the central concepts in electrochemistry is that of the electrochemical cell, which serves as the experimental setup for studying electrochemical reactions [1]. An electrochemical cell typically consists of two electrodes—an anode and a cathode—immersed in an electrolyte solution. The anode is where oxidation occurs (electrons are lost), while the cathode is where reduction occurs (electrons are gained). The two electrodes are connected by an external circuit, allowing the flow of electrons from the anode to the cathode, thereby completing the electrical circuit.

Electrochemical reactions can be classified into two broad categories: galvanic (voltaic) cells and electrolytic cells. In galvanic cells, spontaneous redox reactions generate electrical energy, which can be used to power electronic devices or perform useful work. Batteries and fuel cells are examples of galvanic cells that convert chemical energy into electrical energy. In contrast, electrolytic cells require an external source of electrical energy to drive non-spontaneous redox reactions, enabling processes such as electroplating, electrolysis, and electrochemical synthesis [2, 3]. The thermodynamics and kinetics of electrochemical reactions play a crucial role in determining their feasibility and rates of occurrence. Thermodynamic considerations, governed by concepts such as Gibbs free energy and the Nernst equation, provide insights into the directionality and spontaneity of electrochemical processes. Kinetic considerations, on the other hand, focus on the rates of electron transfer at electrode surfaces, which are influenced by factors such as electrode potential, concentration of reactants, and surface area.

Electrochemistry finds applications across diverse fields, including energy storage and conversion, electroplating, corrosion protection, sensors, biomedical devices, and environmental remediation [4–7]. For example, electrochemical batteries and supercapacitors are critical components of portable electronics and electric vehicles, while electrochemical sensors play a vital role in monitoring pollutants in air and water. Overall, electrochemistry is a rich and interdisciplinary field that spans chemistry, physics, materials science, and engineering. Its principles and applications continue to drive technological advancements and contribute to addressing global challenges in energy, environment, and health. As researchers continue to explore the frontiers of electrochemical science, the potential for innovation and discovery in this field remains vast and promising.

The historical background of electrochemistry is rich and multifaceted, spanning centuries of scientific inquiry and technological advancement. The roots of electrochemistry can be traced back to ancient civilizations, where rudimentary forms of electrochemical phenomena were observed and utilized. However, it was not until the eighteenth and nineteenth centuries that the systematic study of electrochemistry began, laying the groundwork for modern understanding and applications of the field. Figure 1.1 provide a historical layout of electrochemical progress.

The importance and applications of electrochemistry are vast and diverse, permeating numerous fields ranging from energy and materials science to medicine

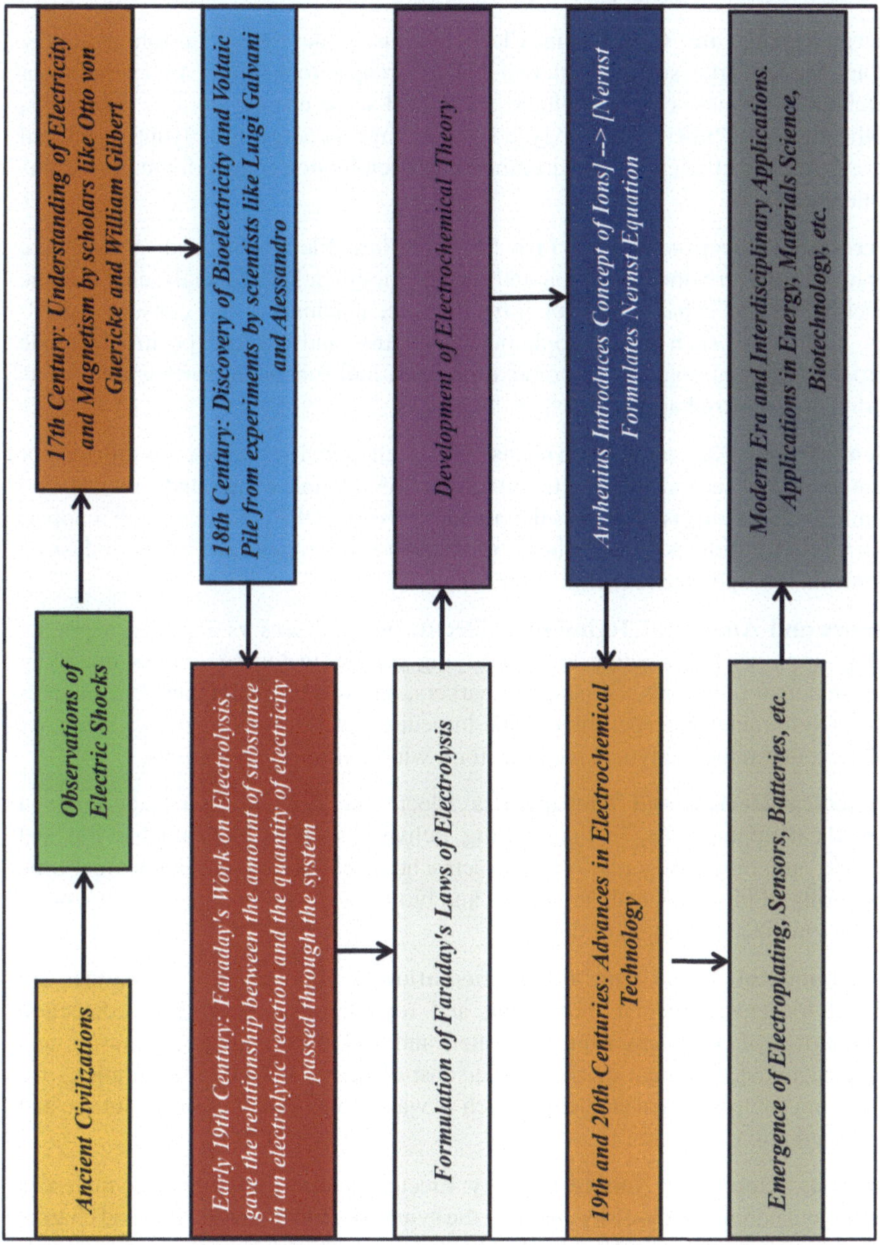

Fig. 1.1 Step by step historical progress of electrochemistry to modern application till today

and environmental protection. Understanding the significance of electrochemistry requires exploring its myriad applications, each contributing to technological advancements and societal progress.

Energy Storage and Conversion: Electrochemistry plays a pivotal role in energy storage technologies such as batteries and supercapacitors, which are essential for portable electronics, electric vehicles, and grid-scale energy storage. Fuel cells, another electrochemical technology, convert chemical energy directly into electrical energy, offering efficient and environmentally friendly power generation for various applications.

Corrosion Protection and Surface Modification: Electrochemical methods are employed for corrosion protection and surface modification of metals, enabling the preservation and enhancement of infrastructure, machinery, and consumer products. Techniques such as electroplating, anodizing, and electrodeposition provide corrosion-resistant coatings, decorative finishes, and functional surface treatments for diverse industrial applications.

Chemical Synthesis and Electrolysis: Electrochemical processes are utilized for chemical synthesis and electrolytic production of valuable compounds, metals, and chemicals. Electrolysis facilitates the production of metals (e.g., aluminum, copper) from ores, as well as the synthesis of chlorine, hydrogen, and other industrial chemicals through electrochemical reactions.

Sensors and Analytical Techniques: Electrochemical sensors and biosensors are widely employed for detecting and quantifying analytes in diverse samples, including environmental pollutants, biomedical markers, and food contaminants. Techniques such as voltammetry, amperometry, and impedance spectroscopy provide sensitive, selective, and rapid analytical methods for a wide range of applications.

Biomedical Devices and Therapeutics: Electrochemistry plays a crucial role in biomedical applications, including drug delivery systems, implantable medical devices, and biosensing platforms. Electrochemical biosensors enable real-time monitoring of physiological parameters and biomarkers, facilitating early diagnosis and personalized treatment of diseases.

Environmental Monitoring and Remediation: Electrochemical techniques are utilized for environmental monitoring and remediation, enabling the detection and removal of pollutants from air, water, and soil. Electrochemical sensors and electrochemical treatment methods offer cost-effective and efficient solutions for addressing environmental challenges such as water contamination, air pollution, and soil remediation.

Materials Science and Nanotechnology: Electrochemistry contributes to materials science and nanotechnology by enabling the synthesis, functionalization, and characterization of nanostructured materials and thin films. Electrodeposition, electrospinning, and electrochemical etching techniques are used to fabricate nanostructures with tailored properties for applications in electronics, catalysis, energy storage, and biomedical devices.

Overall, the importance of electrochemistry lies in its ability to harness the principles of electron transfer and electrochemical reactions to address a wide range of technological challenges and societal needs. From energy storage and environmental protection to healthcare and materials science, electrochemistry continues to drive innovation and impact diverse aspects of modern life.

1.2 Electrochemical Thermodynamics and Kinetics Charge Transfer Mechanisms

1.2.1 Thermodynamic Principles

Thermodynamic principles form the foundation of electrochemistry, providing insights into the energy changes and equilibrium conditions associated with electrochemical reactions. Understanding these principles is essential for predicting the feasibility, directionality, and spontaneity of electrochemical processes.

1.2.1.1 Gibbs Free Energy

Gibbs free energy (ΔG) is a fundamental concept in thermodynamics that provides crucial insights into the spontaneity and directionality of chemical reactions. It represents the maximum amount of reversible work that can be extracted from a system under constant temperature and pressure conditions. In the context of electrochemistry, ΔG serves as a measure of the driving force for a reaction to proceed. When ΔG is negative ($\Delta G < 0$), the reaction is considered spontaneous, meaning it can occur without the need for external energy input. Conversely, when ΔG is positive ($\Delta G > 0$), the reaction is non-spontaneous, requiring an input of energy to proceed. At equilibrium, where the rates of the forward and reverse reactions are equal, ΔG equals zero ($\Delta G = 0$) as shown in Fig. 1.2 [8].

The relationship between Gibbs free energy (ΔG), enthalpy change (ΔH), and entropy change (ΔS) is described by the Gibbs–Helmholtz equation: $\Delta G = \Delta H - T\Delta S$. Here, ΔH represents the enthalpy change (heat absorbed or released) during the reaction, ΔS represents the entropy change (disorder or randomness), and T is the temperature in Kelvin. This equation illustrates the balance between the enthalpic and entropic contributions to the spontaneity of a reaction. A negative ΔH (exothermic reaction) or a positive ΔS (increase in disorder) tends to favor spontaneity, while a positive ΔH (endothermic reaction) or a negative ΔS (decrease in disorder) tends to inhibit spontaneity [9, 10]. The temperature term ($T\Delta S$) accounts for the temperature dependence of the entropy contribution to the overall Gibbs free energy change.

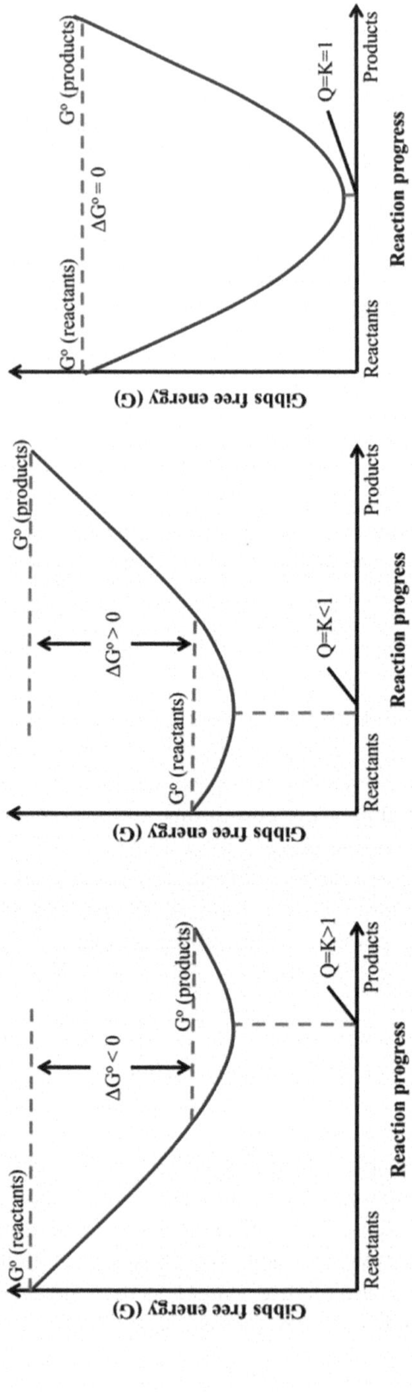

Fig. 1.2 Gibbs free energy versus reaction progress for systems that are **a** negative, **b** positive, and **c** zero

By considering the interplay between ΔH, ΔS, and temperature, scientists can predict whether a reaction will proceed spontaneously under given conditions and optimize reaction conditions for desired outcomes. Thus, Gibbs free energy serves as a valuable thermodynamic parameter for understanding and manipulating chemical reactions in electrochemistry and beyond [11].

1.2.1.2 Nernst Equation

The Nernst equation provides a mathematical relationship between the electrode potential of an electrochemical cell and the concentrations (or activities) of reactants and products involved in the half-cell reactions. For a given cell reaction represented by the equation:

$$aA + bB \rightarrow cC + dD$$

where a, b, c, and d are stoichiometric coefficients, the Nernst equation can be expressed as [12]:

$$E_{Cell} = E^{o}_{Cell} - \frac{RT}{nF}\ln(Q)$$

In this equation:

E_{Cell} represents the cell potential, which is the electromotive force (EMF) or voltage generated by the electrochemical cell.

E^{o}_{Cell} is the standard cell potential, which is the cell potential under standard conditions (usually 1 M concentration for each species, 1 atm pressure, and a specified temperature).

R is the gas constant (8.314 J/(mol K)).

T is the temperature in Kelvin.

n is the number of moles of electrons transferred in the cell reaction.

F is Faraday's constant (96,485 C/mol), representing the charge of one mole of electrons.

Q is the reaction quotient, which is the ratio of the activities (or concentrations) of products to reactants, each raised to the power of its stoichiometric coefficient.

The Nernst equation demonstrates how the cell potential (E_{Cell}) changes with variations in the concentrations of reactants and products. Specifically, as the reaction proceeds and the concentrations of reactants and products change, the value of Q and thus the logarithmic term in the equation change, leading to corresponding changes in the cell potential [13]. This relationship allows for the determination of the cell potential under non-standard conditions, providing valuable insights into the thermodynamics of electrochemical reactions illustrated in Fig. 1.3.

Fig. 1.3 Relation between Gibbs free energy and Nernst equation

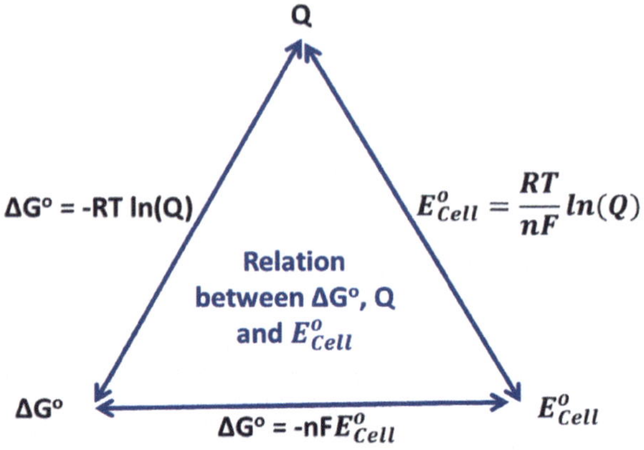

1.2.2 *Kinetic Principles*

Kinetic principles in electrochemistry govern the rates at which electrochemical reactions occur and the mechanisms by which charge transfer processes take place at electrode interfaces. Understanding these principles is essential for optimizing reaction kinetics, enhancing electrochemical performance, and designing efficient electrochemical devices.

Charge transfer at electrode interfaces involves complex processes such as electron transfer, ion transport, and mass transfer. These mechanisms dictate the overall rate of electrochemical reactions. Electron transfer involves the movement of electrons between the electrode and the species in solution, facilitating redox reactions. The rate of electron transfer is influenced by factors such as electrode material, surface structure, and electroactive species concentration. Ion transport refers to the movement of ions in solution towards or away from the electrode surface, which is crucial for maintaining charge neutrality and enabling electrochemical reactions. Diffusion and migration are the primary mechanisms of ion transport in electrolyte solutions.

1.2.2.1 Butler–Volmer Equation

The Butler–Volmer equation is a fundamental expression that describes the kinetics of electrochemical reactions occurring at electrode interfaces. It relates the rate of electron transfer to the overpotential (the deviation of electrode potential from its equilibrium value) and other kinetic parameters such as exchange current density and transfer coefficients. The equation is given by [14, 15]:

$$i = i_o\left[\exp\left(\frac{RT}{\alpha F}\eta\right) - \exp\left(-\frac{RT}{(1-\alpha)F}\eta\right)\right]$$

where

i is the current density,
i_0 is the exchange current density,
α is the charge transfer coefficient,
F is Faraday's constant,
η is the over potential,
R is the gas constant, and
T is the temperature in Kelvin.

This equation illustrates how the rate of electron transfer (current density) depends on the overpotential, with the exponential terms representing the forward and backward reaction rates. The charge transfer coefficient (α) reflects the degree of symmetry in the reaction mechanism, while the exchange current density (i_0) represents the rate of the reaction at equilibrium [16]. By incorporating these parameters, the Butler–Volmer equation provides a quantitative framework for analysing and predicting the kinetics of electrochemical reactions at electrode interfaces.

1.2.2.2 Activation Overpotentials

Activation overpotentials refer to the additional energy required to initiate and sustain electrochemical reactions at a desired rate, arising from the energy barrier that must be overcome for the reaction to proceed. These overpotentials play a crucial role in determining the efficiency and performance of electrochemical devices. Factors influencing activation overpotentials include reaction kinetics, electrode surface properties, electrolyte composition, and temperature. For instance, reactions with slower kinetics or complex reaction mechanisms may exhibit higher activation overpotentials. Additionally, the surface morphology and catalytic activity of electrodes can influence the ease with which reactants can adsorb and participate in the reaction, thereby affecting the magnitude of the overpotential. Minimizing activation overpotentials is essential for optimizing the efficiency and performance of electrochemical devices. Higher overpotentials result in increased energy losses and decreased reaction rates, limiting the overall efficiency of the system. By understanding and controlling the factors contributing to activation overpotentials, researchers can design electrodes, electrolytes, and operating conditions to mitigate these effects and enhance the performance of electrochemical devices.

1.2.3 Integration of Thermodynamics and Kinetics

The integration of thermodynamics and kinetics is essential for a comprehensive understanding of electrochemical processes, allowing researchers to predict and optimize the behavior of electrochemical systems under various conditions.

By combining thermodynamic principles, which govern the feasibility and directionality of reactions, with kinetic considerations, which determine the rates of reaction, researchers can elucidate the underlying mechanisms and dynamics of electrochemical phenomena.

Thermodynamic Considerations: Thermodynamics provides insights into the energy changes and equilibrium conditions associated with electrochemical reactions. Gibbs free energy (ΔG) quantifies the driving force for reactions to occur, with negative ΔG values indicating spontaneity. Understanding thermodynamic principles allows researchers to predict the feasibility of electrochemical reactions and assess the maximum work that can be extracted from a system. The Nernst equation relates the cell potential of an electrochemical reaction to the activities (or concentrations) of reactants and products, providing a thermodynamic framework for analyzing cell potentials under non-standard conditions.

Kinetic Considerations: Kinetics focuses on the rates of reaction and the mechanisms by which reactions occur. Factors such as reaction order, rate constants, and activation energies influence the speed at which electrochemical processes proceed. Activation overpotentials represent the additional energy required to overcome the activation barrier for reactions to proceed at a desired rate. Minimizing activation overpotentials is crucial for optimizing the efficiency and performance of electrochemical devices. The Butler–Volmer equation describes the relationship between the reaction rate and the electrode potential, taking into account factors such as the exchange current density and the concentrations of reactants and products.

Integration of Thermodynamics and Kinetics: The integration of thermodynamics and kinetics allows researchers to predict and understand the behavior of electrochemical systems in real-world applications. By combining thermodynamic driving forces with kinetic rate expressions, researchers can assess the overall performance of electrochemical devices, optimize operating conditions, and design efficient electrode materials. Understanding the interplay between thermodynamic constraints and kinetic limitations enables the rational design and engineering of electrochemical systems with tailored properties and performance characteristics.

The integration of thermodynamics and kinetics is fundamental to advancing the field of electrochemistry, providing a holistic framework for studying and optimizing electrochemical processes. By leveraging both thermodynamic principles and kinetic considerations, researchers can unlock new insights into the behavior of electrochemical systems and develop innovative solutions for a wide range of applications.

1.3 Charge Transfer Mechanisms

Charge transfer mechanisms are fundamental to understanding the dynamics of electrochemical reactions, encompassing processes involving the movement of electrons, ions, and protons across interfaces. By elucidating these mechanisms, researchers can optimize the performance of electrochemical systems and design more efficient devices for various applications.

1.3.1 Electron Transfer Processes

Electron transfer processes are fundamental to electrochemical reactions, governing the flow of charge within electrochemical systems. Understanding electron transfer mechanisms is essential for elucidating reaction kinetics, determining electrode behavior, and optimizing the performance of electrochemical devices. Electron transfer involves the movement of electrons between species participating in redox reactions. These reactions can be categorized as either oxidation (loss of electrons) or reduction (gain of electrons), collectively known as redox reactions. According to Marcus theory, electron transfer occurs through either outer-sphere or inner-sphere mechanisms, depending on the proximity of the reactants and the involvement of solvent molecules or ligands.

The rate of electron transfer is influenced by factors such as the electronic coupling between donor and acceptor, the reorganization energy associated with structural changes, and the driving force for the reaction [16]. Outer-sphere electron transfer involves the transfer of electrons between reactants without direct coordination to the surrounding solvent or ligands. This mechanism is characterized by a relatively low reaction rate and is prevalent in dilute solution environments. Inner-sphere electron transfer involves the formation of a coordination complex between the reactants, facilitating the transfer of electrons through a bridging ligand or solvent molecule. This mechanism is typically observed in condensed phase systems and at electrode surfaces. Electron transfer processes are central to the operation of batteries and supercapacitors, where redox reactions store and release electrical energy. Electron transfer plays a critical role in catalytic processes occurring at electrode surfaces, such as the conversion of reactants in fuel cells, electrolyzers, and electrochemical sensors. Electron transfer reactions are essential for various biological processes, including respiration, photosynthesis, and enzymatic reactions, highlighting the interdisciplinary relevance of electron transfer mechanisms [17].

Voltammetric techniques, such as cyclic voltammetry and differential pulse voltammetry, are widely used to study electron transfer kinetics and mechanisms at electrode interfaces. Spectroscopic methods, such as UV–Vis spectroscopy and infrared spectroscopy, combined with electrochemical techniques, provide insights into the electronic structure and reaction intermediates during electron transfer processes. Computational methods, including density functional theory (DFT) and molecular dynamics (MD) simulations, are employed to elucidate the energetics and dynamics of electron transfer reactions at the atomic level.

1.3.2 Ion Transport Mechanisms

Ion transport mechanisms refer to the processes by which ions move within electrochemical systems, influencing the performance and functionality of various devices such as batteries, fuel cells, sensors, and electrochemical reactors. Understanding ion transport mechanisms is crucial for optimizing the efficiency, stability, and reliability of these systems. This section provides an in-depth exploration of ion transport mechanisms, including diffusion, migration, and convection, and their implications in electrochemistry. Diffusion is the movement of ions through a medium due to their random thermal motion, leading to a net flux of ions from regions of high concentration to low concentration. Fick's first law of diffusion describes the relationship between ion flux (J) and concentration gradient (dc/dx) as: $J = -D\ (dc/dx)$, where D is the diffusion coefficient. In electrochemical systems, diffusion governs the transport of ions within electrolyte solutions, electrode pores, and across interfaces, influencing reaction kinetics and mass transfer rates presented in Fig. 1.4.

Migration refers to the movement of ions under the influence of an electric field, leading to the accumulation of ions with like charges and depletion of ions with opposite charges. Migration is governed by the electromotive force (EMF) exerted on ions in the electric field, which is proportional to the ion mobility and the strength of the electric field. In electrochemical cells, migration contributes to ion transport within the electrolyte and plays a critical role in charge transfer processes at

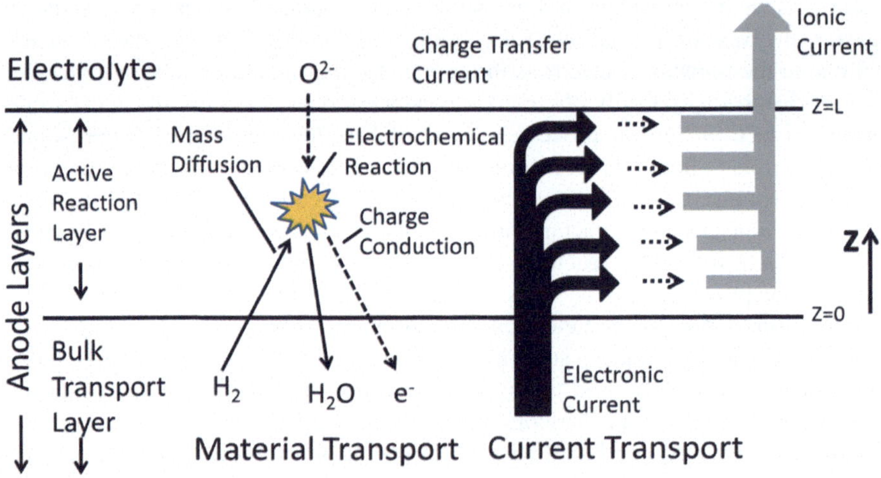

Fig. 1.4 Transport process in electrochemical reactions

electrode surfaces [18]. Convection involves the bulk movement of ions within a fluid medium due to external forces such as fluid flow or stirring. In electrochemical systems, convection can enhance mass transport rates by promoting mixing and reducing concentration gradients, thereby improving reaction kinetics and electrolyte utilization. However, excessive convection may lead to non-uniform ion distribution, concentration polarization, and inefficient utilization of active materials.

Ion transport mechanisms profoundly influence the performance of electrochemical devices by affecting reaction rates, charge transfer kinetics, and mass transport limitations. Optimal ion transport is essential for achieving high energy density, power density, and cycling stability in batteries and fuel cells, as well as high sensitivity and selectivity in electrochemical sensors. Strategies to enhance ion transport include electrode/electrolyte engineering, membrane design, and operating condition optimization, aimed at reducing diffusion/migration resistances and improving overall device performance [19].

Various experimental and computational techniques are employed to study ion transport mechanisms, including electrochemical impedance spectroscopy, cyclic voltammetry, finite element modeling, and molecular dynamics simulations. These techniques provide insights into ion diffusion coefficients, migration velocities, concentration profiles, and transport phenomena at different length and time scales, enabling the design and optimization of electrochemical systems.

1.3.3 Proton Transfer Mechanisms

Proton transfer mechanisms play a pivotal role in various electrochemical processes, particularly in systems where protons are involved in the reaction mechanisms or serve as charge carriers. Understanding the intricacies of proton transfer is essential for optimizing the performance of proton-conducting materials, electrochemical devices, and processes such as fuel cells, electrolyzes, and pH sensors. In-depth knowledge of proton transfer mechanisms encompasses various aspects, including the underlying principles, driving forces, and factors influencing proton mobility and reactivity. Proton transfer reactions involve the transfer of protons (H^+ ions) between species, typically occurring in acidic or aqueous environments. The simplest proton transfer reaction is the self-ionization of water:

$$2H_2O(l) = H_3O^+(aq) + OH^-(aq)$$

Proton transfer reactions can also occur between acids and bases, where a proton is transferred from the acid to the base, forming a conjugate acid–base pair. Proton-coupled electron transfer (PCET) reactions involve the simultaneous transfer of electrons and protons, where the electron transfer is accompanied by a proton transfer. PCET reactions are prevalent in biological systems, redox-active molecules, and electrochemical processes such as fuel cells and water electrolysis. The Marcus theory provides a framework for understanding PCET reactions, describing the relationship between the electron transfer rate and the driving force for proton transfer. Proton mobility refers to the ability of protons to move through a medium, such as an electrolyte or a proton-conducting material [20].

Proton diffusion can occur via mechanisms such as vehicular transport (migration of individual protons) and Grotthuss hopping (proton transfer along hydrogen-bonded networks). Factors influencing proton mobility include the viscosity and composition of the medium, the presence of solvating species, and the strength of hydrogen bonding interactions. Proton transport mechanisms govern the movement of protons across interfaces, membranes, and electrolyte solutions. Proton transport can occur through various pathways, including bulk transport (through the electrolyte), interfacial transport (across electrode–electrolyte interfaces), and proton conduction in solid-state materials (e.g., proton exchange membranes, proton-conducting ceramics). Proton conductivity is a key parameter in proton-conducting materials, with high proton conductivity essential for efficient proton exchange membrane fuel cells and electrolyzers.

The environment, such as pH, temperature, and the presence of solvent molecules, significantly influences proton transfer mechanisms and rates. Electrode materials play a crucial role in facilitating proton transfer reactions at electrode–electrolyte interfaces, with catalysts often employed to enhance proton transport and reaction kinetics.

1.3.4 Impact of Interfaces on Charge Transfer

Interfaces between electrodes, electrolytes, and solution phases play a critical role in determining the efficiency and kinetics of charge transfer processes in electrochemical systems. Understanding the intricate interplay between interfaces and charge transfer is essential for optimizing the performance of electrochemical devices and advancing various applications in energy storage, conversion, sensing, and catalysis. In-depth exploration of the impact of interfaces on charge transfer involves consideration of surface properties, interfacial phenomena, and interface-engineering strategies.

Surface morphology, composition, and reactivity significantly influence charge transfer kinetics at electrode interfaces. High surface area electrodes with nanostructured morphologies provide increased active sites for electrochemical reactions, enhancing reaction rates and efficiency. Surface functionalization techniques, such as surface modification with catalysts or redox-active species, can tailor electrode surfaces to promote specific charge transfer processes. Interfacial phenomena, including adsorption, desorption, and surface reactions, affect the kinetics and thermodynamics of charge transfer at interfaces. Adsorption of reactant molecules or intermediates onto electrode surfaces can facilitate electron transfer reactions and enhance reaction rates. Competitive adsorption of species, such as solvent molecules or impurities, may interfere with charge transfer processes and lead to surface passivation or poisoning [21]. Interface-engineering strategies aim to control and manipulate interfacial properties to optimize charge transfer in electrochemical systems. Surface modification techniques, such as electrodeposition, atomic layer deposition, and self-assembly monolayer formation, enable precise control over surface composition and structure.

Designing electrode–electrolyte interfaces with tailored properties, such as optimized ion conductivity, electron transfer kinetics, and surface wettability, can enhance charge transfer efficiency and device performance. Electrolyte–electrode interactions influence ion transport, solvent dynamics, and charge distribution at the electrode–electrolyte interface [22]. Solvation effects, ion coordination, and ion-surface interactions modulate ion transport kinetics and diffusion coefficients within the electrolyte. Ionic screening and double-layer formation at the interface impact the distribution of charge carriers and the capacitance of electrochemical interfaces. Characterization techniques such as scanning probe microscopy, electrochemical impedance spectroscopy, and surface-sensitive spectroscopies provide insights into interfacial properties and charge transfer mechanisms. In situ and operando techniques allow real-time monitoring of interface dynamics and electrochemical processes under working conditions, enabling a deeper understanding of interface behavior and performance [23] (Fig. 1.5).

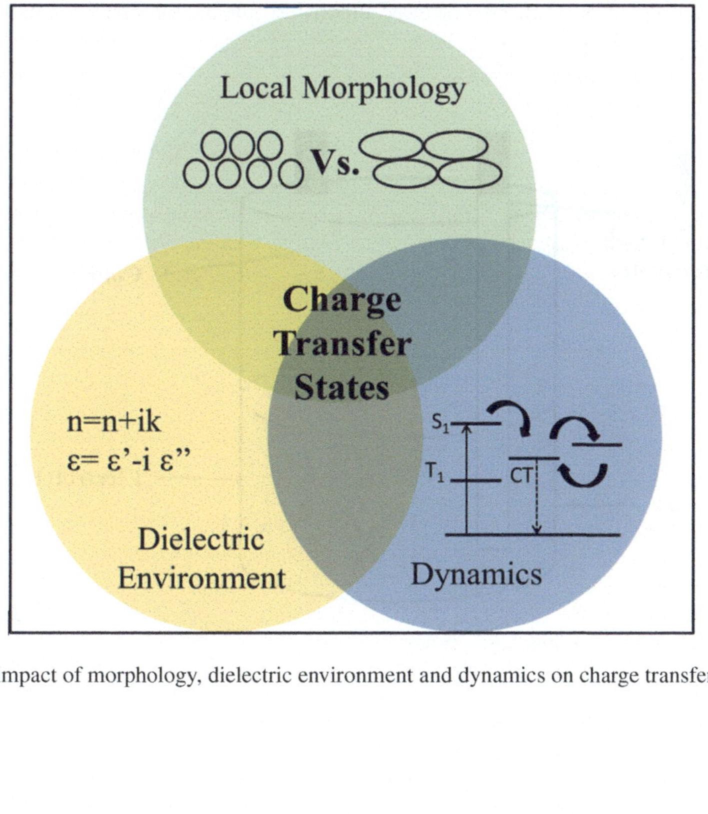

Fig. 1.5 Impact of morphology, dielectric environment and dynamics on charge transfer states

1.4 Electrochemical Cell Configurations

Electrochemical cell configurations play a crucial role in determining the performance and applicability of electrochemical experiments and devices. Different cell designs offer unique advantages and are tailored to specific research needs and applications.

1.4.1 Two-Electrode Cell (Half-Cell Configuration)

The Two-Electrode Cell, also known as the Half-Cell Configuration, represents a foundational setup in electrochemical experimentation, offering simplicity and ease of use. In this configuration, two electrodes, typically a working electrode and a counter/reference electrode, are submerged within an electrolyte solution contained within a single compartment as shown in Fig. 1.6. This setup allows for basic electrochemical measurements without the complexity of additional electrodes or compartments.

The working electrode serves as the site of the electrochemical reaction of interest. It may consist of various materials depending on the experiment's objectives, ranging from noble metals like platinum to conductive substrates coated with catalysts or active materials. The counter/reference electrode, on the other hand, maintains a

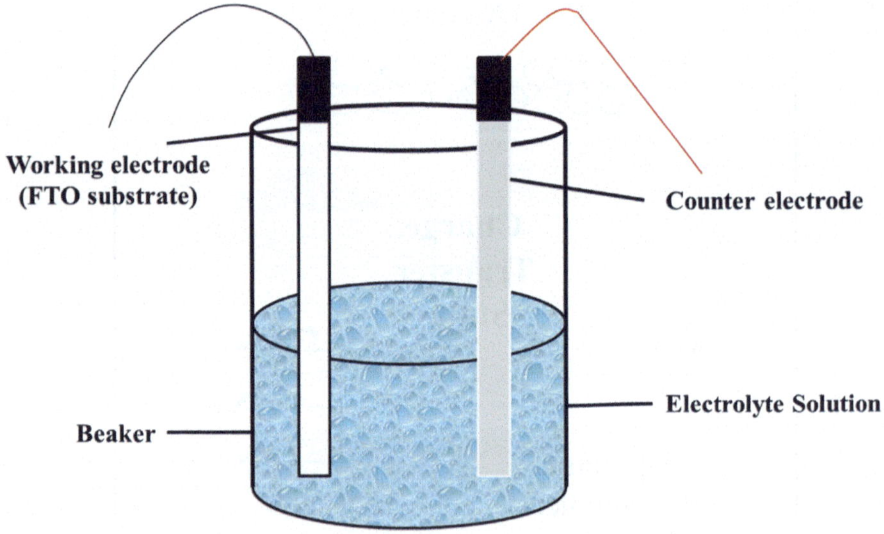

Fig. 1.6 Two electrode configuration system

constant potential to complete the electrical circuit. It serves as both a counter electrode to balance the current flow and a reference electrode to establish a stable reference potential against which the working electrode potential is measured [24].

One of the primary advantages of the two-electrode cell is its simplicity, making it ideal for introductory electrochemistry experiments and rapid screening of electrode materials. This setup allows researchers to perform basic electrochemical techniques, such as cyclic voltammetry or potentiostatic measurements, with minimal setup time and equipment requirements [25]. Additionally, the two-electrode cell provides a straightforward platform for studying electrode kinetics and evaluating the electrochemical behavior of materials in a controlled environment. However, the simplicity of the two-electrode cell also comes with limitations. For instance, the absence of a separate reference electrode can lead to potential drift over time, affecting the accuracy of measurements. Furthermore, the lack of a third electrode for potential control limits the versatility of electrochemical techniques that require precise potential control or separate control of current and potential.

1.4.2 Three-Electrode Cell (Full-Cell Configuration)

The three-electrode cell, often referred to as the full-cell configuration, stands as a cornerstone in electrochemical research and applications due to its versatility and precise control over experimental parameters. Comprising a working electrode, a reference electrode, and a counter electrode, this setup offers a comprehensive platform for investigating electrochemical phenomena with high accuracy and reproducibility as shown in Fig. 1.7.

At the heart of the three-electrode cell lies the working electrode, where the electrochemical reaction of interest occurs. This electrode is carefully chosen to suit the specific application, with materials ranging from metals and metal alloys to conductive polymers and carbon-based materials. The surface of the working electrode plays a pivotal role in dictating reaction kinetics, mass transport, and the overall performance of the electrochemical system [26]. By controlling parameters such as surface area, roughness, and surface chemistry, researchers can tailor the working electrode to optimize electrochemical processes and enhance device performance. The reference electrode serves as a stable reference point for measuring the potential of the working electrode. It maintains a constant and well-defined potential, typically referenced to a standard electrochemical potential such as the standard hydrogen electrode (SHE) or the silver/silver chloride electrode (Ag/AgCl) [27]. The reference electrode's stability and reproducibility are critical for ensuring accurate measurements of the working electrode potential and monitoring changes in electrochemical behavior over time. Common reference electrodes include saturated calomel electrodes (SCE), silver/silver chloride electrodes, and quinhydrone electrodes. The counter electrode, also known as the auxiliary electrode, completes the electrical circuit by balancing the current flow generated at the working electrode. It provides a pathway for the movement of charge carriers, such as electrons or ions, to

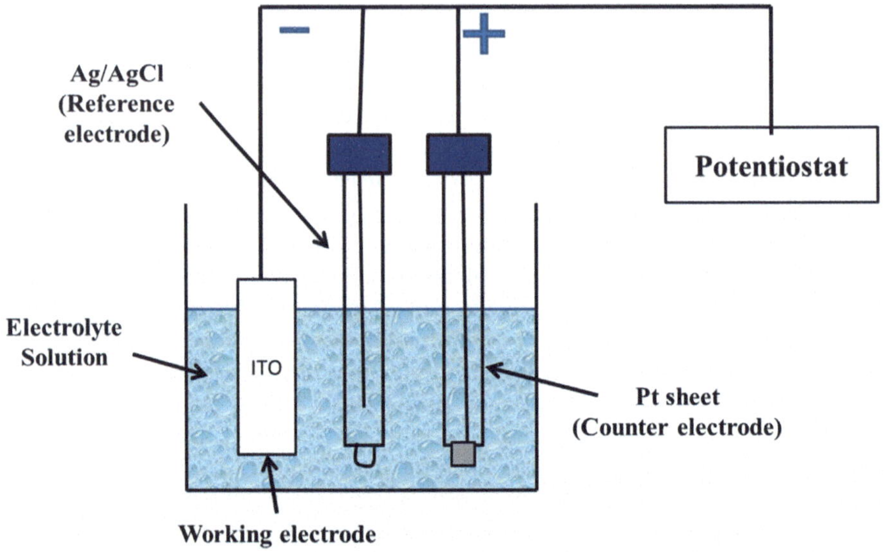

Fig. 1.7 Three electrode configuration system

and from the working electrode during electrochemical reactions. Typically made of inert materials such as platinum, graphite, or carbon, the counter electrode minimizes side reactions and polarization effects, ensuring efficient electron transfer and maintaining a stable electrical potential across the cell. By controlling the geometry and composition of the counter electrode, researchers can optimize current distribution and minimize ohmic losses, thereby improving the accuracy and reproducibility of electrochemical measurements [28].

The three-electrode cell offers unparalleled control over experimental conditions, allowing researchers to manipulate parameters such as electrode potential, current density, and electrolyte composition independently. This enables a wide range of electrochemical techniques and measurements, including cyclic voltammetry, chronoamperometry, and electrochemical impedance spectroscopy. Moreover, the versatility of the three-electrode cell extends to various electrochemical applications, including corrosion studies, electrodeposition, energy storage and conversion, sensor development, and catalysis.

1.4.3 Divided Cell Configuration

The divided cell configuration stands as a pivotal design in electrochemical experimentation, offering a controlled environment for studying ion transport, reaction kinetics, and membrane processes. This setup involves partitioning the electrolyte solution into two compartments using a semi-permeable membrane or barrier,

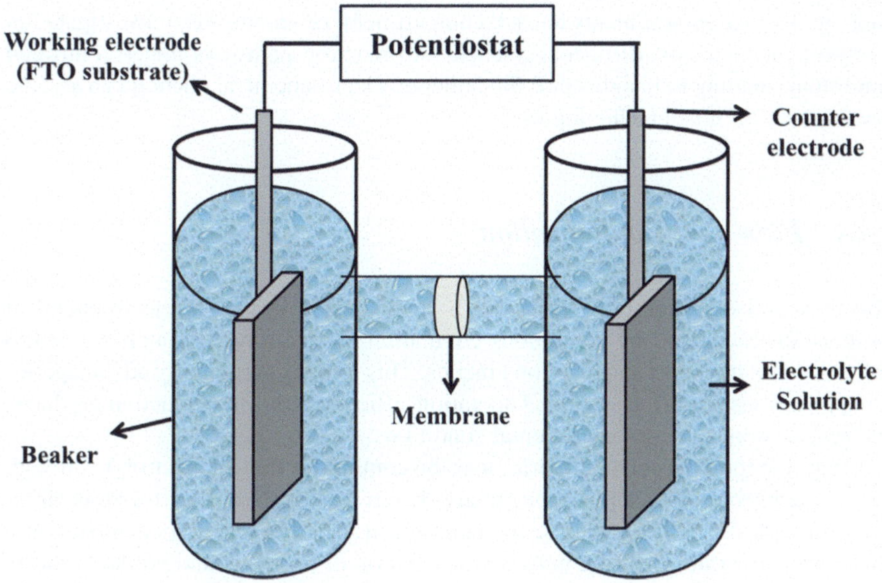

Fig. 1.8 Divided cell configuration system

allowing for selective transport of ions while preventing direct mixing of the solutions depicted in Fig. 1.8.

At the core of the divided cell configuration is the semi-permeable membrane, which acts as a barrier between the two compartments. This membrane selectively permits the passage of specific ions based on their size, charge, and chemical properties, facilitating controlled ion transport across the cell. Common membrane materials include polymer films, glass frits, porous ceramics, and ion-selective membranes, each offering unique characteristics tailored to specific experimental requirements [29]. Within each compartment of the divided cell, electrodes are positioned to facilitate electrochemical reactions and measurements. The working electrode, typically located in one compartment, serves as the site of electrochemical processes of interest, while the counter electrode, situated in the opposite compartment, completes the electrical circuit and balances the current flow. The reference electrode may be placed in either compartment, depending on the experimental setup, to monitor and control the potential of the working electrode relative to a stable reference point. The divided cell configuration enables precise control over experimental conditions, allowing researchers to manipulate parameters such as electrolyte composition, pH, and ion concentration independently in each compartment. This versatility is particularly advantageous for studying electrochemical processes influenced by ionic gradients, such as ion transport phenomena, electrodeposition, redox reactions, and membrane permeation.

Applications of the divided cell configuration span various fields, including electrochemical synthesis, electro dialysis, fuel cells, and ion-selective sensors. By creating distinct environments in each compartment, researchers can investigate ion transport mechanisms, diffusion kinetics, and reaction pathways under controlled conditions, leading to insights into fundamental electrochemical phenomena and the development of novel technologies.

1.4.4 Flow Cell Configuration

The flow cell configuration represents a sophisticated design in electrochemical systems, characterized by continuous circulation of electrolyte through the cell to enhance mass transport and reaction kinetics. This setup is particularly advantageous for applications requiring sustained operation, efficient removal of reaction products, and precise control over experimental conditions.

Central to the flow cell configuration is the continuous flow of electrolyte through the cell, achieved using a pump or gravity-driven flow system. Electrolyte is circulated through the cell, typically containing electrodes and reaction chambers, at a controlled flow rate, ensuring uniform distribution of reactants and products across the electrode surfaces. The flow of electrolyte minimizes concentration gradients and enhances mass transport, facilitating rapid exchange of ions and molecules at the electrode–electrolyte interface. Within the flow cell, electrodes are positioned to facilitate electrochemical reactions and measurements. The working electrode serves as the site of electrochemical processes, while the counter electrode completes the electrical circuit and balances the current flow [30]. The electrolyte flow path may also include additional components such as flow channels, mixing chambers, and sensors to optimize performance and functionality. The continuous flow of electrolyte in the flow cell configuration offers several advantages over static cell designs. It allows for real-time monitoring of electrochemical processes, enabling dynamic measurements and control of experimental parameters. Moreover, the continuous renewal of electrolyte minimizes the accumulation of reaction by-products and electrode fouling, enhancing long-term stability and reproducibility of measurements.

Flow cells find applications across various fields, including fuel cells, batteries, electrochemical sensors, and chemical synthesis. In fuel cell applications, flow cells enable efficient transport of reactants and products to electrode surfaces, enhancing power output and durability. In battery research, flow cells facilitate rapid charging and discharging cycles, enabling high-throughput screening of electrode materials and electrolyte compositions. Additionally, flow cells are utilized in electrochemical sensors for continuous monitoring of analyte concentrations in environmental, biomedical, and industrial settings [31]. The flow cell configuration offers a versatile and efficient platform for studying electrochemical processes and developing innovative technologies. By enabling continuous circulation of electrolyte, this setup enhances mass transport, reaction kinetics, and experimental control, paving the way for advancements in electrochemical research and technology across diverse applications and disciplines.

1.4.5 Customized Cell Configurations

Customized cell configurations represent a tailored approach to electrochemical experimentation, allowing researchers to design specialized setups to address specific research questions and applications. These configurations incorporate unique designs, materials, or additional components to achieve desired functionalities beyond conventional cell designs. Customized cell configurations offer flexibility, adaptability, and enhanced performance for a wide range of electrochemical studies and applications. One common aspect of customized cell configurations is the variation in electrode geometry and design. Researchers may modify electrode shapes, sizes, or arrangements to optimize electrochemical performance and achieve specific experimental goals. For example, microelectrodes with reduced dimensions enable studies at the microscale, offering insights into electrode processes, surface reactions, and transport phenomena with high spatial resolution. Additionally, non-planar electrode geometries, such as three-dimensional (3D) electrodes or nanostructured electrodes, provide increased surface area and improved mass transport, enhancing electrochemical activity and sensitivity for sensing applications.

In addition to electrode design, customized cell configurations may incorporate specialized components or features to enhance functionality. This includes the integration of microfluidic channels, membranes, sensors, and actuators tailored to specific research needs. For instance, microfluidic channels enable precise control over fluid flow and sample delivery, facilitating rapid mixing, reaction kinetics studies, and sample analysis in electrochemical microreactors. Membranes with selective permeability allow for separation of species, concentration gradients, and ion transport studies in membrane-based electrochemical devices such as fuel cells, batteries, and sensors. Furthermore, customized cell configurations enable the integration of advanced measurement techniques and instrumentation for real-time monitoring and characterization of electrochemical processes. This may include in situ or operando techniques such as spectroelectrochemistry, scanning probe microscopy, electrochemical impedance spectroscopy, and surface-enhanced spectroscopies. By combining electrochemical measurements with complementary analytical techniques, researchers can gain deeper insights into reaction mechanisms, interface properties, and dynamic behavior under operando conditions. The versatility and adaptability of customized cell configurations make them invaluable tools for advancing electrochemical research and technology across diverse applications and disciplines. Whether investigating fundamental electrochemical processes, developing novel materials and devices, or addressing complex research challenges, customized cell configurations offer researchers the flexibility to design experiments tailored to their specific needs, ultimately driving innovation and progress in the field of electrochemistry.

1.5 Principles of Light Absorption and Charge Transfer

1.5.1 Fundamentals of Light Absorption

The fundamentals of light absorption delve into the intricate interplay between photons and matter, elucidating the mechanisms by which materials interact with light and absorb its energy. This foundational concept underpins numerous scientific disciplines, including physics, chemistry, materials science, and engineering, and finds widespread applications in optics, spectroscopy, photovoltaics, and photochemistry.

At its core, light absorption occurs when a material absorbs photons, leading to electronic transitions within its atomic or molecular structure. This process is governed by the energy of the incident photons, which must match the energy required to excite electrons from lower energy states to higher energy states. The energy levels available for such transitions are quantized and determined by the electronic structure of the material, including its band structure and electronic configuration. The absorption of light by a material can be described by its absorption spectrum, which represents the wavelength-dependent absorption behavior. Different materials exhibit characteristic absorption spectra, determined by their electronic and molecular properties. For example, semiconductors typically have absorption edges corresponding to the energy bandgap, while molecules may exhibit absorption bands associated with electronic or vibrational transitions.

1.5.2 Photoelectrochemical Processes

Photoelectrochemical (PEC) processes harness the energy of photons to drive electrochemical reactions at the interface of a semiconductor electrode and an electrolyte solution. This synergy between light absorption and charge transfer enables a wide range of applications, including solar energy conversion, water splitting, and pollutant degradation.

PEC processes involve several key steps, including light absorption, generation of electron–hole pairs, charge separation, and redox reactions depicted in Fig. 1.9. Upon absorption of photons, semiconductor materials generate electron–hole pairs through electronic excitation [32]. These charge carriers then migrate to the semiconductor-electrolyte interface, where they participate in electrochemical reactions, such as water oxidation or reduction.

The choice of semiconductor materials is critical for optimizing PEC performance. Criteria such as bandgap energy, band edge positions, charge carrier mobility, and stability influence the efficiency and selectivity of PEC reactions. Common semiconductor materials for PEC applications include metal oxides (e.g., TiO_2, WO_3), metal chalcogenides (e.g., CdS, $CuInSe_2$), and perovskite materials, each offering unique properties and advantages for specific applications. PEC processes find diverse applications in renewable energy generation, environmental remediation, and chemical

Fig. 1.9 Mechanism of photoelectrochemical processes

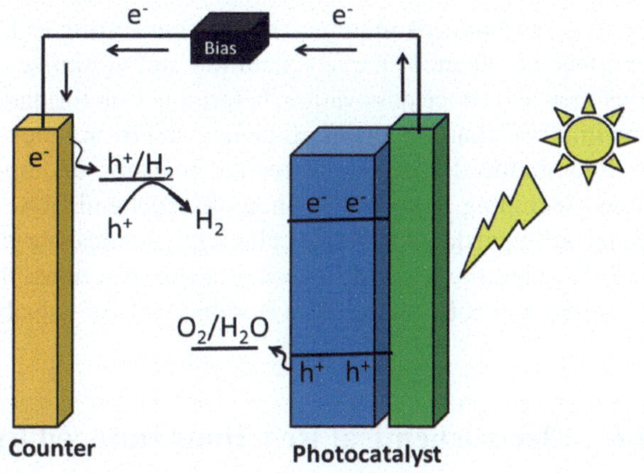

synthesis. Solar water splitting, for instance, utilizes PEC cells to convert solar energy into hydrogen fuel through water electrolysis. PEC sensors enable sensitive and selective detection of analytes in environmental monitoring and biomedical diagnostics [33, 34]. Additionally, PEC reactors facilitate the degradation of organic pollutants and the synthesis of value-added chemicals under mild reaction conditions.

1.5.3 Photoinduced Charge Transfer Mechanisms

Photoinduced charge transfer mechanisms involve the migration of photogenerated charge carriers (electrons and holes) from the semiconductor electrode to redox species in solution or to other electrode surfaces. These mechanisms are central to various photoelectrochemical processes, including solar energy conversion, photocatalysis, and photodetection, and are governed by the electronic structure and interface properties of the semiconductor material. In semiconductor materials, absorption of photons with energies exceeding the bandgap leads to the generation of electron–hole pairs via electronic excitation. These photogenerated charge carriers are characterized by their mobility, lifetime, and recombination rates, which influence their transport and fate within the material. Upon generation, photogenerated electrons and holes migrate to the semiconductor-electrolyte interface, driven by the built-in electric field or concentration gradients. At this interface, charge separation occurs, facilitated by the difference in energy levels between the semiconductor's conduction band and the redox potentials of species in solution. At the semiconductor-electrolyte interface, photogenerated electrons participate in reduction reactions (e.g., reduction of water to hydrogen) by transferring to oxidizable species in solution, while photogenerated holes engage in oxidation reactions (e.g., oxidation of water to oxygen). The kinetics of these charge transfer processes are influenced by factors such as electrode potential, surface states, and solution composition. Optimizing photoinduced charge transfer mechanisms requires careful engineering of the semiconductor-electrolyte interface to enhance charge separation and minimize recombination losses. Strategies such as surface passivation, heterojunction formation, and cocatalyst deposition can improve charge transfer efficiency and overall device performance.

Photoinduced charge transfer mechanisms underpin a wide range of applications, including solar cells, photoelectrochemical water splitting, and photocatalytic pollutant degradation. By efficiently harnessing photogenerated charge carriers and directing them towards desired redox reactions, these technologies enable the conversion of solar energy into storable fuels or valuable chemical products.

1.6 Electrochemical Reactions Induced by Light

Exploring the realm of electrochemical reactions induced by light unveils a fascinating interplay between photon absorption, charge transfer, and chemical transformations. In this section, we delve into the underlying mechanisms and applications of light-driven electrochemical reactions, laying the groundwork for further discussions in subsequent chapters.

Electrochemical reactions induced by light, often referred to as photoelectrochemical (PEC) reactions, represent a unique class of processes wherein light serves as the driving force for chemical transformations at electrode surfaces. This phenomenon harnesses the dual nature of light as both an energy source and a trigger for electron transfer, enabling a wide range of applications in energy conversion, environmental remediation, and chemical synthesis. At the heart of light-induced electrochemical reactions lie the intricate mechanisms of photon absorption, charge carrier generation, and subsequent redox processes. Upon absorption of photons with energies exceeding the semiconductor bandgap, electron–hole pairs are generated within the semiconductor material. These photogenerated charge carriers migrate to the electrode–electrolyte interface, where they participate in electrochemical reactions, such as water oxidation or reduction, facilitated by redox species in solution or immobilized on the electrode surface. Light-driven electrochemical reactions find diverse applications across various fields, ranging from solar energy conversion to environmental remediation and chemical synthesis. Solar water splitting, for instance, utilizes PEC cells to split water into hydrogen and oxygen using sunlight as the sole energy source, offering a sustainable pathway for renewable hydrogen production. Photocatalytic degradation of pollutants leverages PEC processes to oxidize organic contaminants and purify water resources, addressing environmental challenges such as water pollution and wastewater treatment. As we delve deeper into the realm of electrochemical devices and applications, subsequent chapters will further explore the intricacies of light-induced electrochemical reactions and their role in various technologies. Topics such as photoelectrochemical solar cells, photodetectors, and photocatalytic systems will be discussed in detail, highlighting the importance of understanding the fundamental principles of light-driven electrochemistry for advancing sustainable energy technologies and environmental solutions.

In summary, electrochemical reactions induced by light represent a promising frontier in the quest for clean energy and environmental sustainability. By harnessing the power of sunlight to drive chemical transformations, PEC processes offer a pathway towards efficient energy conversion, pollutant remediation, and sustainable chemical synthesis. Through further exploration and understanding of these processes, we can unlock new opportunities for innovation and address pressing global challenges in the transition towards a more sustainable future.

References

1. Hudson, J.L., Tsotsis, T.T.: Electrochemical reaction dynamics: a review. Chem. Eng. Sci. **49**(10), 1493–1572 (1994). https://doi.org/10.1016/0009-2509(94)85063-1
2. Kammona, O., Kotti, K., Kiparissides, C., Celis, J.P., Fransaer, J.: Synthesis of polymeric and hybrid nanoparticles for electroplating applications. Electrochim. Acta **54**(9), 2450–2457 (2009). https://doi.org/10.1016/j.electacta.2008.05.017
3. Rossmeisl, J., Qu, Z.-W., Zhu, H., Kroes, G.-J., Nørskov, J.K.: Electrolysis of water on oxide surfaces. J. Electroanal. Chem. **607**(1), 83–89 (2007). https://doi.org/10.1016/j.jelechem.2006.11.008
4. Lu, C., Chen, X.: Nanostructure engineering of graphitic carbon nitride for electrochemical applications. ACS Nano. **15**(12), 18777–18793 (2021). https://doi.org/10.1021/acsnano.1c06454
5. Murray, P.R.D., et al.: Photochemical and electrochemical applications of proton-coupled electron transfer in organic synthesis. Chem. Rev. **122**(2), 2017–2291 (2022). https://doi.org/10.1021/acs.chemrev.1c00374
6. Xiao, X., Zou, L., Pang, H., Xu, Q.: Synthesis of micro/nanoscaled metal–organic frameworks and their direct electrochemical applications. Chem. Soc. Rev. **49**(1), 301–331 (2020). https://doi.org/10.1039/C7CS00614D
7. Zhao, X., Pachfule, P., Thomas, A.: Covalent organic frameworks (COFs) for electrochemical applications. Chem. Soc. Rev. **50**(12), 6871–6913 (2021). https://doi.org/10.1039/D0CS01569E
8. Campero, A., Díaz Ponce, J.A.: Relationship between the atomic structure and electrochemistry. 1. Electric force, standard reduction potential E°, and standard reaction Gibbs free energy ΔG°. ACS Omega **5**(21), 12046–12056 (2020). https://doi.org/10.1021/acsomega.0c00257
9. Shapiro, N.Z., Shapley, L.S.: Mass action laws and the Gibbs free energy function. J. Soc. Ind. Appl. Math. **13**(2), 353–375 (1965). https://doi.org/10.1137/0113020
10. Doménech, A., Koshevoy, I.O., Montoya, N., Karttunen, A.J., Pakkanen, T.A.: Determination of individual Gibbs energies of anion transfer and excess Gibbs energies using an electrochemical method based on insertion electrochemistry of solid compounds. J. Chem. Eng. Data **56**(12), 4577–4586 (2011). https://doi.org/10.1021/je200514c
11. Hernández-Rizo, S.G., Larios-Durán, E.R., Bárcena-Soto, M.: Frequency response of Gibbs free energy and enthalpy changes of electrochemical systems analyzed as thermometric transfer functions. J. Solid State Electrochem. **27**(11), 3177–3188 (2023). https://doi.org/10.1007/s10008-023-05553-3
12. Feiner, A.-S., McEvoy, A.J.: The Nernst equation. J. Chem. Educ. **71**(6), 493 (1994). https://doi.org/10.1021/ed071p493
13. Vidal-Iglesias, F.J., Solla-Gullón, J., Rodes, A., Herrero, E., Aldaz, A.: Understanding the Nernst equation and other electrochemical concepts: an easy experimental approach for students. J. Chem. Educ. **89**(7), 936–939 (2012). https://doi.org/10.1021/ed2007179

14. Dickinson, E.J.F., Wain, A.J.: The Butler-Volmer equation in electrochemical theory: origins, value, and practical application. J. Electroanal. Chem. **872**, 114145 (2020). https://doi.org/10.1016/j.jelechem.2020.114145
15. Dreyer, W., Guhlke, C., Müller, R.: A new perspective on the electron transfer: recovering the Butler-Volmer equation in non-equilibrium thermodynamics. Phys. Chem. Chem. Phys. **18**(36), 24966–24983 (2016). https://doi.org/10.1039/C6CP04142F
16. Lukács, Z., Kristóf, T.: Linear transformations of the Butler-Volmer equation. Electrochem. Commun. **154**, 107556 (2023). https://doi.org/10.1016/j.elecom.2023.107556
17. Li, X., Zhang, J., Huo, Y., Dai, K., Li, S., Chen, S.: Two-dimensional sulfur- and chlorine-codoped g-C_3N_4/CdSe-amine heterostructures nanocomposite with effective interfacial charge transfer and mechanism insight. Appl. Catal. B Environ. **280**, 119452 (2021). https://doi.org/10.1016/j.apcatb.2020.119452
18. Kim, D., Yong, K.: Boron doping induced charge transfer switching of a C_3N_4/ZnO photocatalyst from Z-scheme to type II to enhance photocatalytic hydrogen production. Appl. Catal. B Environ. **282**, 119538 (2021). https://doi.org/10.1016/j.apcatb.2020.119538
19. Wang, C., Chi, W., Qiao, Q., Tan, D., Xu, Z., Liu, X.: Twisted intramolecular charge transfer (TICT) and twists beyond TICT: from mechanisms to rational designs of bright and sensitive fluorophores. Chem. Soc. Rev. **50**(22), 12656–12678 (2021). https://doi.org/10.1039/D1CS00239B
20. Cheng, C., He, B., Fan, J., Cheng, B., Cao, S., Yu, J.: An inorganic/organic S-scheme heterojunction H_2-production photocatalyst and its charge transfer mechanism. Adv. Mater. **33**(22), 2100317 (2021). https://doi.org/10.1002/adma.202100317
21. Phogat, P., Shreya, Jha, R., Singh, S.: Diffusion controlled features of microwave assisted ZnS/ZnO nanocomposite with reduced band gap. ECS J. Solid State Sci. Technol. **12**(3), 034004 (2023). https://doi.org/10.1149/2162-8777/acc426
22. Azad, M., Hussain, Z., Baig, M.M.: MWCNTs/NiS_2 decorated Ni foam based electrode for high-performance supercapacitors. Electrochim. Acta **345**, 136196 (2020). https://doi.org/10.1016/j.electacta.2020.136196
23. Zhou, B., et al.: Platinum modulates redox properties and 5-hydroxymethylfurfural adsorption kinetics of $Ni(OH)_2$ for biomass upgrading. Angew. Chemie Int. Ed. **60**(42), 22908–22914 (2021). https://doi.org/10.1002/anie.202109211
24. Chu, F., Su, M., Xiao, G., Tan, Z., Yang, G.: Analysis of electrode configuration effects on mass transfer and organic redox flow battery performance. Ind. Eng. Chem. Res. **61**(7), 2915–2925 (2022). https://doi.org/10.1021/acs.iecr.1c04689
25. Shaheen, I., et al.: Recent advancements in metal oxides for energy storage materials: design, classification, and electrodes configuration of supercapacitor. J. Energy Storage **72**, 108719 (2023). https://doi.org/10.1016/j.est.2023.108719
26. Pandey, D., Kumar, K.S., Thomas, J.: Supercapacitor electrode energetics and mechanism of operation: uncovering the voltage window. Prog. Mater. Sci. **141**, 101219 (2024). https://doi.org/10.1016/j.pmatsci.2023.101219
27. Nugroho, F.A., Sani, M.M., Apriyanti, F., Aryanti, P.T.P.: The influence of applied current strength and electrode configuration in laundry wastewater treatment by electrocoagulation. J. Phys. Conf. Ser. **1477**(5), 52018 (2020). https://doi.org/10.1088/1742-6596/1477/5/052018
28. Wu, Y., et al.: Thick-network electrode: enabling dual working voltage plateaus of Zn-ion micro-battery with ultrahigh areal capacity. Adv. Funct. Mater. **34**(5), 2301734 (2024). https://doi.org/10.1002/adfm.202301734
29. Merlo, A., Bò, M.C., Campanini, I.: Electrode size and placement for surface EMG bipolar detection from the brachioradialis muscle: a scoping review. Sensors **21**(21) (2021). https://doi.org/10.3390/s21217322
30. Niu, Z.-Z., Chi, L.-P., Liu, R., Chen, Z., Gao, M.-R.: Rigorous assessment of CO_2 electroreduction products in a flow cell. Energy Environ. Sci. **14**(8), 4169–4176 (2021). https://doi.org/10.1039/D1EE01664D

31. Irkham, Nagashima, S., Tomisaki, M., Einaga, Y.: Enhancing the electrochemical reduction of CO_2 by controlling the flow conditions: an intermittent flow reduction system with a boron-doped diamond electrode. ACS Sustain. Chem. Eng. **9**(15), 5298–5303 (2021). https://doi.org/10.1021/acssuschemeng.0c08955
32. Dong, G., et al.: Cadmium sulfide nanoparticles-assisted intimate coupling of microbial and photoelectrochemical processes: mechanisms and environmental applications. Sci. Total. Environ. **740**, 140080 (2020). https://doi.org/10.1016/j.scitotenv.2020.140080
33. Divyapriya, G., Singh, S., Martínez-Huitle, C.A., Scaria, J., Karim, A.V., Nidheesh, P.V.: Treatment of real wastewater by photoelectrochemical methods: an overview. Chemosphere **276**, 130188 (2021). https://doi.org/10.1016/j.chemosphere.2021.130188
34. Tolbert, C.L., McDonald, D.M., Hill, C.M.: Electrochemical techniques for visualizing photoelectrochemical processes at the nanoscale. Curr. Opin. Electrochem. **37**, 101164 (2023). https://doi.org/10.1016/j.coelec.2022.101164

Chapter 2
Photoelectrochemical Solar Cells

This chapter provides a comprehensive overview of the principles, materials, configurations, and applications of photoelectrochemical solar cells. Beginning with the fundamental principles of photoelectrochemical detection, it explores the design and characteristics of semiconductor photoelectrodes tailored for photoelectrochemical (PEC) solar cell applications. The chapter delves into the mechanisms of analyte detection, highlighting the role of electrolytes in enhancing device performance. Moreover, it discusses various PEC detector configurations and device architectures, elucidating their impact on sensitivity and selectivity. Through a series of case studies and real-world applications, readers gain insights into the diverse range of PEC devices and their potential for renewable energy generation, environmental monitoring, and beyond. Overall, this chapter serves as a valuable resource for researchers, engineers, and students seeking to understand and harness the capabilities of photoelectrochemical solar cells.

2.1 Introduction

Photoelectrochemical (PEC) solar cells stand at the intersection of photovoltaics and photocatalysis, offering a unique approach to solar energy conversion. Unlike traditional photovoltaic cells, which solely rely on the generation of electric current from sunlight, PEC solar cells integrate electrochemical reactions that can enhance the overall efficiency of energy conversion. This dual functionality is particularly advantageous because it allows PEC solar cells to not only capture and convert sunlight into electricity but also to drive chemical processes, such as water splitting or carbon dioxide reduction, simultaneously. This makes PEC solar cells a versatile and promising technology in the quest for renewable energy solutions [1]. Figure 2.1 shows the experimental setup of the PEC solar cell circuit. The core of PEC solar cells lies in their ability to utilize semiconducting materials to absorb sunlight and

P. Phogat et al., *Electrochemical Devices*, Engineering Materials,
https://doi.org/10.1007/978-981-96-0527-9_2

generate electron–hole pairs. These charge carriers are then separated and driven to respective electrodes, where they can either produce electricity directly or participate in electrochemical reactions. The incorporation of photocatalytic elements in PEC cells introduces an additional layer of functionality, where the absorbed solar energy can be used to drive reactions that are otherwise challenging to achieve. For instance, in PEC water-splitting cells, the generated electrons and holes are utilized to split water molecules into hydrogen and oxygen, providing a clean and renewable source of hydrogen fuel [2].

Over the years, the design and optimization of PEC solar cells have evolved significantly, with advancements in materials science playing a crucial role. The choice of semiconductor materials, their bandgap properties, and surface modifications have all been fine-tuned to enhance light absorption, charge separation, and overall cell efficiency. Moreover, the integration of co-catalysts and the development of nanostructured materials have further improved the performance of PEC solar cells by facilitating faster reaction kinetics and reducing recombination losses [3]. PEC solar cells also offer flexibility in terms of design, with various cell architectures being explored to maximize efficiency. Single-junction PEC cells, which consist of a single light-absorbing material, are simpler in design but often limited by their efficiency.

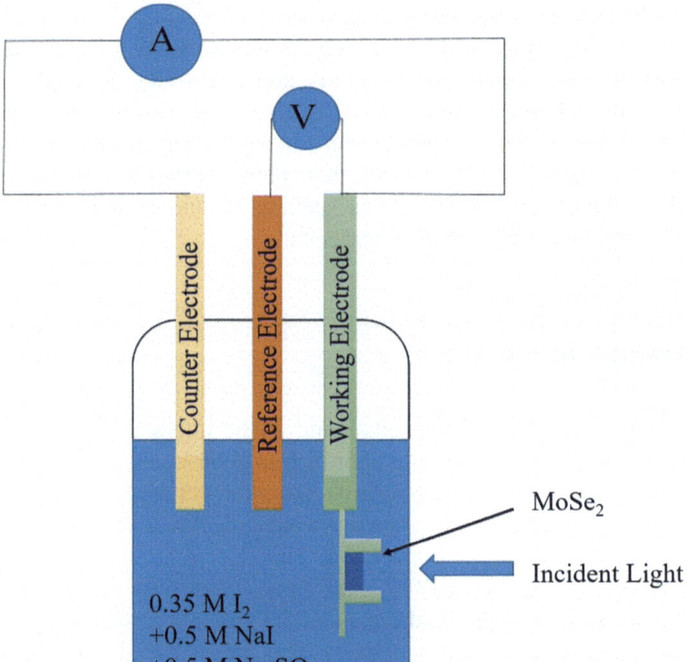

Fig. 2.1 Experimental setup of the PEC solar cell circuit, reproduced from [5] Copyright © 2011–2024 Journal of Emerging Investigators

On the other hand, tandem and multi-junction PEC cells, which incorporate multiple layers of semiconductors with varying bandgaps, have shown promise in achieving higher efficiencies by capturing a broader spectrum of sunlight. These advanced architectures are paving the way for more efficient and commercially viable PEC solar cells as shown in Fig. 2.1 [4]. The significance of PEC solar cells extends beyond their potential for high-efficiency solar energy conversion. Their ability to integrate seamlessly with other renewable energy systems, such as hydrogen production or carbon capture, positions them as a key player in the future of sustainable energy. As research continues to advance, the potential applications of PEC solar cells are expanding, with possibilities ranging from large-scale solar farms to decentralized energy generation systems.

The evolution of PEC solar cells is deeply rooted in the broader history of solar energy conversion technologies. The concept of utilizing sunlight to drive electrochemical reactions has intrigued scientists for decades, beginning with early experiments that laid the groundwork for the modern PEC solar cell. The journey from rudimentary photoelectrodes to sophisticated PEC systems has been marked by significant milestones, each contributing to the advancement of this promising technology. The origins of PEC solar cells can be traced back to the mid-twentieth century when researchers first began exploring the possibility of using semiconductors to harness solar energy for chemical transformations. In 1972, a seminal discovery by Fujishima and Honda demonstrated the potential of titanium dioxide (TiO_2) as a photoanode for water splitting under ultraviolet light. This breakthrough marked the first practical demonstration of a PEC cell and sparked widespread interest in the field. The Fujishima–Honda experiment was not only a pivotal moment in the development of PEC cells but also set the stage for future research into the use of semiconductors for solar energy conversion [6]. Following this discovery, the 1970s and 1980s saw a surge in research aimed at improving the efficiency and stability of PEC cells. Researchers began investigating alternative semiconductor materials with narrower bandgaps to extend light absorption into the visible spectrum, thus enhancing the efficiency of solar energy conversion. Materials such as iron oxide (Fe_2O_3) and tungsten trioxide (WO_3) were explored, but challenges such as poor charge transport and photocorrosion limited their practical application. Despite these hurdles, these early studies were crucial in expanding the understanding of the fundamental processes governing PEC cells and identifying key factors that influence their performance [7]. The 1990s brought about a renewed focus on nanostructuring and surface modification techniques as researchers sought to overcome the limitations of traditional semiconductor materials. The development of nanostructured photoelectrodes, which offered improved charge separation and transport properties, represented a significant advancement in the field. During this period, the integration of co-catalysts into PEC cells also emerged as a critical strategy for enhancing the kinetics of photoelectrochemical reactions, thereby improving overall efficiency. These innovations laid the groundwork for the development of more efficient and durable PEC solar cells [8]. The turn of the twenty-first century witnessed a shift towards more sophisticated PEC cell architectures, including tandem and multi-junction designs. These advanced configurations, which stack multiple layers of semiconductors with varying bandgaps, have the potential to capture a broader spectrum of sunlight and achieve higher solar-to-electricity conversion efficiencies. Research during this era also began to focus on emerging materials, such as perovskites and 2D materials, which offered promising properties for PEC applications. These materials not only extended the absorption range but also provided new avenues for optimizing charge carrier dynamics and improving overall cell performance.

In recent years, the field of PEC solar cells has continued to advance with the introduction of novel materials and innovative design strategies. The exploration of hybrid structures, combining organic and inorganic materials, has opened up new possibilities for enhancing light absorption and charge transport. Additionally, advances in computational modeling and simulation have provided deeper insights into the mechanisms underlying PEC processes, enabling more targeted and efficient material design. The use of advanced characterization techniques, such as in-situ spectroscopy and microscopy, has also allowed researchers to gain a more comprehensive understanding of the dynamic processes occurring within PEC cells. Today, PEC solar cells are recognized as a viable and potentially game-changing technology in the quest for sustainable energy solutions. The historical development of PEC cells, marked by key milestones in material science, cell architecture, and process optimization, has paved the way for the current generation of high-performance PEC systems. As research continues to push the boundaries of what is possible, PEC solar cells are poised to play a significant role in the future of renewable energy, offering a pathway to efficient, clean, and scalable solar energy conversion [9].

The significance of Photoelectrochemical (PEC) solar cells in the renewable energy landscape lies in their unique ability to directly convert solar energy into chemical fuels, in addition to electricity. This dual functionality distinguishes PEC solar cells from conventional photovoltaic (PV) technologies, offering not only a pathway to sustainable electricity generation but also the potential for producing hydrogen—a clean and versatile fuel. As the world transitions toward more sustainable energy systems, the integration of PEC solar cells into the renewable energy mix presents an exciting opportunity to address both energy and environmental challenges [10]. One of the most compelling aspects of PEC solar cells is their ability to produce hydrogen through water splitting, driven by sunlight. Hydrogen, as an energy carrier, has the potential to revolutionize various sectors, including transportation, industry, and power generation, by providing a clean alternative to fossil fuels. Unlike conventional hydrogen production methods, which often rely on natural gas and result in significant carbon emissions, PEC solar cells offer a green and sustainable approach to hydrogen production. By using only water and sunlight as inputs, PEC cells can generate hydrogen with zero carbon emissions, contributing to a reduction in greenhouse gases and helping to mitigate climate change [11]. In addition to their role in hydrogen production, PEC solar cells offer several advantages over traditional PV systems. The ability to directly couple solar energy with electrochemical processes enables PEC cells to achieve higher overall energy conversion efficiencies under certain conditions. This is particularly relevant in integrated energy systems where both electricity and chemical fuels are needed. For instance, PEC solar cells can be used to produce hydrogen during periods of high solar irradiance, which can then be stored and used to generate electricity when sunlight is not available. This capability provides a means of overcoming the intermittency issues associated with solar energy, thereby enhancing the reliability and stability of renewable energy systems [12].

Furthermore, PEC solar cells are versatile in terms of their potential applications. Beyond hydrogen production, they can be tailored to drive other chemical reactions, such as carbon dioxide reduction, nitrogen fixation, or organic synthesis. These additional applications make PEC solar cells a valuable tool in the broader context of sustainable chemistry and green manufacturing. The ability to harness sunlight to drive a variety of chemical processes opens up new possibilities for reducing reliance on fossil fuels and promoting the development of a circular economy, where waste products are converted into valuable resources using renewable energy [13].

The environmental benefits of PEC solar cells extend beyond their operational phase. The materials used in PEC cells, particularly those based on earth-abundant and non-toxic elements, can be more environmentally benign compared to the materials used in conventional PV technologies. Additionally, the potential for using less material in PEC cells due to advances in nanostructuring and surface engineering further reduces their environmental footprint. As the technology matures, the development of more sustainable manufacturing processes and the recycling of PEC materials will be crucial in ensuring that these cells contribute to a truly sustainable energy future [14].

From an economic perspective, PEC solar cells hold the promise of reducing the cost of renewable energy, particularly for hydrogen production. As the technology advances and scales up, the cost of PEC-based hydrogen is expected to decrease, making it competitive with other forms of hydrogen production. This cost reduction, combined with the environmental benefits of PEC technology, positions PEC solar cells as a key component in the future energy market. Moreover, the ability to integrate PEC cells into existing infrastructure, such as solar farms or industrial facilities, could facilitate their adoption and accelerate the transition to a hydrogen economy [15].

The significance of PEC solar cells in renewable energy is multifaceted, encompassing their role in sustainable hydrogen production, their advantages over traditional PV systems, their versatility in driving various chemical processes, and their potential environmental and economic benefits. Figure 2.2 shows a house powered by renewable sources of energy (sun and wind) As the global community continues to seek solutions to the pressing challenges of energy security, climate change, and resource sustainability, PEC solar cells offer a promising pathway to a cleaner, more sustainable energy future.

Fig. 2.2 A house powered by renewable sources of energy (sun and wind) generated by Lexica.art (AI)

2.2 Fundamental Principles of PEC Solar Cells

The operation of Photoelectrochemical (PEC) solar cells is grounded in the interplay between light absorption, charge generation, and electrochemical reactions at the interface of photoelectrodes and electrolytes. Understanding the fundamental principles that govern these processes is essential for the design and optimization of PEC systems. Unlike conventional photovoltaic cells, where the primary goal is to generate electrical power, PEC solar cells are designed to drive specific chemical reactions, such as the production of hydrogen through water splitting. This dual functionality introduces additional layers of complexity in terms of both the materials and mechanisms involved. Figure 2.3 shows a PEC cell uses light to separate water molecules into hydrogen and oxygen. At the heart of PEC solar cells are the photoelectrodes, typically made of semiconducting materials that absorb sunlight and generate charge carriers—electrons and holes. These charge carriers must then be efficiently separated and transported to the electrode surfaces, where they participate in redox reactions with species in the electrolyte. The efficiency of these processes depends on several factors, including the energy band alignment of the photoelectrodes, the kinetics of

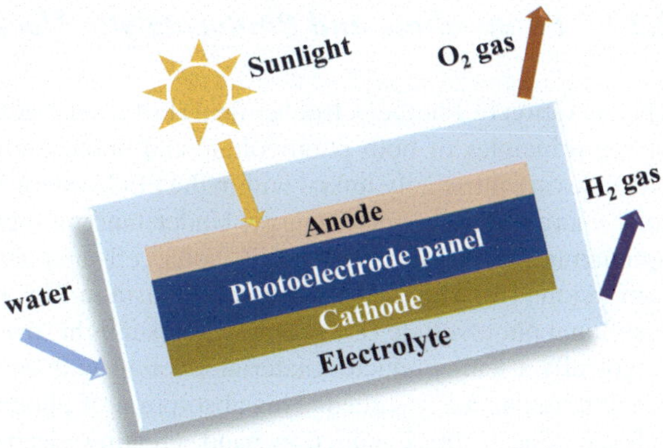

Fig. 2.3 PEC cell uses light to separate water molecules into hydrogen and oxygen

charge transfer at the interfaces, and the minimization of recombination losses, where electrons and holes recombine without contributing to the desired reaction [16].

The fundamental principles governing PEC solar cells also encompass the thermodynamics and kinetics of the electrochemical reactions they are designed to drive. For instance, the water-splitting reaction in PEC cells involves two key half-reactions: the Hydrogen Evolution Reaction (HER) and the Oxygen Evolution Reaction (OER). Each of these reactions has specific energetic and kinetic requirements that must be met to achieve efficient and sustained operation. The ability to tailor the photoelectrode materials and cell design to meet these requirements is a critical aspect of PEC solar cell development [17].

Additionally, the interaction between light and matter plays a crucial role in the performance of PEC solar cells. The absorption of photons by the semiconductor material must be optimized to maximize the generation of charge carriers. This requires careful consideration of the material's bandgap, which determines the portion of the solar spectrum that can be absorbed. Techniques such as bandgap engineering and nanostructuring are often employed to enhance light absorption and improve the efficiency of charge carrier generation and separation. Overall, the operation of PEC solar cells involves a complex interplay of physical and chemical principles. A deep understanding of these fundamentals is necessary to overcome the challenges associated with material selection, cell design, and process optimization, ultimately leading to the development of more efficient and durable PEC systems. The following sections will delve into these fundamental principles in greater detail, exploring the thermodynamics and kinetics of water splitting, the intricacies of energy band alignment and charge carrier dynamics, and the critical electrochemical reactions that define PEC solar cell performance.

2.2.1 *Photovoltaic and Photocatalytic Mechanisms*

The operation of Photoelectrochemical (PEC) solar cells is fundamentally governed by the principles of both photovoltaic and photocatalytic mechanisms. These two processes are intricately linked within the PEC system, working in tandem to convert solar energy into chemical energy. Understanding these mechanisms is crucial for optimizing the performance of PEC solar cells and achieving efficient solar-to-fuel conversion [18]. The photovoltaic mechanism in PEC solar cells is similar to that in traditional photovoltaic (PV) cells. When sunlight strikes the photoelectrode, which is typically a semiconductor material, photons with energy greater than the bandgap of the material are absorbed. This absorption of photons excites electrons from the valence band to the conduction band, creating electron–hole pairs. These charge carriers—electrons in the conduction band and holes in the valence band—are then separated and driven towards opposite electrodes under the influence of an internal electric field, which can be generated by the semiconductor's built-in potential or by an externally applied bias [19]. In a PEC solar cell, the electrons and holes generated through the photovoltaic effect play a central role in driving the photocatalytic reactions at the surface of the photoelectrodes. The photocatalytic mechanism involves the participation of these charge carriers in redox reactions at the interface between the photoelectrode and the electrolyte. For instance, in the water-splitting reaction, the photogenerated electrons are typically directed to the cathode, where they reduce protons (H^+) to produce hydrogen gas (H_2), while the photogenerated holes are directed to the anode, where they oxidize water molecules (H_2O) to produce oxygen gas (O_2) [20].

The efficiency of these photocatalytic reactions depends heavily on the ability of the PEC system to effectively separate and transport the photogenerated charge carriers to the reaction sites without significant recombination losses. Recombination, where electrons and holes recombine before reaching the reaction sites, is a major loss mechanism in PEC systems and can severely limit the overall efficiency of solar-to-fuel conversion. To mitigate this, various strategies are employed, including the use of materials with high charge carrier mobility, the engineering of interfaces to facilitate charge transfer, and the application of surface catalysts that lower the reaction barriers [21]. Another critical aspect of the photocatalytic mechanism is the alignment of energy levels between the semiconductor photoelectrode and the redox potentials of the reactions involved. For efficient photocatalysis, the conduction band edge of the semiconductor must be positioned at a more negative potential than the reduction potential of the reaction (e.g., the reduction of protons to hydrogen), and the valence band edge must be at a more positive potential than the oxidation potential of the reaction (e.g., the oxidation of water to oxygen). This energy band alignment ensures that the photogenerated charge carriers have sufficient energy to drive the desired redox reactions [22]. In addition to energy band alignment, the surface properties of the photoelectrode play a significant role in the photocatalytic mechanism. The presence of surface states, defects, or adsorbed species can influence the charge transfer kinetics and the overall efficiency of the photocatalytic process.

Surface modifications, such as the deposition of co-catalysts or the introduction of passivation layers, are often employed to enhance the reactivity and stability of the photoelectrode surfaces, thereby improving the overall performance of the PEC solar cell.

The integration of photovoltaic and photocatalytic mechanisms within PEC solar cells represents a powerful approach to harnessing solar energy for chemical fuel production. By carefully designing the photoelectrode materials, optimizing the energy band alignment, and engineering the surface properties, it is possible to maximize the efficiency of both the light absorption and the subsequent catalytic reactions. This dual mechanism approach not only allows for the direct conversion of solar energy into chemical fuels like hydrogen but also provides a versatile platform for driving a variety of other solar-driven chemical processes, such as carbon dioxide reduction or nitrogen fixation [23].

2.2.2 *Thermodynamics and Kinetics in PEC Solar Cells*

The efficiency and functionality of Photoelectrochemical (PEC) solar cells are fundamentally rooted in the principles of thermodynamics and kinetics. These principles dictate the feasibility and rate of the chemical reactions that PEC cells are designed to drive, most notably the splitting of water into hydrogen and oxygen. Understanding these principles is crucial for optimizing the design and operation of PEC systems, as they determine the energy requirements and potential efficiency limits of the processes involved. Thermodynamics in PEC solar cells primarily concerns the energy required to drive the water-splitting reaction. This reaction can be broken down into two half-reactions: the Hydrogen Evolution Reaction (HER) and the Oxygen Evolution Reaction (OER). The overall reaction can be expressed as:

$$2H_2O(l) \rightarrow 2H_2(g) + O_2(g)$$

Figure 2.4 shows the oxygen evolution reaction and hydrogen evolution reaction for overall water splitting (under acidic conditions). The Gibbs free energy change (ΔG) for this reaction is approximately + 237.2 kJ/mol under standard conditions, which corresponds to a theoretical voltage of 1.23 V. This is the minimum thermodynamic energy required to split water into hydrogen and oxygen. However, in practical PEC systems, the actual energy required often exceeds this value due to additional losses, including overpotentials and inefficiencies associated with charge transport and recombination. Overcoming these losses is a key challenge in PEC solar cell design [24].

Overpotentials are a critical factor in the kinetics of the HER and OER. Overpotential refers to the additional voltage required beyond the thermodynamic minimum to drive the reaction at a practical rate. The sources of overpotential include activation overpotential, which arises from the intrinsic kinetics of the electron transfer reactions at the electrode surfaces; concentration overpotential, due to the diffusion

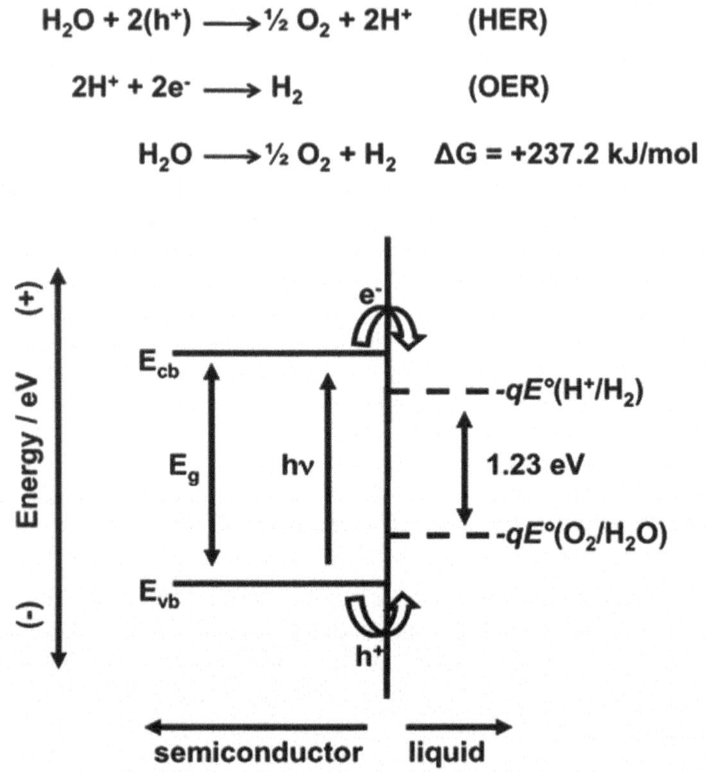

Fig. 2.4 Oxygen evolution reaction and hydrogen evolution reaction for overall water splitting (under acidic conditions) reproduced from Ref. [25] Copyright © 2010 American Chemical Society

limitations of reactants to the electrode surface; and resistance overpotential, which is related to the internal resistance of the cell. Minimizing overpotentials is essential for enhancing the efficiency of PEC cells, as they directly reduce the overall solar-to-hydrogen (STH) conversion efficiency. The kinetics of the HER and OER are influenced by the nature of the electrode materials, their surface properties, and the presence of catalysts. For instance, platinum is widely recognized as an excellent catalyst for the HER due to its low overpotential and high catalytic activity. However, the scarcity and cost of platinum drive the search for alternative, earth-abundant catalysts that can provide similar performance. The OER, being a more complex reaction involving the transfer of four electrons and the formation of an oxygen–oxygen bond, generally exhibits higher overpotentials and slower kinetics. Materials such as transition metal oxides (e.g., RuO_2, IrO_2) are commonly used as OER catalysts, but ongoing research focuses on developing more efficient and stable alternatives [26].

Another important aspect of thermodynamics in PEC solar cells is the energy band alignment between the semiconductor photoelectrode and the redox potential of the electrolyte. For effective water splitting, the conduction band minimum of the photoanode must be more negative than the hydrogen reduction potential (0 V vs. NHE), while the valence band maximum should be more positive than the oxygen oxidation potential (1.23 V vs. NHE). This alignment ensures that the photogenerated electrons and holes have sufficient energy to drive the HER and OER, respectively. In cases where the band alignment is not optimal, external biases or additional materials, such as co-catalysts, are often employed to facilitate the reactions. The interplay between thermodynamics and kinetics also determines the stability of PEC solar cells. Photocorrosion, a common issue in PEC systems, occurs when the photoelectrode materials themselves undergo undesirable oxidation or reduction reactions, leading to degradation. The stability of PEC cells is thus a function of both the thermodynamic stability of the materials under operational conditions and the kinetic barriers to unwanted side reactions. Engineering the surface of photoelectrodes, for instance by applying protective coatings or using passivation layers, can help enhance stability by altering the kinetics of degradation processes [27].

The performance of PEC solar cells is intricately tied to the principles of thermodynamics and kinetics. The energy required to split water, the overpotentials associated with the HER and OER, and the stability of the materials under operational conditions all play pivotal roles in determining the overall efficiency and viability of PEC systems. A deep understanding of these principles is essential for advancing PEC technology, enabling the development of more efficient, stable, and cost-effective solar-driven water-splitting devices. The next section will delve into the critical aspects of energy band alignment and charge carrier dynamics, which further influence the efficiency of PEC solar cells.

2.2.3 Charge Carrier Generation, Separation, and Transport

Charge carrier generation, separation, and transport are fundamental processes that underpin the operation of Photoelectrochemical (PEC) solar cells. These processes determine the efficiency with which solar energy is converted into chemical energy, making them critical to the overall performance of PEC systems. Understanding and optimizing these mechanisms is essential for improving the efficiency of PEC solar cells, particularly in the context of water splitting and other solar-driven chemical reactions. The generation of charge carriers begins when sunlight is absorbed by the semiconductor photoelectrode in the PEC cell. This absorption occurs when photons with energies greater than the semiconductor's bandgap excite electrons from the valence band to the conduction band, creating electron–hole pairs. The efficiency of this process depends on several factors, including the absorption coefficient of the material, the thickness of the photoelectrode, and the spectral match between the incident light and the bandgap of the semiconductor. Materials with high absorption coefficients and bandgaps that are well-aligned with the solar spectrum are typically favored to maximize photon absorption and charge carrier generation.

Once generated, the charge carriers—electrons in the conduction band and holes in the valence band—must be efficiently separated and transported to the reaction sites at the surface of the photoelectrode. The separation of charge carriers is driven by the internal electric field within the semiconductor, which is typically established by the built-in potential at the semiconductor-electrolyte interface or by an externally applied bias. This electric field helps to prevent the recombination of electrons and holes, which is a significant loss mechanism in PEC cells. Recombination can occur both within the bulk of the semiconductor and at the surface, leading to a decrease in the number of charge carriers available for driving the desired electrochemical reactions [28]. To enhance charge carrier separation, various strategies are employed. One common approach is the use of semiconductor materials with high charge carrier mobilities, which facilitate the rapid transport of electrons and holes to the respective electrodes. Additionally, nanostructuring techniques, such as the creation of nanowires, nanotubes, or nanorods, can be used to reduce the distance that charge carriers need to travel before reaching the reaction sites, thereby minimizing recombination losses. These nanostructured architectures provide larger surface areas for the redox reactions and can also improve light absorption by trapping photons within the nanostructures, further enhancing charge carrier generation [29].

The transport of charge carriers to the reaction sites is another critical aspect of PEC solar cell operation. For efficient transport, the semiconductor material must possess high conductivity and minimal defect states that could trap charge carriers and impede their movement. The interface between the semiconductor and the electrolyte also plays a crucial role in charge transport. At this interface, the charge carriers must overcome energy barriers to transfer from the semiconductor to the electrolyte, where they participate in the redox reactions. The presence of surface states or defects at the interface can create additional barriers or recombination centers, further reducing the efficiency of charge carrier transport. Surface modifications are often employed to improve charge carrier transport across the semiconductor-electrolyte interface. For example, the deposition of co-catalysts on the surface of the photoelectrode can lower the activation energy required for the redox reactions, facilitating the transfer of charge carriers from the semiconductor to the electrolyte. Additionally, passivation layers can be applied to the surface to reduce the density of surface states and minimize recombination at the interface. These surface engineering techniques are essential for achieving efficient charge carrier transport and maximizing the overall performance of PEC solar cells [30]. Another important consideration in charge carrier transport is the alignment of energy levels between the semiconductor and the electrolyte. The conduction band edge of the semiconductor must be aligned with the reduction potential of the reaction (such as hydrogen evolution), and the valence band edge must align with the oxidation potential (such as oxygen evolution). Proper energy band alignment ensures that the photogenerated charge carriers have sufficient energy to drive the electrochemical reactions, which is crucial for efficient PEC operation.

2.2.4 Role of Electrolytes and Interfaces

In Photoelectrochemical (PEC) solar cells, the electrolytes and interfaces play pivotal roles in facilitating charge transfer and driving the electrochemical reactions that convert solar energy into chemical fuels. The interaction between the photoelectrode and the electrolyte is crucial for achieving high efficiency in PEC systems. This section explores the role of electrolytes and interfaces, examining their impact on the performance of PEC solar cells and the strategies employed to optimize these components for enhanced efficiency.

Electrolytes serve as the medium through which charge carriers migrate between the photoelectrode and the counter electrode, enabling the redox reactions necessary for solar energy conversion. The choice of electrolyte has significant implications for the overall performance of the PEC cell. Electrolytes can be broadly categorized into liquid, gel, and solid-state types, each with distinct advantages and challenges [31]. Liquid electrolytes are commonly used in PEC systems due to their high ionic conductivity and ease of integration. These electrolytes typically contain aqueous or non-aqueous solutions with dissolved salts or acids, such as sulfuric acid or potassium hydroxide, which provide the necessary ions for the electrochemical reactions. However, liquid electrolytes can pose issues such as corrosion of the photoelectrode and degradation over time. To mitigate these issues, researchers are developing more stable and less corrosive electrolyte formulations [32]. Gel and solid-state electrolytes offer alternatives to liquid electrolytes, addressing some of the challenges associated with liquid systems. Gel electrolytes, which are semi-solid and contain a polymer matrix, combine high ionic conductivity with improved stability and reduced risk of leakage. Solid-state electrolytes, which are entirely solid, provide enhanced stability and durability but often suffer from lower ionic conductivity compared to liquid systems. Innovations in solid-state electrolyte materials and fabrication techniques are continuously being explored to overcome these limitations [33].

The **interface** between the photoelectrode and the electrolyte is another critical component that significantly affects PEC cell performance. This interface is where the charge carriers generated in the photoelectrode must transfer to or from the electrolyte to participate in the redox reactions. The efficiency of this charge transfer process is influenced by several factors, including the nature of the interface, the quality of the contact between the photoelectrode and the electrolyte, and the presence of any intermediate layers or coatings. One key aspect of interface optimization is the reduction of charge transfer resistance. High resistance at the interface can lead to significant losses in efficiency, as it impedes the movement of charge carriers and increases the likelihood of recombination. Techniques such as applying conductive coatings, using co-catalysts, or modifying the surface of the photoelectrode can help reduce this resistance and improve overall performance.

Another important consideration is the **energy band alignment** at the interface. For efficient charge transfer, the energy levels of the semiconductor photoelectrode must be well-matched with the redox potentials of the electrolyte. Proper alignment ensures that the photogenerated charge carriers have sufficient energy to drive the electrochemical reactions, facilitating smooth and efficient charge transfer. Misalignment can create energy barriers that hinder charge movement and reduce the efficiency of the PEC cell.

In addition to optimizing the direct interface between the photoelectrode and electrolyte, researchers are also exploring the use of **interface modifiers** and **passivation layers**. These materials can be applied to improve the stability of the interface, protect the photoelectrode from corrosion, and enhance the overall efficiency of charge transfer. For example, passivation layers can reduce surface states and defects that act as recombination centers, thereby improving the performance of the PEC cell [34].

Co-catalysts are another important component at the interface, particularly for enhancing the kinetics of the electrochemical reactions. Co-catalysts are often deposited onto the surface of the photoelectrode to facilitate the reaction processes by lowering activation energy barriers and improving reaction rates. The choice and deposition method of co-catalysts can significantly impact the performance of the PEC cell, making it a critical area of research and development.

2.3 Materials for PEC Solar Cells

The performance of Photoelectrochemical (PEC) solar cells is profoundly influenced by the materials used in their construction. The selection and optimization of these materials are critical for achieving high efficiency and stability in solar energy conversion processes. PEC solar cells require a careful balance of several material properties, including light absorption, charge carrier generation, separation, and transport, as well as catalytic activity. This section delves into the various materials employed in PEC solar cells, exploring their roles, characteristics, and the advancements made in enhancing their performance. The effectiveness of a PEC solar cell depends on the synergy between its photoelectrode materials, electrolytes, and co-catalysts. Photoelectrode materials are particularly crucial, as they directly absorb sunlight and initiate the charge generation process. These materials must possess a suitable bandgap to efficiently capture solar energy and facilitate the necessary redox reactions. In addition to photoelectrodes, the choice of semiconductor materials and their modification can significantly impact the overall efficiency of the PEC cell. This section will explore the range of materials used, including metal oxides, sulfides, nitrides, and emerging materials such as perovskites and 2D materials, highlighting their respective advantages and challenges. Understanding these materials' properties and their interactions within the PEC system is essential for advancing the field and developing more efficient solar-to-fuel conversion technologies.

2.3.1 *Photoelectrode Materials*

Photoelectrode materials are central to the functioning of Photoelectrochemical (PEC) solar cells, as they directly absorb sunlight and drive the photoelectrochemical reactions required for solar-to-fuel conversion. The choice of photoelectrode material profoundly impacts the efficiency, stability, and overall performance of PEC systems. This section explores the various types of photoelectrode materials, their properties, and the advancements in their design and optimization. Semiconductor Materials form the backbone of photoelectrodes in PEC solar cells. These materials must have a suitable bandgap to absorb a significant portion of the solar spectrum and convert it into usable energy. Among the most widely studied semiconductor materials are metal oxides, sulfides, and nitrides, each offering unique advantages and challenges [35].

2.3.1.1 Metal Oxides

Metal oxides have long been integral to the development of photoelectrochemical (PEC) solar cells due to their stability, versatility, and well-understood properties. As photoelectrode materials, they play a crucial role in absorbing solar energy and facilitating the necessary electrochemical reactions. This section delves into the characteristics, advantages, and challenges associated with metal oxides used in PEC systems, focusing on key examples such as titanium dioxide (TiO_2), zinc oxide (ZnO), and iron oxide (Fe_2O_3).

Titanium Dioxide (TiO_2) is one of the most widely utilized metal oxides in PEC applications, owing to its exceptional chemical stability, low cost, and non-toxic nature. TiO_2 exhibits a wide bandgap (~ 3.2 eV), which limits its absorption to the ultraviolet (UV) portion of the solar spectrum. To overcome this limitation, significant research has focused on enhancing TiO_2's performance through various strategies. Doping with metals or non-metals, creating composite materials, and designing heterojunctions with other semiconductors are effective methods to extend its light absorption range into the visible spectrum. For instance, nitrogen or carbon doping has been employed to narrow the bandgap and improve its light absorption capabilities, making TiO_2 a more versatile material for PEC cells [36].

Zinc Oxide (ZnO) is another prominent metal oxide used in PEC solar cells, known for its suitable bandgap (~ 3.3 eV) and high electron mobility. ZnO's wide bandgap allows it to absorb UV light, but similar to TiO_2, it benefits from modifications to enhance its visible light absorption. ZnO can be easily synthesized and fabricated into various nanostructures, such as nanorods and nanowires, which improve its surface area and charge separation efficiency. Additionally, ZnO can form heterojunctions with other semiconductors to create more efficient photoelectrodes. However, issues such as photocorrosion and poor stability under long-term operational conditions remain challenges that need to be addressed [37].

Iron Oxide (Fe_2O_3), also known as hematite, is notable for its abundance, low cost, and suitable bandgap (~ 2.2 eV) that allows for visible light absorption. Fe_2O_3 has attracted considerable attention due to its potential for achieving high solar-to-hydrogen efficiency. Despite its advantages, Fe_2O_3 suffers from limitations such as low charge carrier mobility and short-lived photoactivity, which can impede its overall performance. Researchers have focused on overcoming these challenges by optimizing Fe_2O_3's surface properties and creating composite materials that enhance charge separation and transport. Strategies such as co-catalyst deposition and creating heterojunctions with other semiconductors have been employed to improve its performance [38].

2.3.1.2 Sulfides

Sulfide-based materials represent a significant class of photoelectrodes for photoelectrochemical (PEC) solar cells, distinguished by their ability to absorb visible light and their promising performance in solar-to-fuel conversion processes. This section explores the key sulfide materials used in PEC applications, focusing on their properties, advantages, and the challenges associated with their use. Key examples include cadmium sulfide (CdS), copper sulfide (Cu_2S), and their various modifications.

Cadmium Sulfide (CdS) is one of the most studied sulfide materials for PEC solar cells due to its suitable bandgap (~ 2.4 eV), which allows it to absorb a significant portion of the visible spectrum. CdS exhibits high photochemical activity, making it effective in driving the photoelectrochemical reactions required for hydrogen evolution. However, its use is limited by environmental and health concerns related to cadmium toxicity. To address these issues, researchers are investigating alternative materials with similar properties but reduced environmental impact. Innovations in material design, such as the development of cadmium-free sulfide analogs, are ongoing to mitigate these concerns while retaining high performance [39].

Copper Sulfide (Cu_2S) is another sulfide material that has garnered attention for its advantageous electronic properties and relatively narrow bandgap (~ 1.2 eV), which allows it to absorb a broad spectrum of visible light. Cu_2S has demonstrated good performance in PEC applications, but its practical use is hindered by stability issues under prolonged exposure to the photoelectrochemical environment. To enhance its stability and performance, researchers have explored various strategies, including the creation of composite materials, the application of protective coatings, and the development of mixed-phase systems. These approaches aim to improve the material's durability while maintaining its high photoelectrochemical activity [40].

Zinc Sulfide (ZnS) is another notable sulfide material used in photoelectrochemical (PEC) solar cells due to its suitable bandgap (~ 3.6 eV), which allows it to absorb a portion of the visible light spectrum. ZnS exhibits good photochemical stability and can be synthesized in various nanostructured forms, such as nanoparticles, nanowires, and nanosheets, which enhance its surface area and improve charge separation and

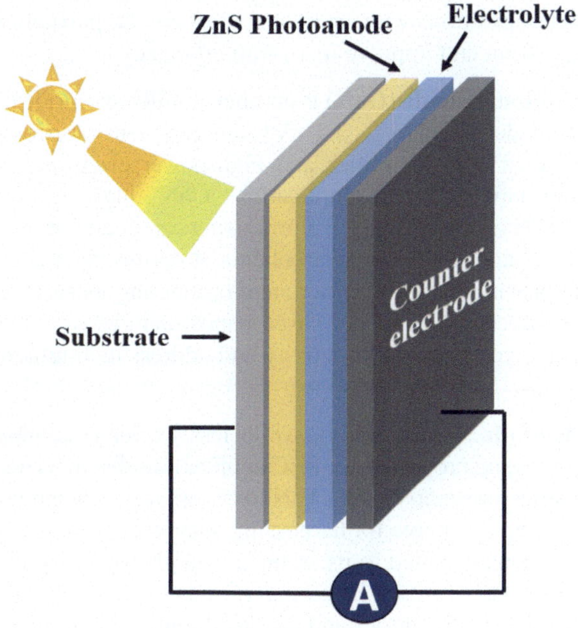

Fig. 2.5 Schematic of a ZnS based PEC solar cell

transport. Figure 2.5 shows the schematic of a ZnS based PEC solar cell. Despite its advantages, ZnS suffers from limitations related to its bandgap, which restricts its absorption to the UV and blue regions of the spectrum. Research efforts are focused on modifying ZnS through doping, alloying, or creating heterojunctions with other materials to extend its light absorption range and enhance its photoelectrochemical performance [41].

2.3.1.3 Nitrides

Nitrides are an emerging class of materials with notable potential for use in photoelectrochemical (PEC) solar cells due to their unique electronic properties and stability. This section explores the characteristics, advantages, and challenges associated with nitride-based photoelectrodes, focusing on key examples such as titanium nitride (TiN), gallium nitride (GaN), and bismuth nitride (BiN).

Titanium Nitride (TiN) is renowned for its high electrical conductivity, chemical stability, and resistance to corrosion, making it an attractive candidate for photoelectrochemical applications. TiN exhibits a metallic conductivity that can be advantageous for facilitating efficient charge transfer. However, its wide bandgap (~ 2.9 eV) limits its light absorption primarily to the UV region. To enhance its performance in PEC cells, TiN is often used in combination with other materials or as a component in composite structures. Strategies such as doping with other elements or creating heterojunctions with semiconductors can help extend its absorption into the visible spectrum and improve its overall efficiency [42].

Gallium Nitride (GaN) is another significant nitride material, particularly noted for its wide bandgap (~ 3.4 eV) and excellent electronic properties. GaN is widely used in optoelectronics and high-power electronic devices due to its robustness and stability. In PEC applications, GaN's high electron mobility and stability under harsh conditions make it a promising candidate for photoelectrode materials. Efforts to enhance its photoelectrochemical performance focus on modifying its bandgap, improving light absorption, and optimizing its surface properties. Techniques such as the creation of GaN-based nanostructures or the integration of co-catalysts are employed to boost its activity and address its limitations related to light absorption [43].

Bismuth Nitride (BiN), while less explored compared to TiN and GaN, offers promising properties for PEC applications due to its narrow bandgap (~ 1.4 eV). This narrow bandgap enables BiN to absorb a significant portion of the visible spectrum, making it suitable for harnessing solar energy more efficiently. However, BiN faces challenges related to its stability and photocorrosion under operational conditions. Researchers are investigating various methods to enhance BiN's stability, such as surface modifications and the development of composite materials that can improve its performance and longevity in PEC systems [44].

2.3.1.4 Emerging Materials

The field of photoelectrochemical (PEC) solar cells is continuously evolving, with emerging materials offering new possibilities for enhancing performance and efficiency. These innovative materials often provide unique properties that can address the limitations of traditional photoelectrodes, such as bandgap limitations, stability issues, and inefficiencies in charge separation and transport. This section explores several promising emerging materials, including perovskites, two-dimensional (2D) materials, and hybrid structures, focusing on their properties, potential applications, and the challenges they face.

Perovskite Materials have gained significant attention in recent years due to their remarkable optoelectronic properties, including tunable bandgaps, high light absorption coefficients, and excellent charge transport characteristics. The general formula for perovskites is ABX_3, where A and B are cations and X is an anion. In the context of PEC solar cells, lead halide perovskites, such as methylammonium lead iodide ($MAPbI_3$), are particularly notable for their high efficiency and relatively simple fabrication processes. The ability to tune the bandgap by varying the composition of the perovskite allows for optimization of light absorption across the solar spectrum. However, challenges such as stability under prolonged light exposure and environmental degradation remain significant. Ongoing research aims to improve the stability

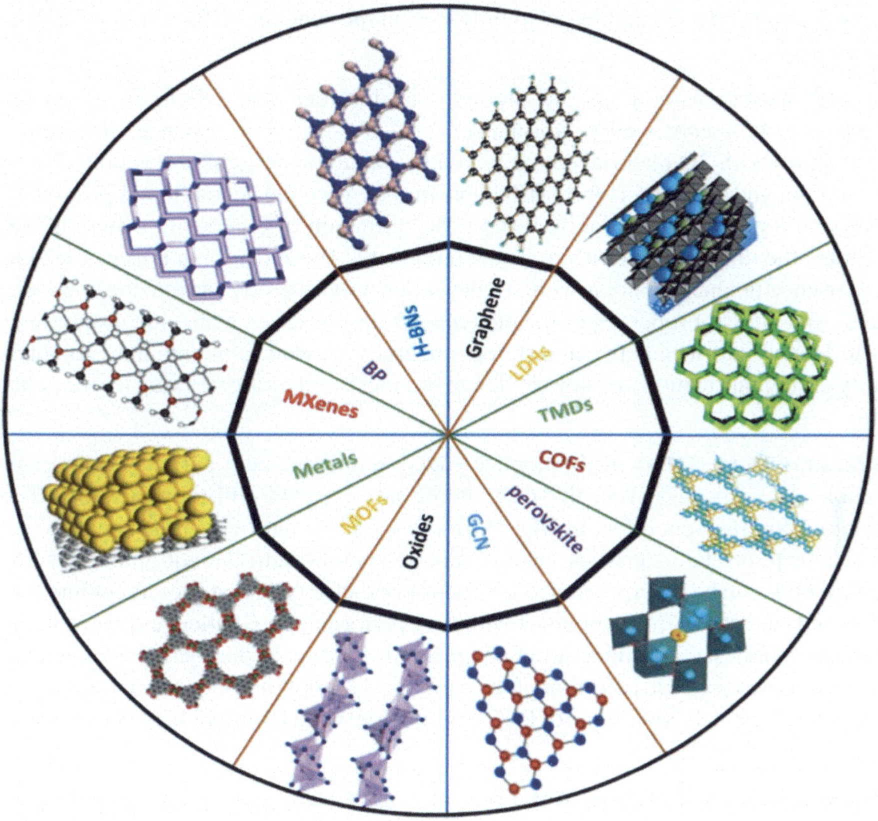

Fig. 2.6 Classification and structure of various 2D materials reproduced from Ref. [47] Copyright © The Royal Society of Chemistry 2023

of perovskite materials by developing new compositions and encapsulation methods, as well as exploring lead-free alternatives to address environmental concerns [45].

Two-Dimensional (2D) Materials, such as graphene and transition metal dichalcogenides (TMDs), represent another class of emerging materials with exceptional properties for PEC applications. Graphene, with its high electrical conductivity and large surface area, facilitates efficient charge transport and can be used to enhance the performance of other photoelectrode materials. Transition metal dichalcogenides (TMDs), including materials like molybdenum disulfide (MoS_2) and tungsten disulfide (WS_2), exhibit unique electronic and optical properties due to their layered structure and tunable bandgaps. Figure 2.6. depicts the classification and structure of various 2D materials. The ability to engineer these materials at the monolayer scale enables high absorption of visible light and effective charge separation. Despite their potential, challenges such as material stability, scalability, and integration with other components need to be addressed to fully exploit the benefits of 2D materials in PEC systems [46].

Hybrid Structures combine the strengths of different materials to create photoelectrodes with enhanced performance characteristics. For instance, integrating perovskites with 2D materials or metal oxides can improve light absorption, charge separation, and stability. Hybrid structures may involve the deposition of perovskite layers on top of graphene or TMDs, or the formation of composite materials that leverage the unique properties of each component. These combinations can lead to improved efficiency in solar-to-fuel conversion processes by optimizing the electronic and optical properties of the photoelectrode. However, achieving a stable and well-defined interface between different materials, as well as maintaining compatibility during fabrication, are key challenges that researchers are working to overcome [48].

Quantum Dots (QDs) are nanometer-sized semiconductor particles with size-tunable electronic properties that offer unique advantages for PEC applications. The quantum confinement effect in QDs allows for precise control over the bandgap and light absorption characteristics, enabling tailored photoelectrodes with enhanced efficiency. QDs can be incorporated into various photoelectrode matrices, including thin films and nanostructured supports, to improve performance. Challenges related to the stability, synthesis, and integration of QDs into functional photoelectrodes remain areas of active research. Efforts are focused on developing robust QD systems that can withstand the harsh conditions of PEC environments and maintain high performance over extended periods [49].

Organic–Inorganic Hybrid Materials combine organic semiconductors with inorganic components to achieve beneficial properties for PEC applications. These materials can offer high light absorption, tunable electronic properties, and flexibility in device fabrication. For example, organic–inorganic halide perovskites have demonstrated impressive efficiency in solar cells, and similar approaches are being explored for PEC systems. The integration of organic materials with inorganic semiconductors can enhance the overall performance by optimizing light absorption and charge transport. However, ensuring the long-term stability and reproducibility of these hybrid materials poses significant challenges that need to be addressed [50].

In conclusion, emerging materials such as perovskites, 2D materials, hybrid structures, quantum dots, and organic–inorganic hybrids hold significant promise for advancing photoelectrochemical solar cells. These materials offer unique properties that can enhance efficiency, stability, and performance in PEC applications. Ongoing research and development efforts aim to address the challenges associated with these materials, including stability, scalability, and integration, to fully realize their potential in future PEC systems.

2.4 Electrolytes in PEC Solar Cells

In photoelectrochemical (PEC) solar cells as depicted in Fig. 2.7, electrolytes play a critical role in facilitating the essential redox reactions at the photoelectrode interfaces. The choice and design of the electrolyte affect the overall efficiency, stability, and performance of the PEC system. This section delves into the various types of electrolytes used in PEC applications, their properties, and the challenges associated with their use.

There are a variety of electrolyte as illustrated in Fig. 2.8. **Liquid Electrolytes** are the most commonly employed type in PEC systems. These electrolytes typically consist of an aqueous or organic solvent containing dissolved salts or acids that facilitate ion conduction. **Aqueous electrolytes**, such as sulfuric acid (H_2SO_4) and sodium sulfate (Na_2SO_4), are popular due to their high ionic conductivity and relatively low cost. They are particularly effective in facilitating the Hydrogen Evolution Reaction (HER) and Oxygen Evolution Reaction (OER). However, aqueous electrolytes often pose challenges related to stability and compatibility with certain photoelectrode materials, especially those sensitive to acidic or basic conditions. **Organic electrolytes**, on the other hand, offer better stability and compatibility with a wider range of photoelectrode materials. Common organic solvents include acetonitrile and propylene carbonate, which can be combined with salts like lithium hexafluorophosphate ($LiPF_6$) to form ionic liquid electrolytes. These electrolytes are particularly

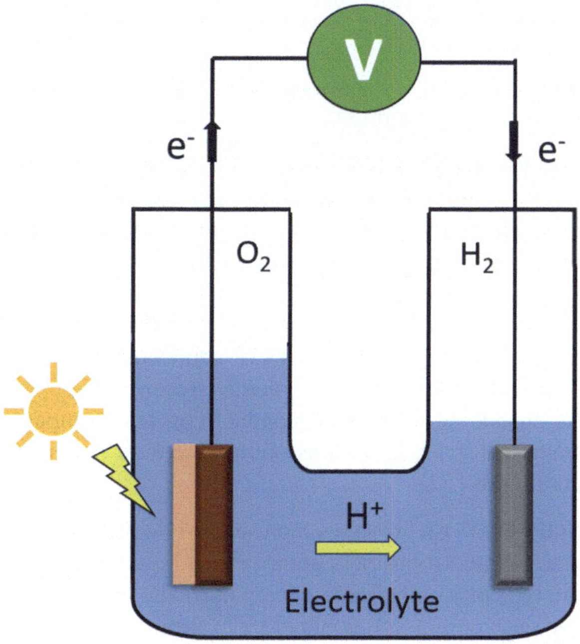

Fig. 2.7 Electrolyte in a PEC solar cell facilitating the transport of charge carriers

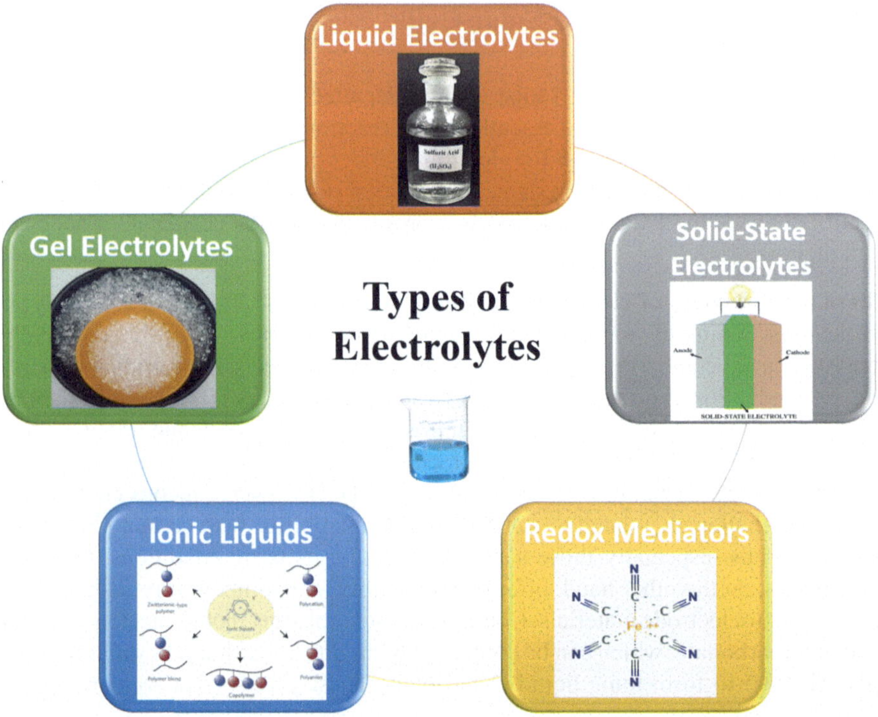

Fig. 2.8 Types of electrolytes in PEC solar cells

useful for PEC systems operating in non-aqueous environments or when higher stability is required [51].

Gel Electrolytes have emerged as a promising alternative to liquid electrolytes due to their improved stability and ease of handling. These electrolytes consist of a gel matrix—often made from polymers such as poly(vinyl alcohol) (PVA) or poly(ethylene glycol) (PEG)—that holds an ionic liquid or aqueous solution. Gel electrolytes combine the high ionic conductivity of liquid electrolytes with the mechanical stability and ease of handling associated with solid materials. They are particularly advantageous in applications where leakage or evaporation of the electrolyte could be problematic. Recent developments in gel electrolytes focus on enhancing their ionic conductivity and compatibility with various photoelectrode materials, as well as optimizing their performance under different operating conditions [52].

Solid-State Electrolytes represent another advanced approach, offering even greater stability and robustness compared to liquid and gel electrolytes. These electrolytes are typically composed of inorganic salts or polymers with high ionic conductivity, such as sodium–ion conductors or lithium–ion conductors. Solid-state electrolytes eliminate the risk of leakage and evaporation, making them ideal for long-term and high-performance PEC applications. However, achieving high ionic conductivity and maintaining good contact with the photoelectrode surfaces are key challenges. Innovations in material science, such as the development of composite solid electrolytes or advanced synthesis techniques, are helping to address these challenges and improve the performance of solid-state electrolytes [53].

Ionic Liquids are a class of electrolytes with unique properties that offer several advantages for PEC systems. These liquid salts, which remain liquid at room temperature or below, exhibit high ionic conductivity and stability. They can be tailored to have specific electrochemical properties and are compatible with a wide range of photoelectrode materials. Ionic liquids also have low volatility and high thermal stability, making them suitable for demanding PEC conditions. Ongoing research focuses on optimizing the ionic conductivity and electrochemical stability of ionic liquids, as well as exploring their interactions with different photoelectrode materials to enhance overall PEC performance [54].

Redox Mediators are additives used in electrolytes to facilitate the transfer of electrons between the photoelectrode and the electrolyte. These mediators can significantly improve the efficiency of PEC systems by enhancing the kinetics of the redox reactions and reducing overpotentials. Common redox mediators include ferrocyanide/ferrocyanide and iodide/iodine couples, which are known for their fast electron transfer rates and compatibility with various photoelectrode materials. Research into new redox mediators aims to identify compounds with even higher efficiencies and stability, as well as to understand their interactions with different electrolyte systems [55].

The choice and design of electrolytes are crucial for the performance of photoelectrochemical solar cells. Liquid, gel, solid-state, and ionic liquid electrolytes each offer unique advantages and challenges. Advances in electrolyte materials and formulations are driving improvements in PEC efficiency, stability, and overall system performance, with ongoing research focused on addressing current limitations and exploring new possibilities for future applications.

2.5 Co-catalysts in PEC Solar Cells

Co-catalysts are crucial components in photoelectrochemical (PEC) solar cells, enhancing the efficiency of photocatalytic reactions by facilitating charge transfer and reducing reaction overpotentials. Their primary role is to improve the kinetics of the Hydrogen Evolution Reaction (HER) and Oxygen Evolution Reaction (OER), which are essential for effective solar energy conversion. This section explores the different types of co-catalysts, their functions, and the challenges associated with their integration into PEC systems.

Noble Metal Co-catalysts such as platinum (Pt), ruthenium (Ru), and iridium (Ir) are widely used in PEC systems due to their high catalytic activity and stability.

These metals are particularly effective for HER, where they significantly lower the overpotential required for hydrogen production. For OER, iridium oxide (IrO_2) and ruthenium oxide (RuO_2) are commonly employed due to their excellent stability and catalytic performance. The high cost and scarcity of noble metals, however, pose significant challenges. Research efforts are focused on developing more cost-effective alternatives or reducing the amount of noble metals required through novel deposition techniques or alloying with less expensive metals [56].

Transition Metal-Based Co-catalysts offer a promising alternative to noble metals due to their abundance and lower cost. Transition metals such as nickel (Ni), cobalt (Co), and iron (Fe) have shown significant potential in enhancing HER and OER activity. These metals are often used in their oxide or sulfide forms, which can exhibit high catalytic activity and stability. For instance, cobalt-based catalysts such as cobalt phosphide (CoP) and cobalt oxide (Co_3O_4) are known for their high activity in HER. Similarly, iron-based catalysts like iron oxide (Fe_2O_3) are effective for OER. The challenge with transition metal-based co-catalysts lies in optimizing their activity and stability to match or exceed that of noble metal counterparts [57].

Metal–Organic Frameworks (MOFs) and **Coordination Polymers (CPs)** are a class of hybrid materials that combine metal centers with organic ligands to create porous structures with high surface areas. These materials offer tunable catalytic properties and can serve as effective co-catalysts in PEC systems. MOFs and CPs can be tailored to enhance charge transfer and improve reaction kinetics, making them suitable for both HER and OER. Research in this area focuses on developing MOFs and CPs with high stability, optimized electronic properties, and compatibility with various photoelectrode materials [58].

Carbon-Based Co-catalysts such as graphene, carbon nanotubes (CNTs), and carbon dots (CDs) have gained attention for their high electrical conductivity, large surface area, and ability to enhance charge transfer processes. These materials can serve as support matrices for other co-catalysts or act as primary catalysts in their own right. For example, graphene and CNTs can improve the electron transfer rate in PEC cells, while carbon dots can enhance the light absorption and catalytic activity. The challenge with carbon-based co-catalysts is optimizing their integration with other materials to achieve synergistic effects without compromising stability or performance [59].

Layered Double Hydroxides (LDHs) are another category of co-catalysts known for their high catalytic activity and stability. LDHs are layered materials composed of metal hydroxides with intercalated anions, which can be tailored to enhance their catalytic properties. These materials can be used as co-catalysts in both HER and OER by modifying their composition and structure to optimize performance. Research is ongoing to improve the synthesis methods for LDHs and to better understand their mechanisms in PEC reactions [60].

The co-catalysts play a vital role in enhancing the efficiency of photoelectrochemical solar cells by improving the kinetics of key reactions. Noble metals, transition metal-based catalysts, MOFs, carbon-based materials, and LDHs each offer unique advantages and challenges. Ongoing research aims to address these challenges and develop new co-catalysts that can further advance PEC technology and contribute to more efficient and cost-effective solar energy conversion.

2.5.1 Bandgap Engineering and Light Absorption

Bandgap engineering is a critical strategy in optimizing the performance of photoelectrochemical (PEC) solar cells. By manipulating the electronic band structure of photoelectrode materials, researchers can enhance light absorption, improve charge carrier dynamics, and increase overall device efficiency. This section delves into the principles of bandgap engineering, its impact on light absorption, and the methods used to tailor the bandgap of photoelectrode materials for optimal PEC performance.

Fundamentals of Bandgap Engineering involve adjusting the energy band structure of materials to absorb a broader spectrum of sunlight. The bandgap of a material, defined as the energy difference between the valence band and the conduction band, determines the range of photon energies that can be absorbed and converted into electrical energy. By modifying the bandgap, either through compositional changes or structural alterations, it is possible to enhance the material's ability to absorb sunlight and improve its photocatalytic activity. Bandgap tuning can be achieved through various approaches, including alloying, doping, and quantum confinement effects [61].

Alloying is one of the most common methods for bandgap engineering. By combining two or more materials with different bandgaps, it is possible to create an alloy with a tunable bandgap that can be optimized for specific light absorption needs. For example, alloying different semiconductors like titanium dioxide (TiO_2) with nitrogen or sulfur can reduce the bandgap, extending the light absorption into the visible region. This method allows for the fine-tuning of the bandgap to match the solar spectrum and improve the efficiency of PEC cells [62].

Doping involves introducing small amounts of specific elements into the photoelectrode material to alter its electronic properties. This process can modify the bandgap, enhance charge carrier concentration, and improve the material's photocatalytic performance. For instance, doping titanium dioxide with metals like platinum or non-metals like nitrogen can shift its bandgap into the visible light region, making it more effective for sunlight absorption. Doping can also help in reducing recombination rates and improving overall charge transport within the material [63].

Quantum Confinement Effects occur in nanomaterials where the dimensions of the material are reduced to the nanoscale, leading to discrete energy levels and altered electronic properties. This effect allows for precise control over the bandgap by adjusting the size and shape of the nanostructures. Quantum dots, for example, can be engineered to have size-dependent bandgaps, enabling them to absorb specific wavelengths of light more efficiently. This method is particularly useful for designing photoelectrodes that can harness a wider spectrum of solar radiation [64].

Bandgap Engineering for Enhanced Light Absorption also involves optimizing the absorption properties of photoelectrodes through structural modifications. Techniques such as **nanostructuring** and **layering** can increase the effective surface area and improve light trapping. For instance, creating nanostructured photoelectrodes with high surface area or using multilayered structures can enhance light absorption and increase the probability of photon capture. Surface plasmon resonance (SPR) effects in metallic nanoparticles, when incorporated into the photoelectrode, can also boost light absorption by enhancing the local electric field [65].

Advanced Materials and Composites play a significant role in bandgap engineering. Emerging materials such as perovskites, two-dimensional (2D) materials, and hybrid organic–inorganic semiconductors offer new opportunities for optimizing bandgap and light absorption. Perovskite materials, for instance, exhibit tunable bandgaps and high absorption coefficients, making them ideal for PEC applications. Similarly, 2D materials like graphene and transition metal dichalcogenides (TMDs) provide unique electronic properties that can be harnessed to enhance light absorption and charge transport [66]. The bandgap engineering is a vital aspect of enhancing the performance of photoelectrochemical solar cells. By modifying the bandgap through alloying, doping, quantum confinement, and structural optimization, researchers can improve light absorption and overall efficiency. Advances in material science and fabrication techniques are driving progress in this field, offering promising avenues for developing more effective and efficient PEC solar cells.

2.5.2 Nanostructured Materials for Enhanced Performance

Nanostructured materials play a pivotal role in advancing the performance of photoelectrochemical (PEC) solar cells. By manipulating materials at the nanoscale, researchers can exploit unique physical and chemical properties that significantly enhance light absorption, charge separation, and overall device efficiency. This section explores the principles of nanostructuring, its impact on PEC performance, and various strategies for designing and utilizing nanostructured materials in PEC systems.

Principles of Nanostructuring involve engineering materials at the nanometer scale to achieve specific electronic, optical, and chemical properties. At the nanoscale, materials exhibit quantum effects, increased surface area, and altered interaction with light, which can lead to improved performance in PEC applications. Nanostructuring can enhance light absorption through increased surface area and light trapping effects, improve charge carrier dynamics by facilitating efficient transport and separation, and boost catalytic activity by providing more active sites for reactions [67].

Nanostructured Photoelectrodes can significantly improve the performance of PEC cells by increasing the surface area available for light absorption and reaction. One common approach is to create nanowires, nanotubes, and nanorods, which provide high surface area-to-volume ratios and enable effective charge separation and transport. For example, titanium dioxide (TiO_2) nanowires have been shown to enhance charge carrier mobility and reduce recombination losses, leading to improved photocatalytic performance. Similarly, zinc oxide (ZnO) nanorods and nanowires offer high surface area and efficient charge transport, making them effective in PEC applications [68].

Quantum Dots are semiconductor nanocrystals with size-dependent electronic properties. By tuning the size of quantum dots, researchers can tailor their bandgap to match specific wavelengths of light, improving absorption and utilization of the solar spectrum. Quantum dots can be incorporated into photoelectrodes to enhance light absorption and charge carrier dynamics. For instance, integrating lead sulfide (PbS) quantum dots into a photoelectrode can extend the absorption range into the near-infrared region, which is often underutilized in conventional photoelectrodes [69].

Nanostructured Composites combine different nanomaterials to leverage their synergistic effects and improve PEC performance. By creating composites of metal nanoparticles, semiconductors, and carbon-based materials, researchers can enhance light absorption, charge transfer, and catalytic activity. For example, integrating gold (Au) nanoparticles with titanium dioxide (TiO_2) can enhance light absorption through surface plasmon resonance and improve photocatalytic activity. Similarly, combining graphene with semiconductor nanomaterials can facilitate efficient charge transfer and reduce recombination losses [70].

2.5.3 Surface Modifications and Catalytic Enhancements

Surface modifications and catalytic enhancements are crucial for optimizing the performance of photoelectrochemical (PEC) solar cells. By altering the surface properties of photoelectrodes and introducing catalytic materials, researchers can significantly improve light absorption, charge transfer, and reaction kinetics. Figure 2.9 shows various techniques of surface modifications of PEC solar cells. This section discusses various surface modification techniques and their impact on enhancing the catalytic activity of photoelectrodes in PEC systems.

Surface Modifications involve altering the surface chemistry or structure of photoelectrode materials to improve their performance in PEC cells. These modifications can enhance the material's interaction with light, optimize charge transfer processes, and increase the number of active sites available for catalytic reactions. Common surface modification techniques include coating, functionalization, and doping. **Coating** is a widely used technique for improving the performance of

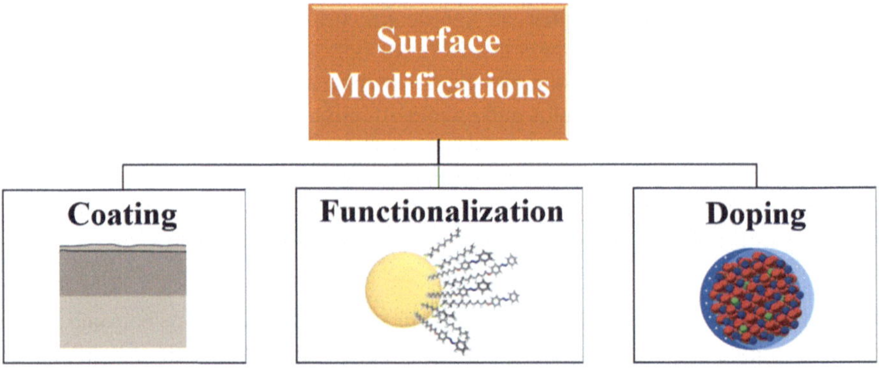

Fig. 2.9 Techniques of surface modifications of PEC solar cells

photoelectrodes. By applying thin layers of additional materials onto the surface of the photoelectrode, researchers can enhance light absorption, improve stability, and increase catalytic activity. For instance, coating a titanium dioxide (TiO_2) photoelectrode with a layer of platinum can enhance its catalytic activity for the hydrogen evolution reaction (HER). Similarly, coating with conductive polymers or carbon-based materials can improve charge transport and stability. **Functionalization** involves attaching specific chemical groups or molecules to the surface of the photoelectrode to enhance its interaction with reactants or improve its stability. Functionalization can introduce additional active sites for catalysis, modify the electronic properties of the surface, or improve the material's compatibility with other components in the PEC cell. For example, functionalizing a photoelectrode with hydroxyl or carboxyl groups can improve its interaction with electrolytes and enhance catalytic performance. **Doping** is another effective surface modification technique that involves introducing small amounts of different elements into the photoelectrode material to alter its electronic properties. Doping can enhance charge carrier concentration, improve charge transport, and increase catalytic activity. For example, doping titanium dioxide with transition metals like iron or copper can shift its bandgap to the visible light region and improve its photocatalytic performance. Similarly, doping with non-metals like nitrogen or sulfur can enhance the material's stability and activity.

Catalytic Enhancements focus on improving the efficiency of the key reactions occurring at the photoelectrode surface. Catalysts play a crucial role in accelerating the hydrogen evolution reaction (HER) and the oxygen evolution reaction (OER), which are essential for the overall efficiency of PEC cells. Various strategies are employed to enhance the catalytic activity of photoelectrodes, including the use of co-catalysts, the optimization of surface morphology, and the introduction of catalytic promoters. **Surface Morphology Optimization** involves designing the surface structure of the photoelectrode to maximize the exposure of active sites and improve the efficiency of catalytic reactions. Nanostructuring techniques, such as creating nanoparticle arrays or hierarchical structures, can increase the surface area and enhance light absorption. Optimizing the surface morphology can also improve the distribution of co-catalysts and enhance their effectiveness in promoting catalytic reactions. **Catalytic Promoters** are substances that are used to improve the efficiency of catalysts by altering their electronic or chemical properties. Promoters can enhance the interaction between the catalyst and the reactants, increase the stability of the catalyst, or modify the reaction pathways to improve efficiency. For example, adding a small amount of a promoter like cobalt or nickel can enhance the activity of a platinum-based catalyst for HER.

The surface modifications and catalytic enhancements are essential for optimizing the performance of photoelectrochemical solar cells. By employing various techniques such as coating, functionalization, and doping, and utilizing co-catalysts and catalytic promoters, researchers can significantly improve the efficiency of PEC cells. Ongoing research and innovation in this field are crucial for advancing PEC technology and achieving higher performance in practical applications [71].

2.6 Design and Optimization of PEC Cell Architectures

The design and optimization of photoelectrochemical (PEC) cell architectures are crucial for achieving high efficiency and performance in solar energy conversion systems. The architecture of a PEC cell influences various factors, including light absorption, charge separation, and reaction kinetics, all of which are essential for optimizing the overall performance of the cell. As such, careful consideration of the cell design is necessary to enhance its functionality and efficiency. The development of PEC cell architectures involves several key aspects, including the selection of appropriate cell configurations, integration of photoelectrodes with co-catalysts, and optimization of electrolyte compositions. Each of these factors plays a significant role in determining the performance of the PEC cell. For instance, the choice between single-junction and tandem cell configurations can impact the cell's ability to absorb and utilize the solar spectrum effectively. Similarly, the integration of photoelectrodes with co-catalysts and the optimization of electrolyte composition can influence the efficiency of charge transfer and catalytic reactions.

Effective light management and photon absorption are also critical components of PEC cell design. The ability to harness and utilize solar energy efficiently requires innovative strategies for managing light within the cell and maximizing photon absorption. This involves designing photoelectrodes and cell structures that enhance light trapping, reduce reflection losses, and improve the overall light-harvesting efficiency. The interplay between these factors requires a comprehensive approach to cell design and optimization. Researchers and engineers must consider how each aspect of the cell architecture interacts with others to achieve the best possible performance. Advances in materials science, nanotechnology, and engineering are driving innovations in PEC cell design, leading to more efficient and effective solar energy conversion systems. In the following subsections, we will delve into specific aspects of PEC cell architecture, including the comparison of single-junction and tandem cells, the role of co-catalysts, the impact of electrolyte composition, and strategies for optimizing light management. Each of these areas plays a vital role in enhancing the performance of PEC cells and advancing the field of solar energy conversion [72].

2.6.1 Single Junction Versus Tandem PEC Solar Cells

The architecture of photoelectrochemical (PEC) cells significantly impacts their efficiency and effectiveness in solar energy conversion. Two prominent configurations in PEC cell design are single-junction and tandem cells. Each configuration has unique advantages and challenges, influencing its suitability for different applications and performance goals. Design considerations for both single-junction and tandem PEC cells include optimizing the choice of photoelectrode materials, ensuring effective charge separation and transport, and minimizing losses due to reflection and absorption. In single-junction cells, selecting materials with appropriate bandgaps and high absorption coefficients is crucial. For tandem cells, the design must ensure that each layer contributes to the overall efficiency without introducing significant losses. Figure 2.10 depicts the efficiency records for perovskite solar cells and monolithic perovskite/Si tandem solar cells compared with other photovoltaic technologies. The choice between single-junction and tandem PEC cells depends on various factors, including efficiency goals, fabrication complexity, and cost considerations. Single-junction cells offer simplicity and lower costs but are limited by their absorption range. Tandem cells provide higher potential efficiencies by utilizing a broader spectrum of sunlight but require more complex design and manufacturing processes. Both configurations represent important approaches in the ongoing development and optimization of PEC technology, with continued advancements in materials and engineering driving improvements in both types of cells [73].

2.6.2 Single Junction PEC Solar Cells

Single-junction photoelectrochemical (PEC) solar cells represent a fundamental and widely studied configuration in PEC technology. These cells consist of a single layer of photoelectrode material that is directly exposed to light and participates in the photoelectrochemical reactions required for solar energy conversion. This section delves into the principles, advantages, challenges, and recent advancements associated with single-junction PEC solar cells. The core principle of a single-junction PEC cell is relatively straightforward: a single photoelectrode layer absorbs sunlight and generates charge carriers that drive the photoelectrochemical reactions. Typically, the photoelectrode material is selected based on its ability to absorb a significant portion of the solar spectrum and to facilitate efficient charge separation and transfer. When

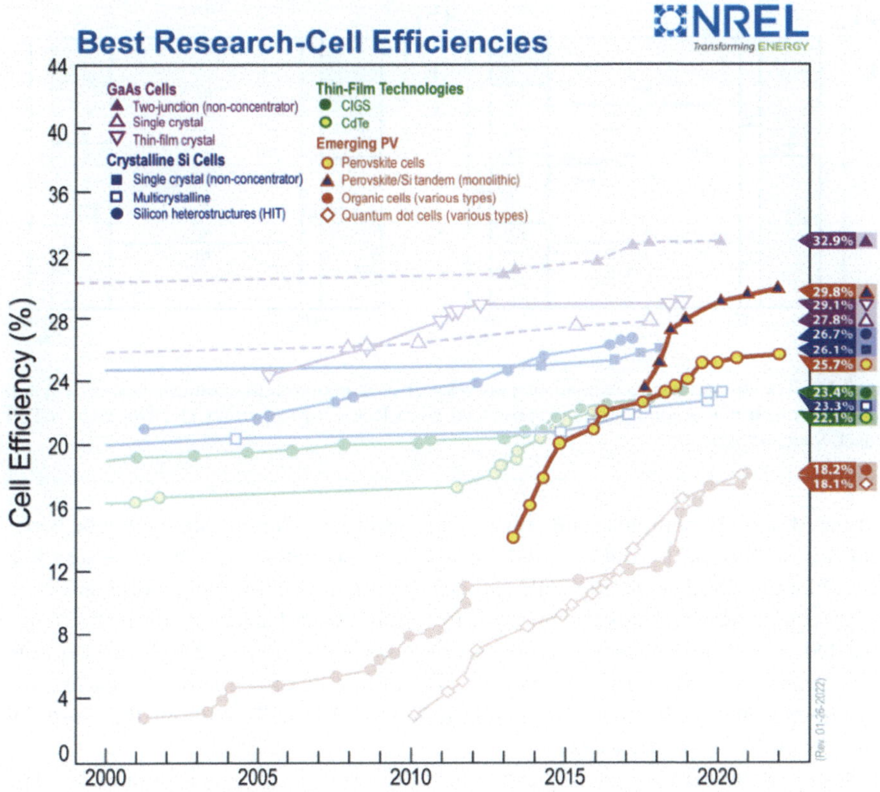

Fig. 2.10 The efficiency records for perovskite solar cells and monolithic perovskite/Si tandem solar cells compared with other photovoltaic technologies reproduced from Ref. [74] copyright Alliance for Sustainable Energy LLC

sunlight strikes the photoelectrode, it excites electrons from the valence band to the conduction band, creating electron–hole pairs. These charge carriers then migrate through the material and participate in the electrochemical reactions occurring at the electrode surface. Figure 2.11 depicts the operating mechanism of perovskite-based photoelectrochemical device (a) n–i–p configuration, (b) p–i–n configuration.

Single-junction PEC cells are characterized by their simplicity in design and construction. They generally consist of a photoelectrode layer deposited onto a conductive substrate, which is often coupled with a counter electrode and an electrolyte. The photoelectrode material must exhibit appropriate bandgap properties to maximize light absorption and ensure effective charge separation. Common materials used in single-junction PEC cells include semiconductors such as titanium dioxide (TiO_2), zinc oxide (ZnO), and various metal chalcogenides. One of the primary advantages of single-junction PEC cells is their relatively simple fabrication process. The straightforward design allows for easier integration and lower production costs

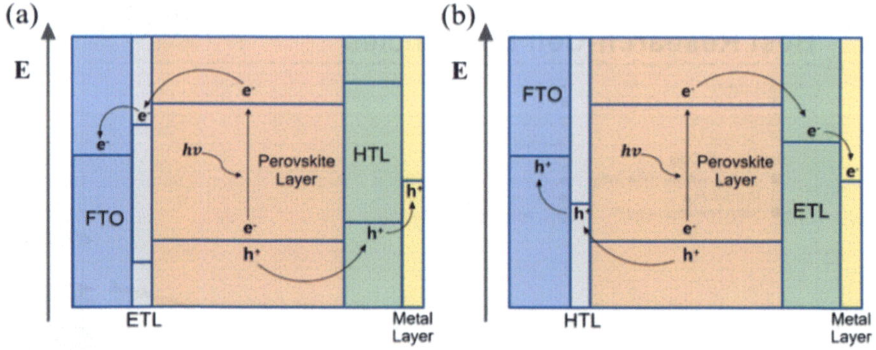

Fig. 2.11 The operating mechanism of perovskite-based photoelectrochemical device **a** n–i–p configuration, **b** p–i–n configuration reproduced from Ref. [76] Copyright © 1996–2024 MDPI (Basel, Switzerland)

compared to more complex multi-layer configurations. This simplicity can be beneficial for scaling up production and for applications where cost-effectiveness is a critical factor. Single-junction cells can also be optimized through careful selection of photoelectrode materials and processing conditions. Advances in material science have led to the development of photoelectrodes with improved light absorption, charge transport, and catalytic properties. By optimizing these materials, researchers can enhance the performance of single-junction PEC cells without the need for complex multi-layer structures.

Despite their advantages, single-junction PEC cells face several limitations. The primary challenge is their inherent efficiency constraint due to the limited spectral range that a single photoelectrode can effectively absorb. The photoelectrode material must be carefully chosen to match the solar spectrum, but even with optimized materials, single-junction cells are constrained by the need to balance light absorption, charge separation, and catalytic activity within a single layer. Another challenge is managing the reflection and transmission losses that occur at the interface between the photoelectrode and the surrounding environment. The efficiency of single-junction PEC cells can be limited by these losses, which reduce the amount of light absorbed and thus the overall energy conversion efficiency.

Recent research has focused on improving the performance of single-junction PEC cells through various strategies. Advances in material science have led to the development of new photoelectrode materials with better light absorption properties and enhanced stability. For example, researchers have explored novel semiconductors, such as perovskite materials and layered transition metal dichalcogenides (TMDs), which offer improved optical and electronic properties compared to traditional photoelectrode materials.

Innovations in surface modification techniques have also contributed to enhanced performance. By applying coatings or functionalizing the surface of the photoelectrode, researchers can improve light absorption, reduce reflection losses, and enhance charge transfer. Additionally, advances in electrode processing and fabrication methods have led to more efficient and reproducible single-junction PEC cells [75].

2.6.3 Tandem PEC Solar Cells

Tandem photoelectrochemical (PEC) solar cells represent an advanced and promising approach to enhancing the efficiency of solar energy conversion. Unlike single-junction PEC cells, which utilize a single layer of photoelectrode material, tandem PEC cells incorporate multiple layers stacked on top of one another. Each layer is designed to absorb different segments of the solar spectrum, thereby improving the overall efficiency of the cell. Figure 2.12 shows the structure of a tandem solar cell. This section explores the principles, advantages, challenges, and recent developments associated with tandem PEC solar cells.

Tandem PEC cells operate on the principle of combining multiple photoelectrode layers, each optimized to absorb specific wavelength ranges of sunlight. By stacking these layers, tandem cells can utilize a broader portion of the solar spectrum compared to single-junction cells. This configuration allows for more efficient

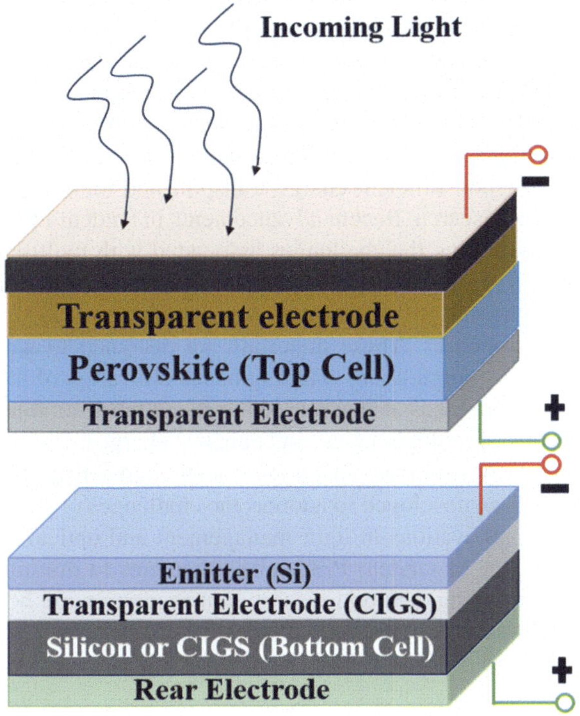

Fig. 2.12 Structure of a tandem solar cell

light absorption and charge generation. Typically, tandem PEC cells consist of two or more photoelectrode materials with different bandgap energies, arranged in a sequence that maximizes the utilization of incoming solar radiation. The structure of a tandem PEC cell usually includes an initial photoelectrode layer that absorbs high-energy photons, followed by additional layers designed to absorb lower-energy photons. This arrangement enables the cell to capture a wider range of solar energy and convert it into usable chemical energy more effectively. Each layer is carefully engineered to ensure proper light transmission and charge transfer between layers, which is crucial for maintaining high efficiency and minimizing losses.

One of the primary advantages of tandem PEC cells is their enhanced efficiency due to the ability to absorb a broader spectrum of sunlight. By stacking multiple photoelectrode layers with complementary absorption properties, tandem cells can achieve higher overall conversion efficiencies than single-junction cells. This increased efficiency arises from the optimized utilization of the solar spectrum and improved charge carrier dynamics. Tandem PEC cells also offer the potential for achieving higher solar-to-hydrogen (STH) conversion efficiencies, which is essential for applications requiring high-performance solar energy systems. The ability to capture and convert more solar energy into chemical energy makes tandem cells particularly attractive for large-scale and high-efficiency solar applications.

Despite their potential advantages, tandem PEC cells face several challenges, primarily related to their complexity and fabrication. The process of fabricating and integrating multiple layers requires precise control and advanced manufacturing techniques. Ensuring proper alignment and bonding of the layers is critical for maintaining efficiency and preventing performance losses. Additionally, managing charge transfer and recombination between different layers can be challenging. Effective interlayer charge transfer is essential to avoid energy losses and ensure that each layer contributes to the overall efficiency of the cell. Developing materials and interfaces that support efficient charge transport and minimize recombination is an ongoing area of research. Recent advancements in tandem PEC cell technology have focused on addressing the challenges associated with multi-layer structures and improving overall performance. Innovations in material science have led to the development of new photoelectrode materials with optimized bandgaps and enhanced light absorption properties. These materials are designed to work synergistically within the tandem configuration, enabling better utilization of the solar spectrum.

Research has also explored advanced fabrication techniques to improve the integration of multiple layers and enhance charge transfer. Techniques such as layer-by-layer deposition, advanced interfacial engineering, and improved bonding methods have been developed to address the challenges of tandem cell construction. Furthermore, innovations in light management and optical design have contributed to the efficiency of tandem PEC cells. Strategies to minimize reflection losses, enhance light trapping, and optimize photon absorption are being explored to maximize the performance of tandem configurations [77].

2.7 Integration of Photoelectrodes with Co-catalysts

The integration of photoelectrodes with co-catalysts is a critical aspect of enhancing the performance of photoelectrochemical (PEC) solar cells. Co-catalysts play a pivotal role in improving the efficiency and effectiveness of PEC systems by facilitating key reactions and optimizing charge transfer processes. This section explores the principles, benefits, challenges, and recent advancements in the integration of photoelectrodes with co-catalysts.

In PEC solar cells, photoelectrodes are responsible for absorbing sunlight and generating charge carriers, which then drive the photoelectrochemical reactions. Co-catalysts are supplementary materials that are typically deposited onto the surface of the photoelectrode or incorporated into the photoelectrode material itself. Their primary role is to enhance the efficiency of the key electrochemical reactions, namely the hydrogen evolution reaction (HER) and the oxygen evolution reaction (OER). Co-catalysts work by providing active sites that facilitate the splitting of water into hydrogen and oxygen. They can significantly lower the overpotentials required for these reactions, thus improving the overall efficiency of the PEC cell. Additionally, co-catalysts can help to mitigate issues related to the slow kinetics of the electrochemical reactions and improve the stability and durability of the photoelectrode materials.

Integrating co-catalysts into PEC systems offers several benefits. One of the main advantages is the enhancement of reaction kinetics. Co-catalysts can lower the activation energy required for the HER and OER, thereby increasing the reaction rates and overall efficiency of the PEC cell. This is particularly important for achieving high solar-to-hydrogen conversion efficiencies. Another benefit is the improved stability of the photoelectrode materials. Co-catalysts can act as protective layers that prevent corrosion and degradation of the photoelectrode surface. This is crucial for maintaining the long-term performance of the PEC cells, especially under harsh operating conditions. Moreover, co-catalysts can also improve the charge transfer efficiency between the photoelectrode and the electrolyte. By providing additional active sites and enhancing the interaction between the photoelectrode and the electrolyte, co-catalysts can reduce charge recombination losses and enhance the overall performance of the PEC cell.

Despite their benefits, integrating co-catalysts into PEC systems presents several challenges. One of the main challenges is the selection and optimization of co-catalyst materials. The co-catalysts must be compatible with the photoelectrode material and the electrolyte, and their properties must be carefully tuned to achieve optimal performance. Another challenge is ensuring uniform deposition and effective interaction between the co-catalyst and the photoelectrode. Inhomogeneous deposition of co-catalysts can lead to uneven reaction rates and reduced overall efficiency. Additionally, the integration process must ensure that the co-catalyst does not introduce additional barriers to charge transfer or interfere with the photoelectrode's light absorption properties.

Recent research has focused on addressing these challenges and improving the integration of co-catalysts in PEC systems. Advances in material science have led to the development of new co-catalyst materials with enhanced catalytic activity and better compatibility with various photoelectrode materials. For example, noble metal nanoparticles such as platinum and gold have been extensively studied for their high catalytic activity in HER and OER. Innovative deposition techniques, such as atomic layer deposition (ALD) and chemical vapor deposition (CVD), have been employed to achieve uniform and controlled coating of co-catalysts on photoelectrode surfaces. These techniques help to ensure optimal interaction between the co-catalyst and the photoelectrode, leading to improved performance.

Furthermore, research has explored the use of hybrid and composite co-catalysts that combine different materials to leverage their synergistic effects. For instance, combining metal nanoparticles with metal oxide supports can enhance the overall catalytic activity and stability. The integration of photoelectrodes with co-catalysts is a crucial aspect of optimizing the performance of PEC solar cells. Co-catalysts enhance reaction kinetics, improve stability, and increase charge transfer efficiency, contributing to higher solar-to-hydrogen conversion efficiencies. While there are challenges related to material selection, deposition, and interaction, ongoing advancements in co-catalyst materials and integration techniques are driving significant improvements in PEC cell performance. By addressing these challenges and leveraging recent innovations, researchers are making strides towards more efficient and durable PEC solar cells [78].

2.7.1 Influence of Electrolyte Composition and pH on Cell Efficiency

The composition and pH of the electrolyte in photoelectrochemical (PEC) solar cells are crucial factors that significantly impact the system's performance. These parameters influence the electrochemical reactions, charge transfer efficiency, and overall stability of the PEC cell. Understanding their effects is essential for optimizing PEC cell design and achieving high solar-to-hydrogen conversion efficiencies.

2.7.1.1 Impact of Electrolyte Composition

The composition of the electrolyte affects various aspects of PEC cell operation, including ion transport, reaction kinetics, and photoelectrode stability. The electrolyte facilitates the transport of ions between the photoelectrode and the counter electrode, which is critical for maintaining charge neutrality and enabling the electrochemical reactions necessary for water splitting. Electrolytes with different compositions can alter the ion transport properties and the interaction between the electrolyte and the photoelectrode. For instance, electrolytes with higher ionic conductivity generally

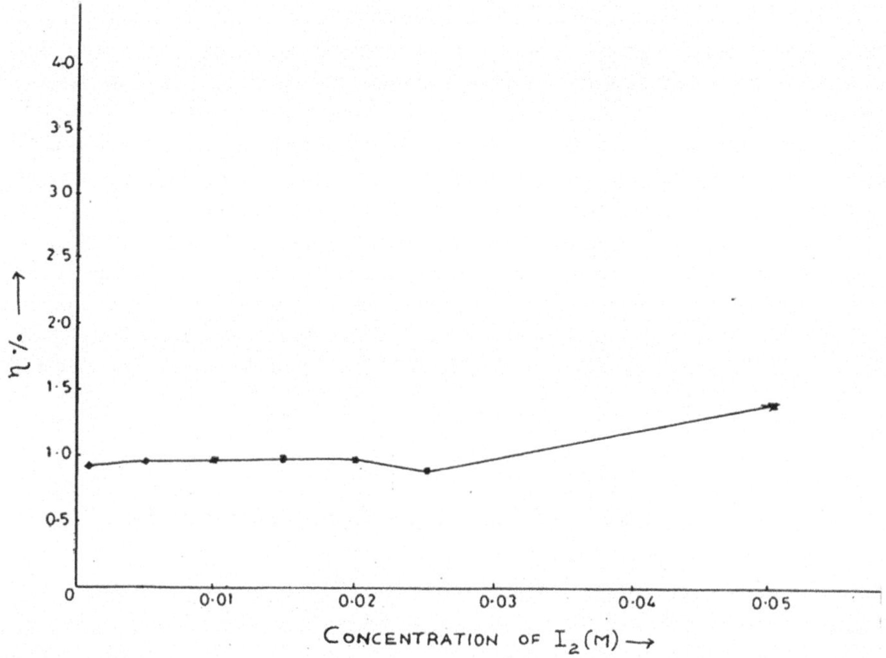

Fig. 2.13 Plot of efficiency versus electrolyte concentrations I_2(M) for PEC solar cells based on $MoTe_2$ reproduced from Ref. [80] Copyright © Copyright 2011–2023 IJSRP

support better charge transport, which can enhance the overall efficiency of the PEC cell. Conversely, electrolytes with lower ionic conductivity might impede ion transport and result in higher internal resistances, reducing cell performance. Figure 2.13. Shows efficiency versus electrolyte concentrations I_2(M) plot for PEC solar cells based on $MoTe_2$. The interaction between the electrolyte and the photoelectrode materials is also influenced by the electrolyte composition. Certain electrolytes may promote or inhibit the formation of intermediates involved in the electrochemical reactions, impacting the efficiency of hydrogen and oxygen production. Additionally, some electrolytes can react with the photoelectrode materials, leading to corrosion or degradation that affects the cell's longevity and stability [79].

2.7.1.2 Influence of pH

The pH of the electrolyte solution plays a critical role in determining the efficiency of PEC cells by affecting the electrochemical reactions and the stability of the photoelectrode materials. The pH influences the surface chemistry of the photoelectrodes and the reaction pathways for the hydrogen evolution reaction (HER) and the oxygen evolution reaction (OER). In acidic conditions, the availability of protons can enhance the HER, making it easier for photoelectrodes to produce hydrogen. However, many photoelectrode materials may suffer from corrosion or instability in acidic environments, which can compromise their performance and durability. Therefore, while acidic conditions can facilitate certain reactions, they may also lead to material degradation.

In contrast, alkaline conditions can promote the OER and enhance the performance of photoelectrodes that are stable in basic environments. Alkaline electrolytes can facilitate the formation of hydroxide ions, which is beneficial for oxygen production. However, alkaline conditions might also lead to the formation of undesired side products or cause stability issues for some photoelectrode materials [81]. Figure 2.14a, b shows the photocurrent density of nanostructured N–ZnO at pH 9–14 in aqueous NaOH and KOH electrolytes. Figure 2.14c, f shows the open-circuit voltage and Fig. 2.14d, e shows the dark current density of N–ZnO at pH 9–14 in aqueous NaOH and KOH electrolytes.

2.7.2 Combined Effects of Electrolyte Composition and pH

The combined effects of electrolyte composition and pH are complex and have a significant impact on PEC cell performance. Optimal electrolyte composition and pH depend on the specific photoelectrode materials used and their compatibility with the electrochemical environment. Adjusting these parameters can help balance the reaction kinetics, enhance charge transfer, and improve overall efficiency. Stability and Corrosion: The stability of photoelectrode materials is a major concern when selecting electrolyte composition and pH. Electrolytes that are not compatible with the photoelectrode material can lead to corrosion and degradation, affecting the cell's performance and lifespan. Careful selection and optimization of these parameters are necessary to ensure long-term stability and reliability of the PEC cell. Efficiency Optimization: Optimizing electrolyte composition and pH can lead to significant improvements in efficiency by enhancing reaction kinetics and reducing energy losses. Fine-tuning these factors helps achieve higher solar-to-hydrogen conversion efficiencies and better overall performance.

The influence of electrolyte composition and pH on PEC cell efficiency is profound and multifaceted. These parameters affect reaction kinetics, material stability, and overall performance. By carefully selecting and optimizing the electrolyte composition and pH, researchers and engineers can enhance the efficiency and durability of PEC solar cells, advancing the development of effective and sustainable solar energy conversion technologies [83].

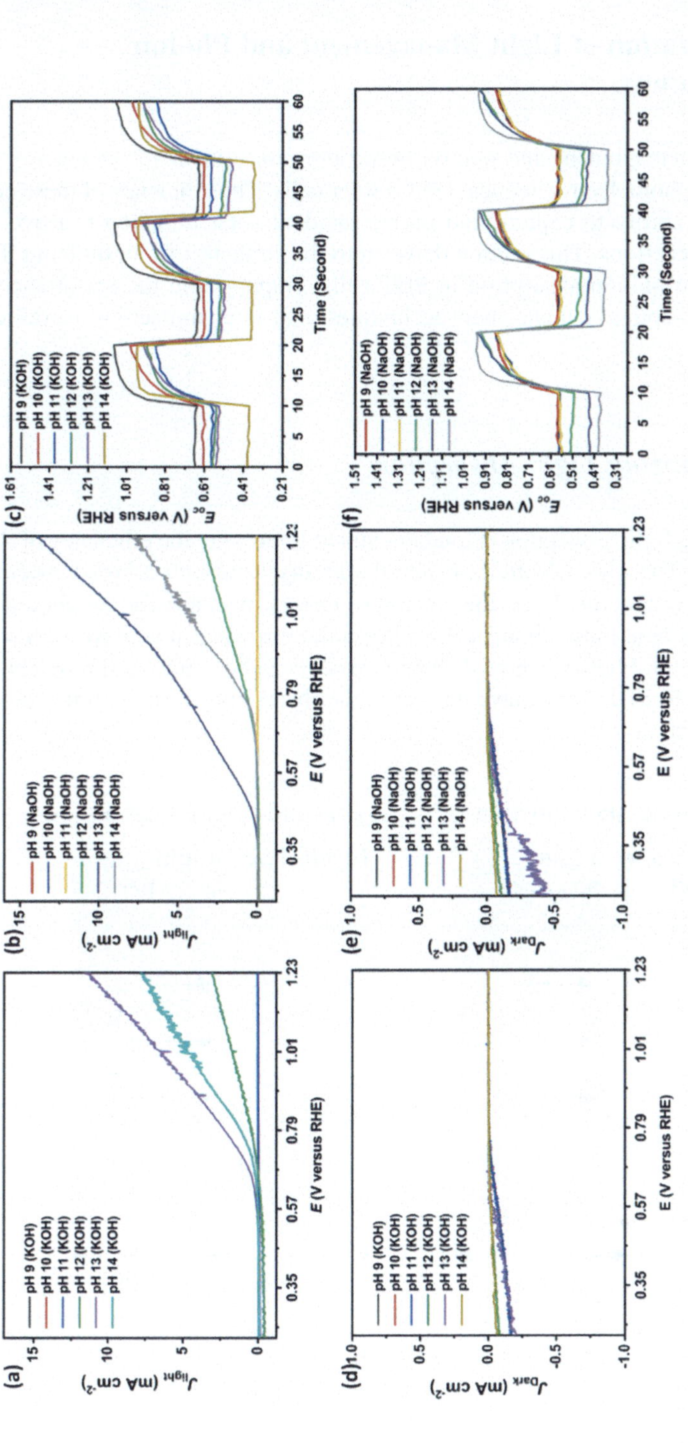

Fig. 2.14 **a**, **b** The photocurrent density of nanostructured N–ZnO at pH 9–14 in aqueous NaOH and KOH electrolytes. **c**, **f** The open-circuit voltage. **d**, **e** The dark current density of N–ZnO at pH 9–14 in aqueous NaOH and KOH electrolytes reproduced from Ref. [82] Copyright © The Royal Society of Chemistry 2023

2.8 Optimization of Light Management and Photon Absorption

Effective light management and photon absorption are critical for enhancing the performance of photoelectrochemical (PEC) solar cells. The efficiency of these cells hinges on their ability to capture and utilize incident solar radiation to drive the water-splitting reactions. This section delves into the strategies for optimizing light management and photon absorption in PEC cells, emphasizing the importance of light harvesting, optical enhancement techniques, and their impact on overall cell performance.

2.8.1 *Maximizing Light Absorption*

(a) Maximizing light absorption is fundamental to improving the efficiency of PEC solar cells. The photoelectrode material's ability to absorb a broad spectrum of solar light directly affects the amount of energy available for the photoelectrochemical reactions. Figure 2.15. Shows the electrical junctions on a solid semiconductor. Photo-generated charge carriers can be extracted through either (a) SSJ or (b) SLJ. To achieve optimal light absorption, several strategies can be employed:

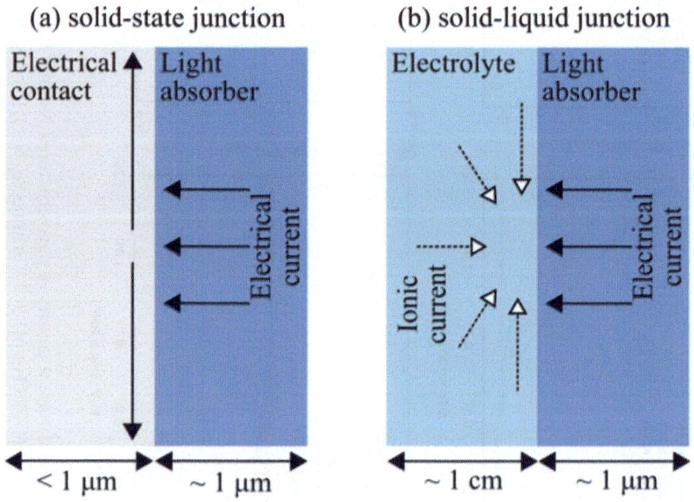

Fig. 2.15 Electrical junctions on a solid semiconductor. Photo-generated charge carriers can be extracted through either **a** SSJ or **b** SLJ reproduced from Ref. [85] Copyright © 2024 Springer Nature

(b) **Bandgap Engineering**: One of the primary methods for enhancing light absorption is bandgap engineering. By tuning the bandgap of the photoelectrode material, it is possible to align its absorption spectrum with the solar spectrum, thereby increasing the amount of absorbed light. Materials with bandgaps that cover a wide range of the solar spectrum can absorb more photons and, consequently, generate more charge carriers.

(c) **Light Trapping Structures**: Incorporating light trapping structures can significantly improve photon absorption by increasing the path length of light within the photoelectrode. Nanostructures, such as photonic crystals, gratings, and textured surfaces, can scatter and trap light, thereby enhancing the effective optical path length and improving the overall light absorption efficiency. These structures can be integrated into the photoelectrode surface to maximize the utilization of incident light.

(d) **Anti-reflective Coatings**: Anti-reflective coatings (ARCs) are applied to the surface of the photoelectrode to minimize light reflection and maximize light absorption. ARCs work by reducing the refractive index mismatch between the photoelectrode and the surrounding medium, thereby minimizing reflective losses. Designing ARCs with optimal thickness and refractive index can significantly enhance light absorption and improve PEC cell efficiency.

(e) **Light Concentration Techniques**: Utilizing light concentration techniques, such as lenses and mirrors, can increase the intensity of incident light on the photoelectrode. Concentrated solar energy leads to higher photon flux and improved absorption efficiency. However, this approach requires careful design to ensure that the concentrated light does not lead to overheating or damage to the photoelectrode material [84].

2.8.2 Photonic Enhancement

Photonic enhancement involves optimizing the interaction of light with the photoelectrode material to improve photon absorption and overall efficiency. Several advanced photonic techniques are employed to achieve this:

(a) **Surface Plasmon Resonance**: Surface plasmon resonance (SPR) occurs when conductive nanostructures, such as metal nanoparticles, interact with incident light to generate localized electromagnetic fields. These fields can enhance the absorption of photons in the photoelectrode material through localized surface plasmon resonance (LSPR). By integrating metal nanoparticles with SPR properties, the effective light absorption of the photoelectrode can be significantly improved.

(b) **Light Scattering and Diffusion**: Light scattering and diffusion techniques can increase the effective interaction time between light and the photoelectrode material. By employing scattering layers or textured surfaces, light can be dispersed within the photoelectrode, increasing the likelihood of photon absorption. This approach helps to overcome limitations associated with direct absorption and improves the overall light management within the PEC cell.

(c) **Optical Filters**: Optical filters can be used to selectively transmit specific wavelengths of light that are most effective for driving the photoelectrochemical reactions. By employing filters that enhance the transmission of wavelengths corresponding to the absorption peaks of the photoelectrode material, it is possible to optimize the light spectrum utilized by the PEC cell and improve its efficiency.

Optimizing light management and photon absorption directly impacts the performance of PEC solar cells. Enhanced light absorption leads to increased photon flux, which results in a higher generation of charge carriers. This, in turn, improves the efficiency of the water-splitting reactions and overall solar-to-hydrogen conversion efficiency. Additionally, effective light management techniques can help minimize energy losses and ensure that the PEC cell operates at its maximum potential.

The optimization of light management and photon absorption is crucial for enhancing the efficiency of PEC solar cells. By employing strategies such as bandgap engineering, light trapping structures, anti-reflective coatings, and photonic enhancement techniques, it is possible to maximize light absorption and improve overall cell performance. Effective light management ensures that a higher proportion of incident solar energy is utilized for the photoelectrochemical reactions, leading to more efficient and effective solar energy conversion [86].

2.9 Performance Metrics and Efficiency Challenges

Evaluating the performance of PEC solar cells involves a thorough understanding of various metrics that define their efficiency and effectiveness. These metrics provide insights into how well the cells convert solar energy into chemical energy and highlight areas where improvements can be made. Performance metrics such as solar-to-electricity conversion efficiency (STEC) are central to assessing the practical viability of PEC cells.

Challenges related to PEC cell efficiency are multifaceted and include issues such as recombination losses, overpotentials, and stability concerns. Understanding and addressing these challenges is essential for optimizing the performance of PEC cells and ensuring their long-term viability. Recombination losses, for instance, can significantly reduce the number of charge carriers available for the desired electrochemical reactions, while overpotentials can increase the energy required to drive these reactions.

Additionally, the scalability of PEC technology and its integration with existing photovoltaic systems are crucial considerations for advancing this technology from the laboratory to real-world applications. Scalability involves not only the ability to produce PEC cells at a larger scale but also their compatibility with current energy systems and infrastructures. Effective integration with photovoltaic systems can enhance overall energy efficiency and facilitate the widespread adoption of PEC technology. This section provides a comprehensive examination of these performance metrics and challenges, setting the stage for a deeper exploration of specific issues and solutions in the subsections that follow [87].

2.9.1 Solar-to-Electricity Conversion Efficiency (STEC)

The Solar-to-Electricity Conversion Efficiency (STEC) is a fundamental metric for evaluating the performance of photoelectrochemical (PEC) solar cells. STEC measures the effectiveness with which a PEC cell converts incident solar energy into electrical energy, providing a direct indication of the cell's capability to perform its intended function in energy conversion applications.

STEC is defined as the ratio of the electrical power output of the PEC cell to the incident solar power, expressed as a percentage. Mathematically, it is represented as:

$$\mathrm{STEC} = \mathrm{P_{out}}/\mathrm{P_{in}} \times 100\%$$

where P_{out} is the electrical power output of the cell, and P_{in} is the incident solar power. This ratio quantifies how efficiently the cell can transform the energy from sunlight into usable electrical power. Figure 2.16 depicts the connection of efficiency with performance. A solar module of total cell area 2 m^2 produces a voltage of 45 V and a current of 5 A at the peak power.

The STEC of PEC cells is crucial for determining their practical viability and competitiveness compared to other energy conversion technologies. High STEC

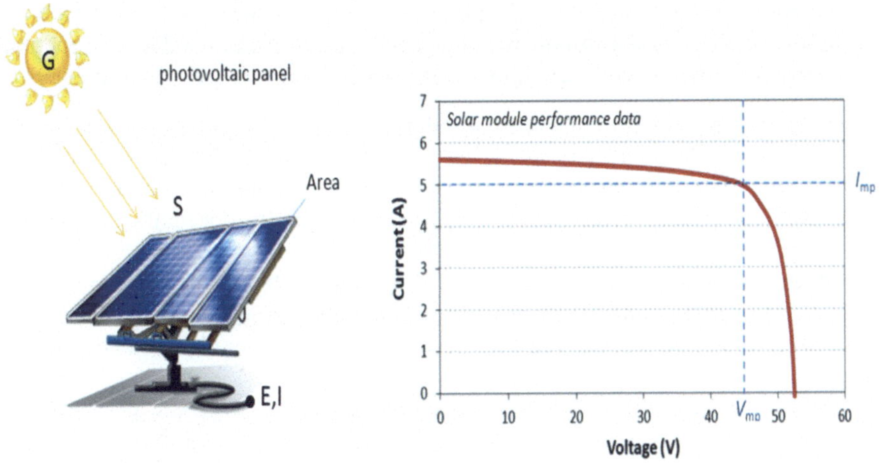

Fig. 2.16 The connection of efficiency with performance. A solar module of total cell area 2 m^2 produces a voltage of 45 V and a current of 5 A at the peak power reproduced from Ref. [88] Copyright The Pennsylvania State University © 2023

values indicate that the PEC cell is effectively harnessing solar energy and converting it into electrical energy, making it a more attractive option for renewable energy applications. Conversely, low STEC values highlight inefficiencies in the cell's performance, signaling the need for further optimization and improvement.

Several factors can influence the STEC of PEC cells, including:

(a) **Photoelectrode Material Properties**: The choice of photoelectrode material significantly affects STEC. Materials with optimal bandgaps for solar absorption, high charge carrier mobility, and robust stability tend to exhibit higher STEC values. Advances in material science, such as the development of new semiconductors and composites, play a crucial role in enhancing STEC.
(b) **Light Absorption Efficiency**: Effective light management strategies, such as bandgap engineering and light trapping, directly impact the amount of solar energy absorbed by the photoelectrode material. Improved light absorption leads to higher generation of charge carriers, thereby increasing the electrical power output and STEC.
(c) **Charge Carrier Dynamics**: Efficient separation and transport of charge carriers are essential for minimizing losses and maximizing electrical output. Enhancements in charge carrier dynamics, achieved through nanostructuring and surface modifications, contribute to improved STEC by reducing recombination losses and facilitating better charge transfer.
(d) **Electrochemical Reactions**: The efficiency of the electrochemical reactions involved in PEC cells, including the hydrogen evolution reaction (HER) and the oxygen evolution reaction (OER), affects STEC. Optimization of reaction kinetics and reduction of overpotentials are critical for maximizing the electrical power output.
(e) **Cell Design and Architecture**: The design and architecture of the PEC cell, including the configuration of the photoelectrodes and the overall cell structure, influence STEC. Innovations in cell design, such as tandem and multi-junction configurations, can enhance light absorption and improve overall efficiency.

Achieving high STEC values involves addressing several challenges. One significant challenge is the balance between light absorption and charge carrier transport. While maximizing light absorption is essential for generating more charge carriers, ensuring efficient transport and minimizing losses are equally important for achieving high STEC.

Additionally, stability and durability of PEC cells under prolonged exposure to sunlight and operating conditions are critical. Degradation of materials and performance over time can impact the long-term STEC and overall efficiency of the cells.

The Solar-to-Electricity Conversion Efficiency (STEC) is a key performance indicator for photoelectrochemical solar cells, reflecting their ability to convert solar energy into electrical power. Enhancing STEC involves optimizing material properties, improving light absorption, and addressing challenges in charge carrier dynamics and cell design. Achieving high STEC values is essential for advancing PEC technology and its practical application in renewable energy systems [89].

2.9.2 Factors Affecting PEC Cell Performance: Recombination, Overpotentials and Stability

The performance of photoelectrochemical (PEC) cells is influenced by several critical factors that impact their efficiency and longevity. Among these factors, recombination losses, overpotentials, and stability are pivotal in determining how effectively a PEC cell converts solar energy into chemical energy and how well it performs under real-world conditions. Understanding and addressing these factors are essential for optimizing PEC cell performance and achieving sustainable energy solutions.

2.9.2.1 Recombination Losses

Recombination losses refer to the phenomenon where charge carriers—electrons and holes—recombine before they can contribute to the desired electrochemical reactions. In PEC cells, recombination can occur at various interfaces, including within the photoelectrode material and at the surface of the electrode. This loss of charge carriers significantly reduces the number of carriers available for the water-splitting reactions, thereby diminishing the overall efficiency of the cell. Figure 2.17 illustrates the typical quantum efficiency in an ideal and actual solar cell, illustrating the impact of optical and recombination losses Recombination losses can be categorized into several types, including:

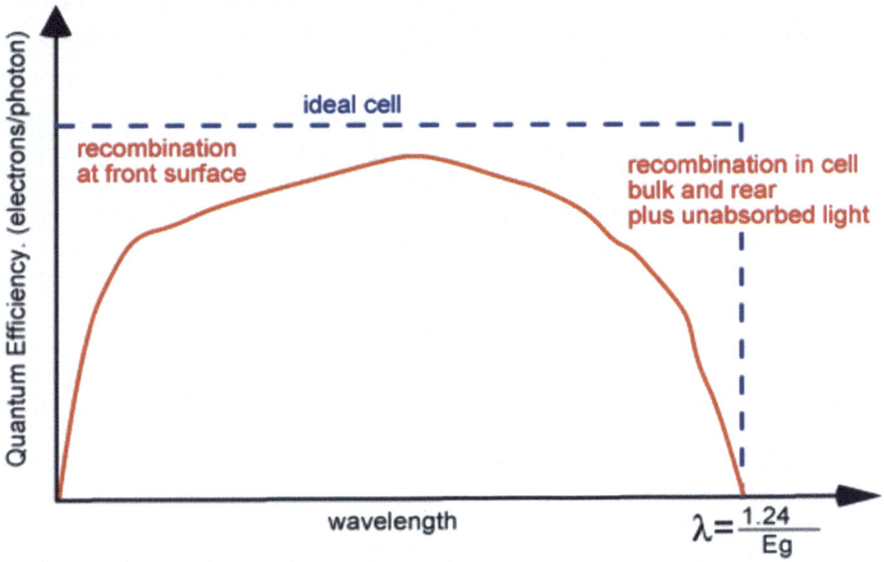

Fig. 2.17 Typical quantum efficiency in an ideal and actual solar cell, illustrating the impact of optical and recombination losses reproduced from Ref. [90] Copyright PVEducation

(a) **Radiative Recombination**: This occurs when electrons and holes recombine and emit photons. While this type of recombination is less common in PEC cells compared to other types of recombination, it still contributes to energy losses.
(b) **Non-radiative Recombination**: This type involves the loss of energy through mechanisms such as defects, traps, and impurities within the photoelectrode material. Non-radiative recombination is typically more detrimental to PEC cell performance and is a primary focus for improvement.
(c) **Surface Recombination**: Charge carriers that reach the surface of the photoelectrode can recombine at the surface instead of participating in the electrochemical reactions. Surface recombination can be minimized through surface passivation techniques and the use of protective coatings.

Addressing recombination losses involves optimizing material properties, including reducing defects and traps, improving charge carrier mobility, and implementing effective surface passivation strategies. Advanced nanostructuring and compositional modifications can also help in mitigating recombination losses and enhancing cell performance.

2.9.2.2 Overpotentials

Overpotentials are the additional voltages required beyond the theoretical potential to drive electrochemical reactions at a reasonable rate. In PEC cells, overpotentials are encountered during the hydrogen evolution reaction (HER) and the oxygen evolution reaction (OER), and they result from kinetic barriers and resistance to charge transfer. Figure 2.18 depicts the overpotential in a PEC cell. Several factors contribute to overpotentials in PEC cells:

(a) **Electrode Material**: The intrinsic properties of the electrode material, such as its catalytic activity and electronic conductivity, influence the overpotential. Materials with poor catalytic performance require higher overpotentials to achieve the desired reaction rates.
(b) **Electrolyte Composition**: The composition of the electrolyte and its interaction with the electrode surface affect the overpotential. Optimizing the electrolyte can help in reducing the resistance and improving reaction kinetics.
(c) **Reaction Kinetics**: The kinetics of the HER and OER are crucial in determining the overpotential. Slow reaction kinetics lead to higher overpotentials and reduced cell efficiency.

Reducing overpotentials involves enhancing the catalytic activity of the electrode materials, improving charge transfer efficiency, and optimizing reaction conditions. Strategies such as using advanced co-catalysts, improving electrode surface area, and employing innovative cell designs can contribute to lower overpotentials and better performance.

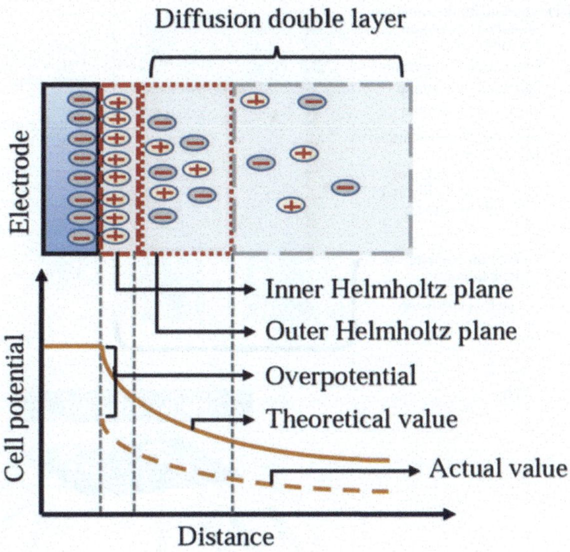

Fig. 2.18 Depiction of overpotential in a PEC cell reproduced Ref. [91] Copyright © 2023 The Authors. Published by Elsevier Ltd.

2.9.2.3 Stability

Stability is a critical factor in the long-term performance of PEC cells. Stability issues can arise from various factors, including degradation of materials, degradation of the photoelectrode surface, and degradation of the electrolyte. Figure 2.19 shows an overview of various degradation mechanisms. Figure 2.19a illustrates the degradation mechanisms occurring in the different components and at the interfaces and Fig. 2.19b depicts the interplay between the degradation effects of different components.

(a) **Material Degradation**: Exposure to harsh operating conditions, such as high temperatures and corrosive environments, can lead to the degradation of photoelectrode materials. Stability concerns can be mitigated by using more robust materials and protective coatings.
(b) **Surface Degradation**: The photoelectrode surface can degrade over time due to chemical reactions with the electrolyte and prolonged exposure to light. Surface degradation can be minimized by employing protective layers and optimizing surface treatments.
(c) **Electrolyte Stability**: The stability of the electrolyte is also crucial for maintaining PEC cell performance. Electrolytes that undergo chemical changes or degradation can negatively impact cell efficiency and longevity.

Enhancing stability involves selecting durable materials, implementing protective strategies, and optimizing operating conditions. Regular monitoring and maintenance of PEC cells are also essential to ensure sustained performance and reliability. Recombination losses, overpotentials, and stability are critical factors affecting the

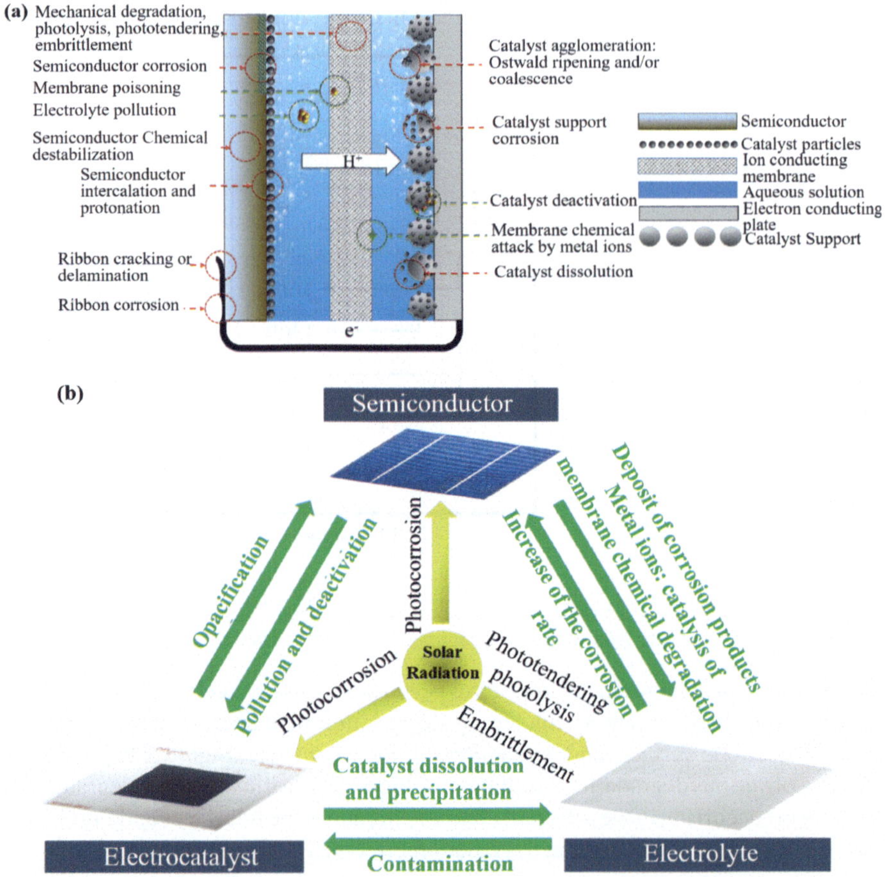

Fig. 2.19 Overview of various degradation mechanisms. **a** Degradation mechanisms occurring in the different components and at the interfaces. **b** Interplay between the degradation effects of different components reproduced from Ref. [93] Copyright T. Bosserez et al., published by IFP Energies nouvelles, 2015

performance of photoelectrochemical cells. Addressing these issues through material optimization, advanced cell design, and effective operational strategies is essential for improving the efficiency and longevity of PEC cells. By focusing on these factors, researchers and engineers can enhance the performance of PEC technology and advance its application in renewable energy systems [92].

2.10 Techniques for Reducing Charge Carrier Losses

Minimizing charge carrier losses is crucial for enhancing the efficiency of photoelectrochemical (PEC) cells. Charge carriers—electrons and holes—are generated when the photoelectrode absorbs solar energy. For the PEC cell to function effectively, these charge carriers must be efficiently separated and transported to the respective electrodes to drive the desired electrochemical reactions. Several techniques and strategies can be employed to reduce charge carrier losses and improve the overall performance of PEC cells.

2.10.1 Material Optimization

The intrinsic properties of the photoelectrode material significantly influence charge carrier dynamics. Optimizing material properties such as bandgap, charge carrier mobility, and carrier lifetime is essential for reducing losses. Bandgap Engineering: Tailoring the bandgap of photoelectrode materials ensures that they absorb a broader spectrum of solar light, leading to more efficient charge carrier generation. Materials with a well-matched bandgap can absorb more photons and generate a higher number of charge carriers, which is crucial for effective PEC cell operation. High Charge Carrier Mobility: Materials with high charge carrier mobility allow for faster transport of electrons and holes to their respective electrodes. This reduces the likelihood of recombination losses and enhances the overall efficiency of the PEC cell. Advanced materials such as nanostructured semiconductors and composites can offer improved mobility. Long Carrier Lifetime: Extending the lifetime of charge carriers before they recombine is vital for efficient PEC cell performance. Materials with reduced defect densities and improved crystal quality contribute to longer carrier lifetimes and lower recombination rates.

2.10.2 Nanostructuring Techniques

Nanostructuring involves the design and fabrication of materials at the nanoscale to enhance their properties. Figure 2.20 illustrates the advantages of nanostructuring of samples. This approach can significantly improve charge carrier dynamics by:

Increasing Surface Area: Nanostructuring increases the surface area of photoelectrodes, providing more active sites for light absorption and electrochemical reactions. This enhances the efficiency of charge carrier generation and reduces losses due to recombination at the surface.

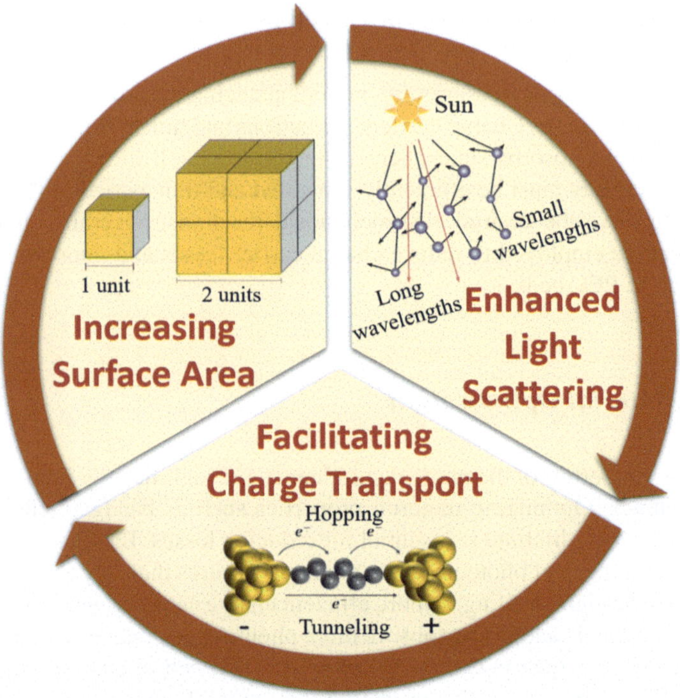

Fig. 2.20 Advantages of nanostructuring of samples

Facilitating Charge Transport: Nanostructures, such as nanowires, nanotubes, and nanorods, can provide efficient pathways for charge transport. These structures reduce the distance that charge carriers must travel, minimizing the chance of recombination and improving overall efficiency.

Enhanced Light Scattering: Nanostructures can also improve light trapping by scattering light more effectively within the photoelectrode material. This increases the optical path length and enhances light absorption, leading to more efficient charge carrier generation.

2.10.3 Surface Passivation

Surface passivation involves applying a coating or treatment to the photoelectrode surface to reduce surface recombination and protect against degradation. Effective passivation techniques include: **Protective Coatings**: Applying thin layers of materials such as metal oxides or polymers can shield the photoelectrode surface from harsh environmental conditions and chemical reactions with the electrolyte. These coatings help to maintain the integrity of the photoelectrode and reduce recombination losses. **Surface Functionalization**: Functionalizing the surface of the photoelectrode with specific chemical groups or molecules can improve charge carrier separation and reduce recombination. For example, certain surface treatments can create favorable energy level alignments and enhance the interaction between the photoelectrode and electrolyte.

2.10.4 Interface Engineering

Optimizing the interfaces between different components of the PEC cell can significantly impact charge carrier dynamics. Key strategies include:

Improving Contact Quality: Ensuring good electrical contact between the photoelectrode and other cell components, such as the counter electrode and conductive substrates, reduces resistance and facilitates efficient charge transfer.

Minimizing Interfacial Resistance: Interfacial resistance between the photoelectrode and electrolyte can hinder charge transfer. Techniques such as optimizing electrolyte composition, using interfacial layers, and improving surface wettability can help minimize this resistance and enhance performance.

2.11 Scalability and Integration with Photovoltaic Systems

The scalability of photoelectrochemical (PEC) systems and their integration with existing photovoltaic technologies are pivotal factors in realizing their potential for large-scale energy production. As PEC cells continue to advance, addressing these aspects is crucial for transitioning from laboratory-scale demonstrations to practical, commercial applications. Effective scalability and integration involve multiple considerations, including system design, material compatibility, and economic feasibility.

2.11.1 Scaling Up PEC Systems

Scaling up PEC systems from laboratory prototypes to large-scale production involves overcoming several technical and engineering challenges. One of the primary concerns is the uniformity of materials and processes across larger areas. In laboratory settings, PEC cells are often tested on small, controlled substrates. However, for commercial applications, it is essential to develop techniques for fabricating large-area photoelectrodes with consistent quality and performance. This requires advancements in deposition techniques, such as roll-to-roll processing or large-area coating methods, to ensure that the properties of the photoelectrode materials remain uniform over extensive surfaces.

Additionally, the stability and longevity of PEC cells must be addressed when scaling up. Laboratory cells are typically operated under ideal conditions, but real-world applications expose them to varying environmental factors such as temperature fluctuations, humidity, and mechanical stresses. Developing robust materials and protective coatings that can withstand these conditions without significant degradation is critical for ensuring the long-term performance of large-scale PEC systems.

2.11.2 Integration with Photovoltaic Technologies

Integrating PEC cells with existing photovoltaic systems presents both opportunities and challenges. One approach is to combine PEC cells with conventional photovoltaic panels to create hybrid systems that leverage the strengths of both technologies. For example, while traditional solar cells efficiently convert sunlight into electricity, PEC cells can be designed to produce hydrogen fuel through water splitting, offering a dual-functionality system. This hybrid approach can enhance the overall efficiency and utility of solar energy systems by providing both electrical power and chemical energy storage.

To achieve successful integration, compatibility between PEC cells and photovoltaic panels is essential. This includes matching the electrical output characteristics and ensuring that the physical and chemical interactions between the two components do not adversely affect performance. For instance, the design of the PEC cell must account for the need to interface with existing photovoltaic systems without causing damage or interference. Additionally, the integration should consider the spatial arrangement and structural support to accommodate both technologies within a cohesive system.

2.11.3 Economic Feasibility and Market Viability

The economic feasibility of scaling up PEC systems and integrating them with photovoltaic technologies is a crucial consideration for their commercial adoption. The costs associated with large-scale manufacturing, material supply, and system installation must be balanced against the potential benefits of improved energy production and storage. Economies of scale can help reduce costs as production volumes increase, but initial investments in research, development, and infrastructure are significant. Therefore, it is essential to conduct comprehensive cost-benefit analyses to evaluate the financial viability of large-scale PEC systems and their integration with existing technologies. Figure 2.21 depicts the large scale manufacturing of solar cells. Moreover, market incentives and policies play a role in promoting the adoption

Fig. 2.21 Large scale manufacturing of solar cells generated by Lexica.art (AI)

of PEC and hybrid systems. Government subsidies, research grants, and supportive regulations can help accelerate the development and deployment of these technologies. By fostering a favorable environment for innovation and commercialization, stakeholders can facilitate the transition from laboratory research to widespread application.

Scalability and integration with photovoltaic systems are critical factors for the successful deployment of photoelectrochemical technologies. Addressing challenges related to material uniformity, system durability, and economic feasibility is essential for advancing PEC systems from experimental stages to large-scale applications. By exploring hybrid solutions and ensuring compatibility with existing photovoltaic technologies, researchers and engineers can pave the way for more efficient and versatile solar energy solutions, ultimately contributing to the broader adoption of renewable energy technologies.

2.12 Advanced Characterization Techniques for PEC Solar Cells

The advancement of photoelectrochemical (PEC) solar cells hinges on the ability to accurately characterize and understand the underlying material properties and performance dynamics. As PEC technology evolves, sophisticated characterization techniques become essential for gaining insights into the mechanisms driving cell operation and for optimizing material and device design. This section explores the role of advanced characterization methods in unraveling the complexities of PEC solar cells, providing a foundation for both fundamental research and practical application. Effective characterization of PEC solar cells involves examining materials at various scales and under different conditions to elucidate their behavior and performance. These techniques encompass a broad range of approaches, from spectroscopic and microscopic methods that analyze material properties to in-situ and operando techniques that monitor cell performance in real-time. Additionally, computational modeling and simulation offer invaluable insights into the theoretical and practical aspects of PEC cells, guiding experimental efforts and facilitating design improvements. Spectroscopic and microscopic methods are essential for the detailed analysis of materials used in photoelectrochemical (PEC) solar cells. These techniques provide valuable insights into the chemical composition, structural properties, and surface characteristics of materials, which are critical for optimizing their performance in PEC applications. This section discusses various spectroscopic and microscopic techniques and their applications in analyzing PEC materials.

2.12.1 Spectroscopic Techniques

Spectroscopy encompasses a range of techniques used to study the interaction between electromagnetic radiation and matter. In the context of PEC solar cells, spectroscopic methods are employed to investigate the electronic structure, chemical composition, and optical properties of materials. Figure 2.22 illustrates the various spectroscopic techniques utilised for material used in PEC solar cells.

(a) **X-ray Photoelectron Spectroscopy (XPS)**: XPS is a powerful technique used to analyze the surface chemistry of materials. It provides information on the elemental composition, chemical state, and electronic environment of the elements present in a sample. In PEC solar cells, XPS is often used to investigate the oxidation states of metal components, the presence of surface contaminants, and the effectiveness of surface modifications. This information is crucial for understanding how surface chemistry impacts charge transfer and overall cell performanc [94].

(b) **Ultraviolet–Visible Spectroscopy (UV–Vis)**: UV–Vis spectroscopy is employed to measure the absorption spectra of materials, which reveals their optical properties and bandgap energies. For PEC solar cells, this technique

Fig. 2.22 Various spectroscopic techniques utilised for material used in PEC solar cells

helps in determining the light absorption capabilities of photoelectrode materials. By analyzing the absorption spectrum, researchers can assess the efficiency of light capture and the suitability of materials for specific wavelengths, which is fundamental for enhancing solar absorption and improving cell efficiency [95].

(c) **Raman Spectroscopy**: Raman spectroscopy provides information on vibrational modes within a material, which is useful for identifying molecular structures and analyzing material phases. In PEC cells, Raman spectroscopy can be used to investigate the structural integrity of photoelectrode materials, detect phase transitions, and study the effects of doping or compositional changes on material properties [96].

2.12.2 Microscopic Techniques

Microscopy techniques offer high-resolution imaging of material surfaces and interfaces, allowing for detailed examination of structural and morphological features. Figure 2.23 shows various microscopic techniques utilised for material used in PEC solar cells.

(a) **Scanning Electron Microscopy (SEM)**: SEM is widely used to observe the surface morphology and topography of materials at the nanometer scale. For

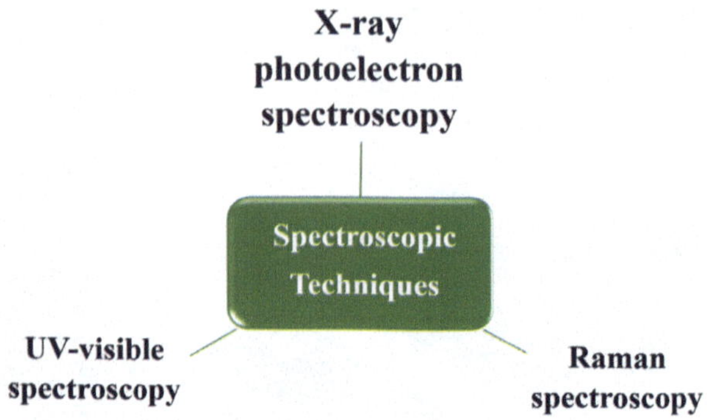

Fig. 2.23 Various microscopic techniques utilised for material used in PEC solar cells

PEC solar cells, SEM provides critical insights into the surface roughness, porosity, and structural uniformity of photoelectrodes. By analyzing SEM images, researchers can evaluate the impact of nanostructuring or surface treatments on material performance and identify potential issues such as defects or non-uniformities that could affect charge transport.

(b) **Transmission Electron Microscopy (TEM)**: TEM offers even higher resolution imaging than SEM and is capable of providing information about the internal structure of materials. TEM is used to investigate the crystallographic properties, grain boundaries, and interfaces within photoelectrodes. This technique is particularly useful for studying nanoscale features and understanding the interactions between different material components in composite or hybrid photoelectrodes.

(c) **Atomic Force Microscopy (AFM)**: AFM is employed to measure the surface topography and mechanical properties of materials at the atomic scale. In the context of PEC cells, AFM can be used to analyze surface roughness, evaluate film thickness, and investigate the mechanical properties of thin films and nanostructures. This information is important for optimizing the surface characteristics of photoelectrodes and ensuring effective interaction with light and electrolytes.

The spectroscopic and microscopic methods are invaluable tools for the analysis of materials used in PEC solar cells. By providing detailed information on chemical composition, optical properties, and structural characteristics, these techniques enable researchers to optimize material properties, enhance cell performance, and advance the development of efficient and durable PEC solar cells.

2.13 In-Situ and Operando Techniques for Performance Monitoring

In-situ and operando techniques are essential for understanding the real-time behavior and performance of photoelectrochemical (PEC) solar cells under operational conditions. Unlike ex-situ methods, which analyze materials in a static or pre-prepared state, in-situ and operando techniques monitor the dynamic processes occurring within a PEC cell as it functions. These methods provide critical insights into the mechanisms driving cell operation, reveal performance degradation factors, and guide the development of more efficient and stable PEC systems. Figure 2.24 depicts the testing a PEC solar cell.

Fig. 2.24 Testing a PEC solar cell generated by Lexica.art (AI)

2.13.1 *In-Situ Techniques*

In-situ techniques involve observing and measuring the properties of materials while they are in contact with or subjected to the conditions they will experience during actual use. For PEC solar cells, in-situ methods can be employed to study various aspects of cell performance, including charge transfer, material stability, and reaction kinetics.

(a) **In-Situ Spectroscopy**: In-situ spectroscopy techniques, such as in-situ UV–Vis and in-situ Raman spectroscopy, enable the real-time monitoring of changes in the optical properties and vibrational modes of materials during PEC cell operation. For instance, in-situ UV–Vis spectroscopy can track the changes in light absorption characteristics of photoelectrodes as they interact with the electrolyte and light. Similarly, in-situ Raman spectroscopy can provide insights into the evolution of material phases and surface chemistry during the PEC process. These techniques help in understanding how operational conditions affect material properties and overall cell performance [97].

(b) **In-Situ Electrochemical Methods**: Techniques such as in-situ electrochemical impedance spectroscopy (EIS) and in-situ cyclic voltammetry (CV) are used to monitor the electrochemical behavior of PEC cells in real-time. In-situ EIS provides information on charge transfer resistance, ion diffusion, and interface properties, while in-situ CV allows for the evaluation of redox reactions and electrode kinetics. These methods are crucial for analyzing the efficiency of charge transfer processes and identifying factors that may lead to performance losses [98].

2.13.2 *Operando Techniques*

Operando techniques involve monitoring PEC cell performance while the cell is actively performing its intended function, typically under realistic operational conditions. These methods provide valuable information on the cell's behavior during actual use, including the effects of external factors such as light intensity, temperature, and electrolyte composition.

(a) **Operando Spectroscopy**: Operando spectroscopy methods, such as operando X-ray absorption spectroscopy (XAS) and operando photoluminescence spectroscopy, are used to investigate the electronic and structural changes occurring within a PEC cell during operation. For example, operando XAS can provide insights into the oxidation states and local environments of metal centers in photoelectrodes, while operando photoluminescence spectroscopy can reveal information about charge carrier dynamics and recombination processes. These techniques help in understanding the real-time interactions between materials and their impact on cell performance [99].

(b) **Operando Microscopy**: Techniques such as operando scanning electron microscopy (SEM) and operando atomic force microscopy (AFM) enable the observation of material surface changes and structural dynamics during PEC cell operation. Operando SEM can reveal morphological changes, such as surface corrosion or film degradation, while operando AFM provides information on surface roughness and mechanical properties. These observations are essential for identifying performance degradation mechanisms and optimizing material properties [100].

(c) **Operando Spectroelectrochemistry**: Combining electrochemical techniques with spectroscopy, operando spectroelectrochemistry provides simultaneous information on the electrochemical and optical properties of PEC cells. This method allows for the real-time monitoring of charge transfer processes and material changes, providing a comprehensive understanding of the factors influencing cell performance [101].

The in-situ and operando techniques are vital for gaining a deeper understanding of the real-time behavior and performance of PEC solar cells. By providing insights into charge transfer, material stability, and reaction kinetics under operational conditions, these methods help in identifying performance limitations, optimizing cell design, and advancing the development of efficient and durable PEC solar cells.

2.14 Computational Modeling and Simulation Approaches

Computational modeling and simulation approaches are invaluable tools for advancing the design, optimization, and understanding of photoelectrochemical (PEC) solar cells. By employing computational methods, researchers can predict material behaviors, optimize cell architectures, and evaluate performance metrics without the need for extensive experimental work. These techniques offer insights into complex interactions within PEC systems, guiding the development of more efficient and reliable solar cells.

2.14.1 Modeling of Photocatalytic Processes

Computational models of photocatalytic processes simulate the behavior of materials under light irradiation and electrochemical conditions. These models are essential for understanding the fundamental mechanisms driving PEC reactions and predicting the performance of different materials.

(a) **Density Functional Theory (DFT)**: DFT is widely used to study the electronic structure and properties of materials at the atomic level. In the context of PEC solar cells, DFT calculations provide insights into the band structure, electronic density of states, and charge distribution of photoelectrode materials. By analyzing these properties, researchers can predict how materials will interact with light, how charge carriers will behave, and how different modifications will impact performance [102].

(b) **Molecular Dynamics (MD) Simulations**: MD simulations model the behavior of materials over time, providing insights into dynamic processes such as adsorption, diffusion, and reaction kinetics. In PEC cells, MD simulations can be used to study the interaction of molecules with photoelectrode surfaces, investigate the stability of materials under operational conditions, and explore the effects of structural changes on performance.

2.14.2 *Optimization of PEC Cell Design*

Computational approaches are also employed to optimize PEC cell designs and architectures. By simulating various design parameters and configurations, researchers can identify the most effective strategies for enhancing cell efficiency and performance.

(a) **Finite Element Analysis (FEA)**: FEA is used to model the physical and electrochemical processes occurring within PEC cells. This technique allows for the simulation of electric fields, charge transport, and thermal effects within the cell. By optimizing design parameters such as electrode configurations, light management strategies, and electrolyte compositions, FEA helps in enhancing overall cell efficiency and performance [103].

(b) **Multiscale Modeling**: Multiscale modeling integrates information from different length scales, from atomic to macroscopic levels. In PEC cells, multiscale models combine insights from molecular dynamics, DFT, and continuum models to provide a comprehensive understanding of material behavior and cell performance. This approach enables the simulation of complex interactions between materials, light, and electrolytes, guiding the development of more effective cell designs.

2.14.3 *Performance Prediction and Analysis*

Computational simulations are used to predict the performance of PEC solar cells under various operating conditions. These predictions help in identifying potential issues, optimizing design parameters, and assessing the impact of different materials and configurations on cell efficiency.

(a) **Performance Modeling**: Performance models simulate the overall efficiency of PEC cells based on input parameters such as light intensity, electrolyte composition, and material properties. These models help in predicting key performance metrics, such as solar-to-hydrogen efficiency and current–voltage characteristics, and can guide the selection of materials and designs that maximize performance.

(b) **Sensitivity Analysis**: Sensitivity analysis involves varying model parameters to assess their impact on performance outcomes. This approach helps in identifying critical factors that influence cell efficiency and stability, allowing researchers to focus on optimizing the most impactful aspects of the PEC cell design.

The computational modeling and simulation approaches are essential for advancing the development of PEC solar cells. By providing insights into material properties, optimizing cell designs, and predicting performance metrics, these techniques enable researchers to design more efficient and reliable PEC systems. The integration of computational methods with experimental work accelerates the advancement of PEC technology and facilitates the development of innovative solutions for sustainable energy production.

2.15 Future Directions and Prospects

The future of photoelectrochemical (PEC) solar cells holds exciting potential as research and development efforts continue to advance the technology. As we look ahead, several key areas promise to drive significant progress in enhancing PEC cell performance, integration with other renewable systems, and practical applications. This section explores the pathways to achieving higher solar-to-electricity conversion efficiency, integrating PEC cells with other renewable technologies, exploring their potential applications, and evaluating their long-term stability and commercialization prospects. Achieving high solar-to-electricity conversion efficiency is a primary goal in the development of PEC solar cells. Several strategies are being explored to push the boundaries of efficiency and performance. Advances in material science are central to these efforts, particularly through the development of new photoelectrode materials with optimized band gaps and improved light absorption capabilities. Emerging materials such as perovskites, quantum dots, and 2D materials hold promise for significantly enhancing the efficiency of PEC cells by better utilizing the solar spectrum and minimizing energy losses. Another key area is the optimization of cell architectures. Tandem PEC cells, which stack multiple photoelectrodes with complementary absorption spectra, have shown potential for higher efficiency compared to single-junction cells. Enhancing light management through advanced photonic structures, such as photonic crystals and plasmonic nanostructures, can also improve the efficiency of light absorption and reduce reflection losses. Combining these approaches with innovations in catalytic materials and surface modifications can further enhance the overall performance of PEC solar cells [104].

The integration of PEC solar cells with other renewable energy systems and storage technologies is crucial for maximizing their utility in sustainable energy solutions. PEC cells can be integrated with photovoltaic (PV) systems to create hybrid solar energy systems that combine the strengths of both technologies. This hybrid approach can enable the simultaneous generation of electricity and hydrogen, providing a versatile solution for energy production and storage. Coupling PEC cells

Fig. 2.25 **a** Renewable ecosystem combining solar and wind energy generated by Lexica.art (AI) **b** solar cell coupled with batteries generated by Lexica.art (AI)

with energy storage technologies, such as batteries and supercapacitors, is another promising avenue. By integrating PEC cells with energy storage systems, it is possible to store excess energy generated during peak sunlight hours and use it during periods of low sunlight or high demand. This integration enhances the reliability and stability of renewable energy systems and supports the transition to a more sustainable energy infrastructure. Figure 2.25a shows renewable ecosystem combining solar and wind energy and Fig. 2.25b depicts the solar cell coupled with batteries.

PEC solar cells have significant potential in various sustainable energy applications. Their ability to directly convert solar energy into hydrogen makes them an attractive option for clean hydrogen production. Hydrogen, as a versatile and clean energy carrier, can be used in fuel cells, combustion engines, and industrial processes, offering a wide range of applications across different sectors. In addition to hydrogen production, PEC cells can be used in water purification and desalination processes. By harnessing solar energy for these applications, PEC cells can contribute to addressing global challenges related to water scarcity and access to clean water. The development of PEC cells for such applications can improve the sustainability of water treatment technologies and provide an eco-friendly alternative to conventional methods. For PEC solar cells to become a mainstream technology, addressing long-term stability and commercialization challenges is essential. Ensuring that PEC cells maintain their performance and efficiency over extended periods is critical for their practical application. This involves developing materials and designs that resist degradation from environmental factors such as light, moisture, and temperature fluctuations. Figure 2.26 shows the simulation of the new exPEC cell operation showing the homogeneous buildup of the hydrogen concentration through water splitting in vapor phase. Commercialization prospects also depend on reducing production costs and scaling up manufacturing processes. Advances in fabrication techniques, such as roll-to-roll processing and scalable deposition methods, can help lower production costs and make PEC cells more commercially viable. Collaborations between

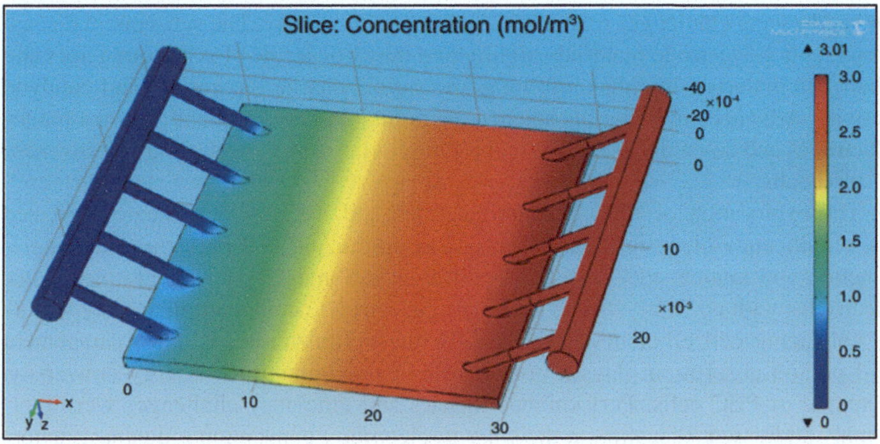

Fig. 2.26 Simulation of the new exPEC cell operation showing the homogeneous buildup of the hydrogen concentration through water splitting in vapor phase reproduced from Ref. [93] Copyright T. Bosserez et al., published by IFP Energies nouvelles, 2015

researchers, industry stakeholders, and policymakers will be crucial in accelerating the commercialization of PEC technology and facilitating its adoption in the energy market.

The future of PEC solar cells is promising, with ongoing research and development addressing key challenges and exploring new opportunities. By focusing on enhancing efficiency, integrating with other renewable systems, and addressing stability and commercialization issues, PEC cells have the potential to play a significant role in the transition to a more sustainable and clean energy future. The continued advancement of PEC technology will be vital in meeting global energy needs and contributing to a more sustainable world.

2.16 Conclusion

In this chapter, we have explored the multifaceted landscape of PEC solar cells, delving into their fundamental principles, material requirements, cell configurations, performance metrics, and future prospects. PEC cells represent a promising technology for harnessing solar energy directly to drive chemical transformations, particularly in the production of hydrogen from water. As a clean and renewable energy solution, PEC cells offer the potential to contribute significantly to the global transition towards sustainable energy systems. We began by examining the fundamental principles underlying PEC water splitting, including the thermodynamics and kinetics of the process, energy band alignment, and charge carrier dynamics. Understanding these core principles is essential for optimizing PEC cell performance and addressing challenges related to efficiency and stability. The subsequent discussion on photoelectrode materials highlighted the critical role of semiconductors and emerging materials in enhancing light absorption, charge separation, and catalytic activity. Advances in materials science, including the development of novel photoelectrodes and co-catalysts, are pivotal for improving the efficiency and effectiveness of PEC cells.

The exploration of PEC cell architectures and integration strategies revealed the significance of design considerations in optimizing cell performance. Single-junction and tandem PEC cells offer different advantages, and integrating photoelectrodes with co-catalysts and optimizing electrolyte composition are key factors in enhancing cell efficiency. Additionally, the optimization of light management and photon absorption plays a crucial role in maximizing the energy conversion potential of PEC cells. Performance metrics and efficiency challenges were thoroughly addressed, focusing on the solar-to-electricity conversion efficiency, factors affecting cell performance such as recombination and overpotentials, and techniques for reducing charge carrier losses. The scalability of PEC technology and its integration with existing photovoltaic systems were also discussed, emphasizing the importance of addressing commercialization and practical deployment challenges.

Advanced characterization techniques are critical for understanding and optimizing PEC cell performance. Spectroscopic and microscopic methods, along with in-situ and operando techniques, provide valuable insights into material properties and cell behavior. Computational modeling and simulation approaches further enhance our ability to predict and optimize PEC cell performance. Looking to the future, several key directions and prospects for PEC solar cells were identified. Achieving high solar-to-electricity conversion efficiency, integrating PEC cells with other renewable energy systems, and exploring their potential applications in sustainable energy solutions are crucial areas for ongoing research. Addressing long-term stability and commercialization challenges will be essential for realizing the full potential of PEC technology. The continued advancement of PEC solar cells offers significant promise for addressing global energy and environmental challenges. By leveraging innovations in materials science, optimizing cell designs, and exploring new applications, PEC technology has the potential to make a meaningful impact on the development of sustainable energy systems. As research progresses and technology matures, PEC cells are poised to play a key role in the transition to a cleaner, more sustainable energy future.

References

1. Photoelectrochemical Solar Cells: Present Status on JSTOR. [Online]. Available: https://www.jstor.org/stable/24098805. Accessed: 03 Sept 2024
2. Chen, Q., Fan, G., Fu, H., Li, Z., Zou, Z.: Tandem photoelectrochemical cells for solar water splitting. Adv. Phys. X **3**(1), 863–884 (2018). https://doi.org/10.1080/23746149.2018.1487267

3. Chen, X., Zhang, Z., Chi, L., Nair, A.K., Shangguan, W., Jiang, Z.: Recent advances in visible-light-driven photoelectrochemical water splitting: catalyst nanostructures and reaction systems. Nanomicro. Lett. **8**(1), 1–12 (2016). https://doi.org/10.1007/S40820-015-0063-3/FIGURES/11
4. Pandey, R.N., Chandra Babu, K.S., Srivastava, O.N.: High conversion efficiency photoelectrochemical solar cells. Prog. Surf. Sci. **52**(3), 125–192 (1996). https://doi.org/10.1016/0079-6816(96)00009-3
5. Patel, S.B., Thakar, B.A.: Impact of light intensity and electrolyte volume on performance of photo-electrochemical (PEC) solar cell. J. Emerg. Investig. (2022). https://doi.org/10.59720/21-146
6. Sankir, N.D., Mehmet, S.: Photoelectrochemical Solar Cells. Advances in Solar Cell Materials and Storage Series, p. 482. Wiley (2019)
7. Müller, A., Kondofersky, I., Folger, A., Fattakhova-Rohlfing, D., Bein, T., and Scheu, C.: Dual absorber Fe_2O_3/WO_3 host-guest architectures for improved charge generation and transfer in photoelectrochemical applications. Mater. Res. Express **4**(1) (2017). https://doi.org/10.1088/2053-1591/AA570F
8. Mishra, P.R., Shukla, P.K., Srivastava, O.N.: Study of modular pec solar cells for photoelectrochemical splitting of water employing nanostructured TiO_2 photoelectrodes. Int. J. Hydrogen Energy **32**(12), 1680–1685 (2007). https://doi.org/10.1016/J.IJHYDENE.2006.10.002
9. Singh, P.K., Nagarale, R.K., Pandey, S.P., Rhee, H.W., Bhattacharya, B.: Present status of solid state photoelectrochemical solar cells and dye sensitized solar cells using PEO-based polymer electrolytes. Adv. Nat. Sci. Nanosci. Nanotechnol **2**(2), 023002 (2011). https://doi.org/10.1088/2043-6262/2/2/023002
10. Wick, R., Tilley, S.D.: Photovoltaic and photoelectrochemical solar energy conversion with Cu_2O. J. Phys. Chem. C **119**(47), 26243–26257 (2015). https://doi.org/10.1021/ACS.JPCC.5B08397/ASSET/IMAGES/MEDIUM/JP-2015-08397H_0013.GIF
11. May, M.M., Rehfeld, K.: Negative emissions as the new frontier of photoelectrochemical CO_2 reduction. Adv. Energy Mater. **12**(21), 2103801 (2022). https://doi.org/10.1002/AENM.202103801
12. Tembhurne, S., Nandjou, F., Haussener, S.: A thermally synergistic photo-electrochemical hydrogen generator operating under concentrated solar irradiation. Nat. Energy **4**(5), 399–407 (2019). https://doi.org/10.1038/s41560-019-0373-7
13. van de Krol, R., Parkinson, B.A.: Perspectives on the photoelectrochemical storage of solar energy. MRS Energy Sustain. **4**(1), E13 (2017). https://doi.org/10.1557/MRE.2017.15
14. Liu, R., Zheng, Z., Spurgeon, J., Yang, X.: Enhanced photoelectrochemical water-splitting performance of semiconductors by surface passivation layers. Energy Environ. Sci. **7**(8), 2504–2517 (2014). https://doi.org/10.1039/C4EE00450G
15. Perathoner, S., Centi, G., Su, D.: Turning perspective in photoelectrocatalytic cells for solar fuels. ChemSusChem **9**(4), 345–357 (2016). https://doi.org/10.1002/CSSC.201501059
16. Li, W., Fu, H.-C., Li, L., Cabán-Acevedo, M., He, J.-H., Jin, S.: Integrated photoelectrochemical solar energy conversion and organic redox flow battery devices. Angew. Chem. **128**(42), 13298–13302 (2016). https://doi.org/10.1002/ANGE.201606986
17. Dong, G., Yan, L., Bi, Y.: Advanced oxygen evolution reaction catalysts for solar-driven photoelectrochemical water splitting. J. Mater. Chem. A Mater. **11**(8), 3888–3903 (2023). https://doi.org/10.1039/D2TA09479G
18. Li, R.: Latest progress in hydrogen production from solar water splitting via photocatalysis, photoelectrochemical, and photovoltaic-photoelectrochemical solutions. Chin. J. Catal. **38**(1), 5–12 (2017). https://doi.org/10.1016/S1872-2067(16)62552-4
19. Jacobsson, T.J., Fjällström, V., Edoff, M., Edvinsson, T.: Sustainable solar hydrogen production: from photoelectrochemical cells to PV-electrolyzers and back again. Energy Environ. Sci. **7**(7), 2056–2070 (2014). https://doi.org/10.1039/C4EE00754A

20. Hodes, G., Howell, I.D.J., Peter, L.M.: Nanocrystalline photoelectrochemical cells: a new concept in photovoltaic cells. J. Electrochem. Soc. **139**(11), 3136–3140 (1992). https://doi.org/10.1149/1.2069045/XML
21. Li, J., Wu, N.: Semiconductor-based photocatalysts and photoelectrochemical cells for solar fuel generation: a review. Catal. Sci. Technol. **5**(3), 1360–1384 (2015). https://doi.org/10.1039/C4CY00974F
22. Wu, H., Tan, H.L., Toe, C.Y., Scott, J., Wang, L., Amal, R., Ng, Y.H.: Photocatalytic and photoelectrochemical systems: similarities and differences. Adv. Mater. **32**(18), 1904717 (2020). https://doi.org/10.1002/ADMA.201904717
23. Wang, R., Liu, H., Zhang, Y., Sun, K., Bao, W.: Integrated photovoltaic charging and energy storage systems: mechanism, optimization, and future. Small **18**(31), 2203014 (2022). https://doi.org/10.1002/SMLL.202203014
24. Landman, A., Dotan, H., Shter, G.E., Wullenkord, M., Houaijia, A., Maljusch, A., Grader, G.S., Rothschild, A.: Photoelectrochemical water splitting in separate oxygen and hydrogen cells. Nat. Mater. **16**(6), 646–651 (2017). https://doi.org/10.1038/nmat4876
25. Walter, M.G., Warren, E.L., McKone, J.R., Boettcher, S.W., Mi, Q., Santori, E.A., Lewis, N.S.: Solar water splitting cells. Chem. Rev. **110**(11), 6446–6473 (2010). https://doi.org/10.1021/CR1002326/ASSET/IMAGES/LARGE/CR-2010-002326_0024.JPEG
26. Ager, J.W., Shaner, M.R., Walczak, K.A., Sharp, I.D., Ardo, S.: Experimental demonstrations of spontaneous, solar-driven photoelectrochemical water splitting. Energy Environ. Sci. **8**(10), 2811–2824 (2015). https://doi.org/10.1039/C5EE00457H
27. Guo, L.J., Luo, J.W., He, T., Wei, S.H., Li, S.S.: Photocorrosion-limited maximum efficiency of solar photoelectrochemical water splitting. Phys. Rev. Appl. **10**(6), 064059 (2018). https://doi.org/10.1103/PHYSREVAPPLIED.10.064059/FIGURES/5/MEDIUM
28. Yang, X., Wang, D.: Photophysics and photochemistry at the semiconductor/electrolyte interface for solar water splitting. In: Semiconductors and Semimetals, vol. 97, pp. 47–80 (2017). https://doi.org/10.1016/BS.SEMSEM.2017.03.001
29. Tang, S., Qiu, W., Xiao, S., Tong, Y., Yang, S.: Harnessing hierarchical architectures to trap light for efficient photoelectrochemical cells. Energy Environ. Sci. **13**(3), 660–684 (2020). https://doi.org/10.1039/C9EE02986A
30. Guijarro, N., Prévot, M.S., Sivula, K.: Surface modification of semiconductor photoelectrodes. Phys. Chem. Chem. Phys. **17**(24), 15655–15674 (2015). https://doi.org/10.1039/C5CP01992C
31. Ampelli, C., Centi, G., Passalacqua, R., Perathoner, S.: Electrolyte-less design of PEC cells for solar fuels: prospects and open issues in the development of cells and related catalytic electrodes. Catal. Today **259**, 246–258 (2016). https://doi.org/10.1016/J.CATTOD.2015.07.020
32. Kozytskiy, A.V., Stroyuk, O.L., Raevskaya, A.E., Kuchmy, S.Y.: Photoelectrochemical solar cells with semiconductor nanoparticles and liquid electrolytes: a review. Theor. Exp. Chem. **53**(3), 145–179 (2017). https://doi.org/10.1007/S11237-017-9512-Z
33. Bandara, T.M.W.J., Jayasundara, W.J.M.J.S.R., Dissanayake, M.A.K.L., Fernando, H.D.N.S., Furlani, M., Albinsson, I., Mellander, B.E.: Quasi solid state polymer electrolyte with binary iodide salts for photo-electrochemical solar cells. Int. J. Hydrogen Energy **39**(6), 2997–3004 (2014). https://doi.org/10.1016/J.IJHYDENE.2013.05.163
34. Cao, S., Zhang, Z., Liao, Q., Kang, Z., Zhang, Y.: Interface engineering for high-performance photoelectrochemical cells via atomic layer deposition technique. Energ. Technol. **9**(2), 2000819 (2021). https://doi.org/10.1002/ENTE.202000819
35. Zhang, D., Shi, J., Zi, W., Wang, P., Liu, S.F.: Recent advances in photoelectrochemical applications of silicon materials for solar-to-chemicals conversion. ChemSusChem **10**(22), 4324–4341 (2017). https://doi.org/10.1002/CSSC.201701674

36. Kusior, A., Wnuk, A., Trenczek-Zajac, A., Zakrzewska, K., Radecka, M.: TiO_2 Nanostructures for Photoelectrochemical Cells (PECs). Int. J. Hydrogen Energy **40**(14), 4936–4944 (2015). https://doi.org/10.1016/J.IJHYDENE.2015.01.103
37. Roza, L., Rahman, M.Y.A., Umar, A.A., Salleh, M.M.: Direct growth of oriented ZnO nanotubes by self-selective etching at lower temperature for photo-electrochemical (PEC) solar cell application. J. Alloys Compd. **618**(1), 153–158 (2015). https://doi.org/10.1016/J.JALLCOM.2014.08.113
38. Shinde, S.S., Bansode, R.A., Bhosale, C.H., Rajpure, K.Y.: Physical properties of hematite α-Fe_2O_3 thin films: application to photoelectrochemical solar cells. J. Semiconductors **32**(1), 013001 (2011). https://doi.org/10.1088/1674-4926/32/1/013001
39. Yadav, A.A., Masumdar, E.U.: Photoelectrochemical investigations of cadmium sulphide (CdS) thin film electrodes prepared by spray pyrolysis. J. Alloys Compd. **509**(17), 5394–5399 (2011). https://doi.org/10.1016/J.JALLCOM.2011.02.061
40. Bai, J., Jiang, S., Liu, D., Tao, R., Fan, X., Xu, C.: The CuS-coated CuO nanostructures grown by ion exchange reaction for efficient photoelectrochemical water splitting. J. Alloys Compd. **1005**, 176036 (2024). https://doi.org/10.1016/J.JALLCOM.2024.176036
41. Mana, P.M., Bhujbal, P.K., Pathan, H.M.: Fabrication and characterization of ZnS based photoelectrochemical solar cell. ES Energy Environ. **12**, 77–85 (2021). https://doi.org/10.30919/ESEE8C1021
42. Li, C.T., Li, S.R., Chang, L.Y., Lee, C.P., Chen, P.Y., Sun, S.S., Lin, J.J., Vittal, R., Ho, K.C.: Efficient titanium nitride/titanium oxide composite photoanodes for dye-sensitized solar cells and water splitting. J. Mater. Chem. A Mater. **3**(8), 4695–4705 (2015). https://doi.org/10.1039/C4TA05606J
43. Jung, Y., Baik, K.H., Ren, F., Pearton, S.J., Kim, J.: Effects of photoelectrochemical etching of N-polar and Ga-polar gallium nitride on sapphire substrates. J. Electrochem. Soc. **157**(6), H676 (2010). https://doi.org/10.1149/1.3384713/XML
44. Sonar, S.M., Guo, Z., Ibrahim, M.M., Mersal, G.A.M., Tarkas, H.S., Ahirrao, P.B.: Study of photoelectrochemical (PEC) solar cell properties of n-type bismuth sulfide (n-Bi_2S_3)/p-type copper (I) iodide (p-CuI) heterojunction synthesized by SILAR. J. Mater. Sci. Mater. Electron. **35**(24), 1–13 (2024). https://doi.org/10.1007/S10854-024-13310-Z/TABLES/2
45. Zhang, H., Lu, Y., Han, W., Zhu, J., Zhang, Y., Huang, W.: Solar energy conversion and utilization: towards the emerging photo-electrochemical devices based on perovskite photovoltaics. Chem. Eng. J. **393**, 124766 (2020). https://doi.org/10.1016/J.CEJ.2020.124766
46. Jun, S.E., Lee, J.K., Jang, H.W.: Two-dimensional materials for photoelectrochemical water splitting. Energy Adv. **2**(1), 34–53 (2023). https://doi.org/10.1039/D2YA00231K
47. Uddin, Md.M., Kabir, M.H., Ali, Md.A., Hossain, Md.M., Khandaker, M.U., Mandal, S., Arifutzzaman, A., Jana, D.: Graphene-like emerging 2D materials: recent progress, challenges and future outlook. RSC Adv. **13**(47), 33336–33375 (2023). https://doi.org/10.1039/D3RA04456D
48. Li, X., Zang, X., Li, X., Zhu, M., Chen, Q., Wang, K., Zhong, M., Wei, J., Wu, D., Zhu, H.: Hybrid heterojunction and solid-state photoelectrochemical solar cells. Adv. Energy Mater. **4**(14), 1400224 (2014). https://doi.org/10.1002/AENM.201400224
49. Buhbut, S., Itzhakov, S., Oron, D., Zaban, A.: Quantum dot antennas for photoelectrochemical solar cells. J. Phys. Chem. Lett. **2**(15), 1917–1924 (2011). https://doi.org/10.1021/JZ200518Q/SUPPL_FILE/JZ200518Q_SI_001.PDF
50. Singh, S., Chen, H., Shahrokhi, S., Wang, L.P., Lin, C.H., Hu, L., Guan, X., Tricoli, A., Xu, Z.J., Wu, T.: Hybrid organic-inorganic materials and composites for photoelectrochemical water splitting. ACS Energy Lett. **5**(5), 1487–1497 (2020). https://doi.org/10.1021/ACSENERGYLETT.0C00327/ASSET/IMAGES/MEDIUM/NZ0C00327_0004.GIF

51. Ileperuma, O.A., Dissanayake, M.A.K.L., Somasunderam, S., Bandara, L.R.A.K.: Photoelectrochemical solar cells with polyacrylonitrile-based and polyethylene oxide-based polymer electrolytes. Sol. Energy Mater. Sol. Cells **84**(1–4), 117–124 (2004). https://doi.org/10.1016/J.SOLMAT.2004.02.040
52. Bandara, T.M.W.J., Fernando, H.D.N.S., Furlani, M., Albinsson, I., Dissanayake, M.A.K.L., Ratnasekera, J.L., Mellander, B.E.: Effect of the alkaline cation size on the conductivity in gel polymer electrolytes and their influence on photo electrochemical solar cells. Phys. Chem. Chem. Phys. **18**(16), 10873–10881 (2016). https://doi.org/10.1039/C6CP00013D
53. Ileperuma, O.A., Dissanayake, M.A.K.L., Somasundaram, S.: Dye-sensitised photoelectrochemical solar cells with polyacrylonitrile based solid polymer electrolytes. Electrochim. Acta **47**(17), 2801–2807 (2002). https://doi.org/10.1016/S0013-4686(02)00166-4
54. Ohno, H.: Electrochemical aspects of ionic liquids. In: Electrochemical Aspects of Ionic Liquids, pp. 1–392 (2005). https://doi.org/10.1002/0471762512
55. Boschloo, G., Hagfeldt, A.: Characteristics of the iodide/triiodide redox mediator in dye-sensitized solar cells. Acc. Chem. Res. **42**(11), 1819–1826 (2009). https://doi.org/10.1021/AR900138M/ASSET/IMAGES/MEDIUM/AR-2009-00138M_0007.GIF
56. Cai, Q., Liu, Z., Han, C., Tong, Z., Ma, C.: $CuInS_2/Sb_2S_3$ heterostructure modified with noble metal Co-catalyst for efficient photoelectrochemical water splitting. J. Alloys Compd. **795**, 319–326 (2019). https://doi.org/10.1016/J.JALLCOM.2019.04.312
57. Bielan, Z., Siuzdak, K.: Transition metal oxides in solar-to-hydrogen conversion. In: Materials for Hydrogen Production, Conversion, and Storage, pp. 1–39 (2023). https://doi.org/10.1002/9781119829584.CH1
58. Rurack, K., Martinez-Manez, R.: The Supramolecular Chemistry of Organic-Inorganic Hybrid Materials, p. 818 (2010)
59. Ali, M., Pervaiz, E., Sikandar, U., Khan, Y.: A review on the recent developments in zirconium and carbon-based catalysts for photoelectrochemical water-splitting. Int. J. Hydrogen Energy **46**(35), 18257–18283 (2021). https://doi.org/10.1016/J.IJHYDENE.2021.02.202
60. Naseem, S., Gevers, B.R., Labuschagné, F.J.W.J., Leuteritz, A.: Preparation of photoactive transition-metal layered double hydroxides (LDH) to replace dye-sensitized materials in solar cells. Materials **13**(19), 4384 (2020). https://doi.org/10.3390/MA13194384
61. Butler, M.A., Ginley, D.S.: Principles of photoelectrochemical, solar energy conversion. J. Mater. Sci. **15**(1), 1–19 (1980). https://doi.org/10.1007/BF00552421/METRICS
62. Fiechter, S., Bogdanoff, P., Bak, T., Nowotny, J.: Basic concepts of photoelectrochemical solar energy conversion systems. Adv. Appl. Ceram. **111**(1–2), 39–43 (2012). https://doi.org/10.1179/1743676111Y.0000000041
63. Chu, D., Yuan, X., Qin, G., Xu, M., Zheng, P., Lu, J., Zha, L.: Efficient carbon-doped nanostructured TiO_2 (anatase) film for photoelectrochemical solar cells. J. Nanopart. Res. **10**(2), 357–363 (2008). https://doi.org/10.1007/S11051-007-9241-7/FIGURES/6
64. Jacobsson, T.J., Edvinsson, T.: Quantum confined stark effects in ZnO quantum dots investigated with photoelectrochemical methods. J. Phys. Chem. C **118**(22), 12061–12072 (2014). https://doi.org/10.1021/JP503098Q/SUPPL_FILE/JP503098Q_SI_001.PDF
65. Chen, Y., Zheng, W., Murcia-López, S., Lv, F., Morante, J.R., Vayssieres, L., Burda, C.: Light management in photoelectrochemical water splitting—from materials to device engineering. J Mater. Chem. C Mater. **9**(11), 3726–3748 (2021). https://doi.org/10.1039/D0TC06071B
66. Low, F.W., Lai, C.W.: Recent developments of graphene-TiO_2 composite nanomaterials as efficient photoelectrodes in dye-sensitized solar cells: a review. Renew. Sustain. Energy Rev. **82**, 103–125 (2018). https://doi.org/10.1016/J.RSER.2017.09.024
67. Maçaira, J., Andrade, L., Mendes, A.: Review on nanostructured photoelectrodes for next generation dye-sensitized solar cells. Renew. Sustain. Energy Rev. **27**, 334–349 (2013). https://doi.org/10.1016/J.RSER.2013.07.011
68. Zhang, Q., Cao, G.: Nanostructured photoelectrodes for dye-sensitized solar cells. Nano Today **6**(1), 91–109 (2011). https://doi.org/10.1016/J.NANTOD.2010.12.007

69. Lee, W., Kwak, W.C., Min, S.K., Lee, J.C., Chae, W.S., Sung, Y.M., Han, S.H.: Spectral broadening in quantum dots-sensitized photoelectrochemical solar cells based on CdSe and Mg-doped CdSe nanocrystals. Electrochem. Commun. **10**(11), 1699–1702 (2008). https://doi.org/10.1016/J.ELECOM.2008.08.025
70. Kongkanand, A., Domínguez, R.M., Kamat, P.V.: Single wall carbon nanotube scaffolds for photoelectrochemical solar cells. Capture and transport of photogenerated electrons. Nano Lett. **7**(3), 676–680 (2007). https://doi.org/10.1021/NL0627238/ASSET/IMAGES/MEDIUM/NL0627238N00001.GIF
71. Yaseen, M., Khalid, K., Bibi, S., Khan, A., Tuzen, M., Saleh, T.A.: Recent trends in photoelectrocatalysts: types, influencing factors, and versatile applications: a comprehensive review. Sustain. Mater. Technol. **41**, e01067 (2024). https://doi.org/10.1016/J.SUSMAT.2024.E01067
72. Tilley, S.D.: Recent advances and emerging trends in photo-electrochemical solar energy conversion. Adv. Energy Mater. **9**(2), 1802877 (2019). https://doi.org/10.1002/AENM.201802877
73. Raza, A., Zhang, X., Ali, S., Cao, C., Rafi, A.A., Li, G.: Photoelectrochemical energy conversion over 2D materials. Photochem **2**(2), 272–298 (2022). https://doi.org/10.3390/PHOTOCHEM2020020
74. Best Research-Cell Efficiency Chart | Photovoltaic Research | NREL. [Online]. Available: https://www.nrel.gov/pv/cell-efficiency.html. Accessed: 02 Sept 2024
75. Harmon, M., Gamba, I.M., Ren, K.: Numerical algorithms based on galerkin methods for the modeling of reactive interfaces in photoelectrochemical (PEC) solar cells. J. Comput. Phys. **327**, 140–167 (2016). https://doi.org/10.1016/J.JCP.2016.08.026
76. Kim, D., Lee, D.K., Kim, S.M., Park, W., Sim, U.: Photoelectrochemical water splitting reaction system based on metal-organic halide perovskites. Materials **13**(1), 210 (2020). https://doi.org/10.3390/MA13010210
77. Prévot, M.S., Sivula, K.: Photoelectrochemical tandem cells for solar water splitting. J. Phys. Chem. C **117**(35), 17879–17893 (2013). https://doi.org/10.1021/JP405291G/ASSET/IMAGES/MEDIUM/JP-2013-05291G_0013.GIF
78. Wang, J., Liu, Z.: Recent advances in two-dimensional layered materials for photoelectrochemical sensing. TrAC Trends Anal. Chem. **133**, 116089 (2020). https://doi.org/10.1016/J.TRAC.2020.116089
79. Licht, S.: Electrolyte modified photoelectrochemical solar cells. Sol. Energy Mater. Sol. Cells **38**(1–4), 305–319 (1995). https://doi.org/10.1016/0927-0248(94)00229-0
80. Joshi, R.M.: Effect of concentration of the electrolyte on the performance of photoelectrochemical (PEC) solar cells using $Mote_2$ single crystals. Int. J. Techn. Res. Appl. **2**(4), 142–146. [Online]. Available: www.ijtra.com. Accessed: 03 Sept 2024
81. Sofia, T., Lopes, T.: Characterization and phenomenological photoelectrochemical production from
82. Biswas, N.K., Srivastav, A., Saxena, S., Verma, A., Dutta, R., Srivastava, M., Upadhyay, S., Satsangi, V.R., Shrivastav, R., Dass, S.: The impact of electrolytic ph on photoelectrochemical water oxidation. RSC Adv. **13**(7), 4324–4330 (2023). https://doi.org/10.1039/D2RA07271H
83. Singh, A.P., Kodan, N., Mehta, B.R., Held, A., Mayrhofer, L., Moseler, M.: Band edge engineering in $BiVO_4/TiO_2$ heterostructure: enhanced photoelectrochemical performance through improved charge transfer. ACS Catal. **6**(8), 5311–5318 (2016). https://doi.org/10.1021/ACSCATAL.6B00956/SUPPL_FILE/CS6B00956_SI_001.PDF
84. Shen, S., Chen, J., Wang, M., Sheng, X., Chen, X., Feng, X., Mao, S.S.: Titanium dioxide nanostructures for photoelectrochemical applications. Prog. Mater. Sci. **98**, 299–385 (2018). https://doi.org/10.1016/J.PMATSCI.2018.07.006
85. Moon, C., Shin, B.: Review on light absorbing materials for unassisted photoelectrochemical water splitting and systematic classifications of device architectures. Discover Mater. **2**(1), 1–16 (2022). https://doi.org/10.1007/S43939-022-00026-2

86. Licht, S., Peramunage, D.: Efficient photoelectrochemical solar cells from electrolyte modification. Nature **345**(6273), 330–333 (1990). https://doi.org/10.1038/345330a0
87. Schleuning, M., Ahmet, I.Y., van de Krol, R., May, M.M.: The role of selective contacts and built-in field for charge separation and transport in photoelectrochemical devices. Sustain. Energy Fuels **6**(16), 3701–3716 (2022). https://doi.org/10.1039/D2SE00562J
88. 1.2 Efficiency of Conversion | EME 812: Utility Solar Power and Concentration. [Online]. Available: https://www.e-education.psu.edu/eme812/node/4. Accessed: 03 Sept 2024
89. Sivula, K., Sivula, K.: Solar-to-chemical energy conversion with photoelectrochemical tandem cells. Chimia (Aarau) **67**(3), 155 (2013). https://doi.org/10.2533/chimia.2013.155
90. Current Losses Due to Recombination | PVEducation. [Online]. Available: https://www.pveducation.org/pvcdrom/current-losses-due-to-recombination. Accessed: 03 Sept 2024
91. Mamun, A.A., Billah, A., Anisuzzaman Talukder, M.: Effects of activation overpotential in photoelectrochemical cells considering electrical and optical configurations. Heliyon **9**(6), e17191 (2023). https://doi.org/10.1016/J.HELIYON.2023.E17191
92. He, L., Zhang, Q., Gong, C., Liu, H., Hu, F., Zhong, F., Wang, G., Su, H., Wen, S., Xiang, S., Zhang, B.: The dual-function of hematite-based photoelectrochemical sensor for solar-to-electricity conversion and self-powered glucose detection. Sens. Actuators B Chem. **310**, 127842 (2020). https://doi.org/10.1016/J.SNB.2020.127842
93. Fontecave, M., Fécant, A., Uzio, D., Passalacqua, R., Centi, G., Perathoner, S., Balantseva, E., Camino, B., Ferrari, A.M., Berlier, G., Udani, P.P.C., Rønning, M., Kawamura, S., Ahmed, N., Carja, G., Izumi, Y., Ogura, Y., Yoshiba, M., Rongé, J., Bosserez, T., Huguenin, L., Dumortier, M., Haussener, S., Martens, J.A., Van Humbeeck, J., Martens, J., Vaiano, V., Iervolino, G., Sarno, G., Sannino, D., Rizzo, L., Mesa, J.J.M., Hidalgo, M.C., Navío, J.A.: Design of compact photoelectrochemical cells for water splitting. Oil Gas Sci. Technol. Rev. d'IFP Energ. Nouv. **70**(5), 877–889 (2015). https://doi.org/10.2516/OGST/2015015
94. Ma, Q.B., Ziegler, J., Kaiser, B., Fertig, D., Calvet, W., Murugasen, E., Jaegermann, W.: Solar water splitting with p-SiC film on p-Si: photoelectrochemical behavior and XPS characterization. Int. J. Hydrogen Energy **39**(4), 1623–1629 (2014). https://doi.org/10.1016/J.IJHYDENE.2013.11.042
95. Van Nguyen, C., Thanh Hai, N., Olejnicek, J., Ksirova, P., Kohout, M., Dvorakova, M., Van Hao, P., Ngoc Hong, P., Cuong Tran, M., Hoang Tung, D., Van Thanh, D.: Preparation and photoelectrochemical performance of porous TiO_2/graphene nanocomposite films. Mater. Lett. **213**, 109–113 (2018). https://doi.org/10.1016/J.MATLET.2017.11.008
96. Favaro, M., Kong, H., Gottesman, R.: In situ and operando Raman spectroscopy of semiconducting photoelectrodes and devices for photoelectrochemistry. J. Phys. D Appl. Phys. **57**(10), 103002 (2023). https://doi.org/10.1088/1361-6463/AD10D3
97. Cen, J., Wu, Q., Liu, M., Orlov, A.: Developing new understanding of photoelectrochemical water splitting via in-situ techniques: a review on recent progress. Green Energy Environ. **2**(2), 100–111 (2017). https://doi.org/10.1016/J.GEE.2017.03.001
98. Freitas, D.V., González-Moya, J.R., Soares, T.A.S., Silva, R.R., Oliveira, D.M., Mansur, H.S., Machado, G., Navarro, M.: Enhanced visible-light photoelectrochemical conversion on TiO_2 nanotubes with Bi_2S_3 quantum dots obtained by in situ electrochemical method. ACS Appl. Energy Mater. **1**(8), 3636–3645 (2018). https://doi.org/10.1021/ACSAEM.8B00375/SUPPL_FILE/AE8B00375_SI_001.PDF
99. Yang, W., Moehl, T., Service, E., Tilley, S.D.: Operando analysis of semiconductor junctions in multi-layered photocathodes for solar water splitting by impedance spectroscopy. Adv. Energy Mater. **11**(9), 2003569 (2021). https://doi.org/10.1002/AENM.202003569
100. Yu, W., Fu, H.J., Mueller, T., Brunschwig, B.S., Lewis, N.S.: Atomic force microscopy: emerging illuminated and operando techniques for solar fuel research. J. Chem. Phys. **153**(2) (2020). https://doi.org/10.1063/5.0009858/76255
101. Nguyen, D.N., Fadel, M., Chenevier, P., Artero, V., Tran, P.D.: Water-splitting artificial leaf based on a triple-junction silicon solar cell: one-step fabrication through photoinduced deposition of catalysts and electrochemical operando monitoring. J. Am. Chem. Soc. **144**(22), 9651–9660 (2022). https://doi.org/10.1021/JACS.2C00666/SUPPL_FILE/JA2C00666_SI_001.PDF

102. Nasir, S.N.F.M., Ullah, H., Ebadi, M., Tahir, A.A., Sagu, J.S., Teridi, M.A.M.: New insights into Se/$BiVO_4$ heterostructure for photoelectrochemical water splitting: a combined experimental and DFT study. J. Phys. Chem. C **121**(11), 6218–6228 (2017). https://doi.org/10.1021/ACS.JPCC.7B01149/SUPPL_FILE/JP7B01149_SI_001.PDF
103. Zandi, S., Razaghi, M.: Finite element simulation of perovskite solar cell: a study on efficiency improvement based on structural and material modification. Sol. Energy **179**, 298–306 (2019). https://doi.org/10.1016/J.SOLENER.2018.12.032
104. He, L., Yang, Z., Gong, C., Liu, H., Zhong, F., Hu, F., Zhang, Y., Wang, G., Zhang, B.: The dual-function of photoelectrochemical glucose oxidation for sensor application and solar-to-electricity production. J. Electroanal. Chem. **882**, 114912 (2021). https://doi.org/10.1016/J.JELECHEM.2020.114912

Chapter 3
Photoelectrochemical (PEC) Detectors

This chapter explores the advanced field of photoelectrochemical (PEC) detectors, focusing on their principles, configurations, and diverse applications. It begins with an overview of the fundamental principles of PEC detection, including signal generation through the interaction of light with semiconductor photoelectrodes. The chapter then delves into various PEC detector configurations, such as flat-band electrodes, nanostructured electrodes, photoelectrochemical cells with reflectors, and integrated devices. Each configuration is analyzed for its advantages and limitations in enhancing PEC sensing performance. The chapter also covers the wide range of applications for PEC detectors, including environmental monitoring, biosensing, food safety, medical diagnostics, and energy harvesting. The conclusion highlights the current state of PEC detector technology and offers perspectives on future advancements, emphasizing material development, nanotechnology, integration with emerging technologies, sustainability, and data connectivity. This comprehensive overview underscores the potential of PEC detectors to revolutionize sensing technologies across multiple domains.

3.1 Introduction

Photoelectrochemical (PEC) detectors have emerged as a pivotal technology in the field of analytical devices, leveraging the synergistic principles of electrochemistry and photochemistry to facilitate the detection of various analytes. At their core, PEC detectors utilize semiconductor materials that, when exposed to light, generate electron–hole pairs. These charge carriers are then separated and transported to the electrode–electrolyte interface, where they participate in redox reactions, producing a measurable electrical signal. This unique mechanism allows PEC detectors to offer high sensitivity and selectivity, making them suitable for a wide range of applications [1] (Fig. 3.1).

P. Phogat et al., *Electrochemical Devices*, Engineering Materials,
https://doi.org/10.1007/978-981-96-0527-9_3

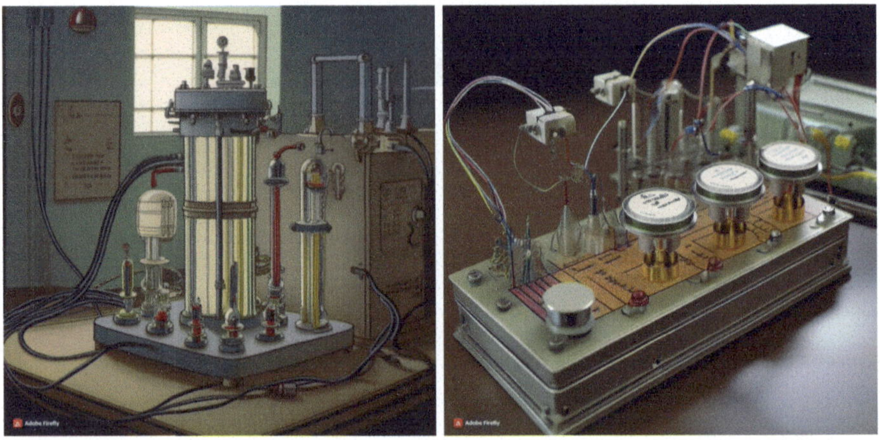

Fig. 3.1 Photoelectrochemical (PEC) detectors created by AI software firefly.adobe.com

The significance of PEC detectors lies in their versatility and efficiency. These devices are capable of operating under ambient conditions and often do not require external power sources, particularly in the case of self-powered PEC detectors. This autonomy enhances their applicability in remote and resource-limited settings. Moreover, the ability to tailor the semiconductor materials and modify the surface properties of photoelectrodes enables the fine-tuning of PEC detectors for specific applications [2]. This adaptability is particularly beneficial in environmental monitoring, where PEC detectors can be used to detect trace levels of pollutants, such as heavy metals and organic contaminants, in water and air [3]. In biomedical diagnostics, PEC detectors are instrumental in identifying biomarkers and pathogens, offering rapid and accurate results that are crucial for early disease detection and management. The integration of PEC detectors into point-of-care diagnostic devices has the potential to revolutionize healthcare, providing accessible and reliable diagnostic solutions in both developed and developing regions [4]. Additionally, in the realm of chemical sensing, PEC detectors are employed to quantify various chemical substances, from glucose in clinical settings to toxic chemicals in industrial environments [5]. The wide-ranging applications of PEC detectors underscore their importance in advancing analytical technologies and addressing critical challenges in environmental protection, healthcare, and industrial safety.

The chapter "Photoelectrochemical (PEC) Detectors" aims to provide a comprehensive understanding of the fundamental principles, materials, configurations, and applications of PEC detectors. The primary objective is to elucidate the underlying mechanisms of photoelectrochemical detection, highlighting the interplay between light and semiconductor materials in generating electrical signals. Additionally, the chapter seeks to guide the selection and design of semiconductor photoelectrodes optimized for PEC detector performance, including surface modification techniques and fabrication methods. Another key objective is to explore the concept of self-powered PEC detectors, emphasizing their potential for autonomous operation and energy conversion efficiency. By examining various configurations and device architectures, the chapter aims to offer insights into design considerations and performance metrics essential for developing efficient PEC detectors. Finally, through detailed case studies and real-world applications, the chapter intends to demonstrate the versatility and utility of PEC detectors in fields such as environmental monitoring, biomedical diagnostics, and chemical sensing, providing readers with a holistic view of current advancements and future possibilities in PEC detector technology [6].

PEC detectors represent a significant advancement in the field of analytical devices due to their high sensitivity, selectivity, and versatility. The underlying photoelectrochemical principles allow for the precise detection of a wide array of analytes, making these detectors invaluable tools in various sectors. The environmental monitoring capabilities of PEC detectors are particularly noteworthy. They can detect low concentrations of pollutants, providing crucial data for maintaining environmental health and safety. This capability is essential in addressing global challenges such as water pollution and air quality monitoring. In the healthcare sector, the rapid and accurate detection of biomarkers and pathogens using PEC detectors can significantly improve patient outcomes. Early diagnosis facilitated by these devices allows for timely intervention, which is vital for managing diseases effectively. The portability and ease of use of PEC detectors make them ideal for point-of-care applications, potentially transforming diagnostic practices and making healthcare more accessible, especially in resource-limited settings. Chemical sensing is another critical application area for PEC detectors. In industrial environments, these detectors can monitor the presence of hazardous chemicals, ensuring worker safety and compliance with environmental regulations. The ability to quantify substances like glucose also has significant implications for diabetes management, providing patients with reliable tools for monitoring their condition. The future of PEC detectors is promising, with ongoing research focused on enhancing their performance and expanding their applications. Advances in semiconductor materials, including the development of novel nanomaterials and composites, are expected to improve the efficiency and sensitivity of PEC detectors. Innovations in device architecture and fabrication techniques will further optimize their functionality and broaden their utility [7].

The development of self-powered PEC detectors is a particularly exciting area of research. These devices harness light energy to operate autonomously, eliminating the need for external power sources and making them ideal for deployment in remote and off-grid locations. Improvements in energy conversion efficiency and the integration of advanced materials will enhance the practicality and effectiveness of self-powered PEC detectors as shown in Fig. 3.2. Photoelectrochemical (PEC) detectors are at the forefront of analytical technology, offering unparalleled sensitivity, selectivity, and versatility. Their applications in environmental monitoring, biomedical diagnostics, and chemical sensing highlight their importance in addressing some of the most pressing challenges of our time. This chapter aims to provide a detailed exploration of PEC detectors, from their fundamental principles to their practical applications, equipping readers with the knowledge to understand and utilize this powerful technology. Through case studies and real-world examples, the chapter will showcase the

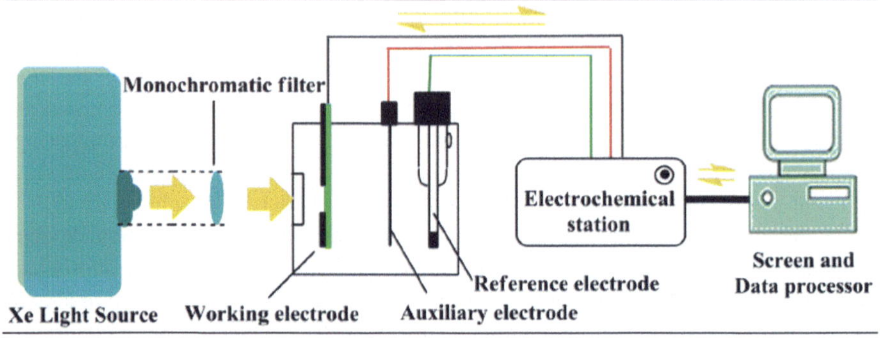

Fig. 3.2 Schematic diagram of the detection process of the photoelectrochemical sensor reproduced from Ref. [9] copyright © 1996–2024 MDPI (Basel, Switzerland)

potential of PEC detectors to make significant contributions to various fields, paving the way for future innovations and advancements [8].

3.2 Principles of Photoelectrochemical Detection

The principle of PEC detection is based on three main phenomenon's. First is photocatalysis and charge separation, second being the electrochemical reaction and third being the signal generation. The explanation of each phenomenon is given below,

3.2.1 Photocatalysis and Charge Separation

Light absorption is the fundamental process that initiates the photocatalytic activity in photoelectrochemical (PEC) detectors. This process begins when the semiconductor photoelectrode is exposed to photons with energy equal to or greater than its bandgap. The bandgap is the energy difference between the valence band, where electrons are normally present, and the conduction band, where electrons can move freely and participate in conduction. When a semiconductor material absorbs light, photons with sufficient energy excite electrons from the valence band to the conduction band, leaving behind positively charged holes in the valence band. This excitation process generates electron–hole pairs, which are crucial for the subsequent photocatalytic reactions. The efficiency of light absorption depends on several factors, including the nature of the semiconductor material, its bandgap, and the intensity and wavelength of the incident light as shown in Fig. 3.3. The choice of semiconductor material is critical for optimizing light absorption. Common materials used in PEC detectors include titanium dioxide (TiO_2), zinc oxide (ZnO), cadmium sulfide (CdS), and others, each with specific bandgap energies. For instance, TiO_2 has a bandgap of

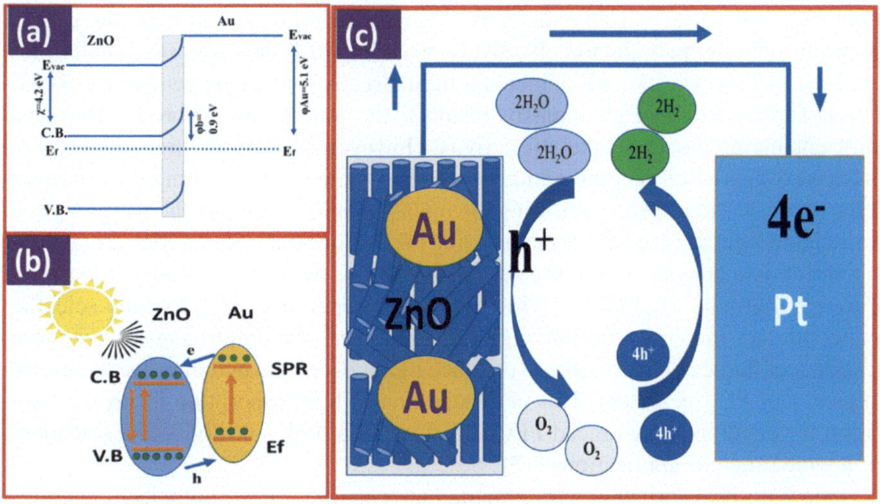

Fig. 3.3 Mechanism of PEC by Au/ZnO photoelectrode; **a** band bending, **b** energy diagram, and **c** PEC water splitting schematics reproduced from Ref. [14] copyright © 1996–2024 MDPI (Basel, Switzerland)

approximately 3.2 eV, making it suitable for absorbing ultraviolet light. In contrast, CdS has a narrower bandgap of around 2.4 eV, allowing it to absorb visible light more efficiently [10, 11].

The bandgap of a semiconductor determines the range of the electromagnetic spectrum it can absorb. Materials with wider bandgaps are typically limited to absorbing higher-energy photons, such as those in the ultraviolet range, while those with narrower bandgaps can absorb lower-energy photons in the visible range. This selection of materials enables the design of PEC detectors tailored for specific applications, depending on the desired range of light absorption. The absorption coefficient of a material indicates how effectively it can absorb light at a particular wavelength. A higher absorption coefficient means that the material can absorb more light over a shorter distance, which is desirable for creating efficient PEC detectors. The thickness of the semiconductor layer also plays a role in light absorption. Thicker layers can absorb more light, but if the layer is too thick, it can hinder the transport of charge carriers, leading to recombination losses. Optimizing the thickness of the semiconductor layer is thus essential to balance light absorption and charge transport. In some designs, nanostructured semiconductors, such as nanowires, nanotubes, or thin films, are employed to enhance light absorption while maintaining efficient charge carrier separation and transport [12].

Surface modification techniques, such as doping and sensitization, are often used to improve light absorption. Doping involves introducing foreign atoms into the semiconductor to create additional energy states within the bandgap, facilitating the absorption of a broader range of wavelengths. Sensitization, on the other hand, involves attaching light-absorbing molecules, such as dyes or quantum dots, to the semiconductor surface. These molecules can absorb light and transfer the excited electrons to the semiconductor, effectively extending its light absorption capabilities. The intensity and wavelength of incident light directly impact the generation of electron–hole pairs. Higher light intensity results in the generation of more electron–hole pairs, enhancing the photocatalytic activity. However, excessively high intensity can also lead to increased recombination of electron–hole pairs, reducing the efficiency of the process. Therefore, controlling the intensity and optimizing the wavelength of the incident light are crucial for maximizing the performance of PEC detectors. Light absorption is a critical step in the operation of PEC detectors, setting off the chain of events that lead to photocatalysis and charge separation. By carefully selecting semiconductor materials, optimizing their properties, and employing surface modification techniques, it is possible to enhance light absorption and improve the overall efficiency of PEC detectors. This understanding of light absorption forms the foundation for developing advanced PEC detectors with high sensitivity and selectivity for a wide range of applications [13].

Once light is absorbed by the semiconductor material in a photoelectrochemical (PEC) detector, the energy from the photons excites electrons from the valence band to the conduction band, creating electron–hole pairs. These pairs are known as excitons. An exciton is a bound state of an electron and a hole attracted to each other by Coulombic forces. The formation of excitons is crucial because they are the primary entities that drive the photocatalytic reactions in PEC detectors. The efficiency of exciton generation depends on the bandgap energy of the semiconductor and the energy of the incident photons. If the photon's energy is greater than or equal to the bandgap energy, it can excite an electron to the conduction band, leaving behind a hole in the valence band. This process creates excitons with energies corresponding to the excess energy of the photons over the bandgap. Excitons can be classified into two types based on their binding energy: tightly bound excitons and loosely bound excitons. Tightly bound excitons, or Frenkel excitons, typically occur in materials with high binding energies, such as organic semiconductors and some inorganic nanostructures. Loosely bound excitons, or Wannier-Mott excitons, are common in materials with lower binding energies, such as bulk inorganic semiconductors [15].

The successful operation of PEC detectors relies on the efficient separation and transport of the generated excitons into free charge carriers—electrons and holes—that can participate in redox reactions at the electrode–electrolyte interface. This process of charge carrier separation is critical for the conversion of absorbed light into a measurable electrical signal. Efficient charge carrier separation involves several key factors:

(a) **Built-In Electric Field**: In many PEC detectors, a built-in electric field at the semiconductor-electrolyte interface aids in the separation of electron–hole pairs. This electric field can be generated by creating a p–n junction or through the application of an external bias. The field drives electrons towards the conductive electrode and holes towards the counter electrode, minimizing recombination and facilitating charge carrier transport [16].

(b) **Nanostructuring**: Nanostructuring the semiconductor material can significantly enhance charge carrier separation. Nanostructures, such as nanowires, nanorods, and quantum dots, provide a large surface area for light absorption and create shorter paths for charge carriers to travel to the electrodes. This reduces the likelihood of recombination and increases the efficiency of charge separation [17].
(c) **Surface Passivation**: Surface passivation techniques are employed to reduce surface recombination of charge carriers. By passivating surface defects and dangling bonds, which often act as recombination centers, the lifespan of the charge carriers can be extended, enhancing their separation and transport. Methods such as coating the semiconductor surface with a thin insulating layer or using chemical treatments can effectively passivate surface states [18].
(d) **Doping and Heterostructures**: Doping the semiconductor with specific impurities can introduce additional energy states that facilitate charge separation. For instance, n-type doping adds extra electrons, while p-type doping adds extra holes, improving the charge carrier density and mobility. Additionally, constructing heterostructures—interfaces between different semiconductor materials—can create favorable energy band alignments that enhance charge separation. In such structures, the difference in bandgaps can create a staggered alignment, known as a Type-II heterojunction, which spatially separates electrons and holes into different materials, reducing recombination [19].
(e) **External Bias**: Applying an external bias can also assist in charge carrier separation. The bias provides an additional driving force that helps in the movement of electrons and holes towards their respective electrodes. However, for self-powered PEC detectors, the goal is to achieve efficient charge separation without the need for an external power source [20].

3.2.2 Electrochemical Reactions

Redox reactions at the electrode–electrolyte interface are central to the functionality of photoelectrochemical (PEC) detectors, as they enable the conversion of absorbed light into a measurable electrical signal. In these systems, semiconductor photoelectrodes absorb photons, generating electron–hole pairs that migrate to the surface and participate in electrochemical reactions with the electrolyte. At the anode, oxidation reactions occur, where the semiconductor or another species loses electrons, resulting in the formation of positive ions. Conversely, at the cathode, reduction reactions take place, where species in the electrolyte gain electrons, forming negative ions or neutral molecules. Transition metals, known for their multiple oxidation states, play a pivotal role in these redox reactions due to their ability to facilitate electron transfer processes. For instance, iron (Fe) can switch between Fe^{2+} and Fe^{3+} states, cobalt (Co) between Co^{2+} and Co^{3+}, and nickel (Ni) between Ni and Ni^{2+}, each undergoing oxidation and reduction as part of the PEC detector's operation.

The overall redox reaction efficiency depends on the material properties of the electrodes, the nature of the electrolyte, and the kinetics of the electron transfer process. Efficient charge separation and minimal recombination at the interface are crucial for optimizing the detector's sensitivity and selectivity. Moreover, the electrode–electrolyte interface can be engineered to enhance redox reaction efficiency through surface modifications, doping, and the use of nanostructured materials. These modifications can create more active sites for redox reactions, reduce energy barriers for electron transfer, and improve the overall charge transfer dynamics. Understanding and controlling these redox processes at the interface is essential for developing high-performance PEC detectors, enabling their application in diverse fields such as environmental monitoring, biomedical diagnostics, and chemical sensing. Redox reactions involve the transfer of electrons between chemical species, resulting in the reduction of one species and the oxidation of another [21].

The redox reactions at the electrode–electrolyte interface can be represented as follows:

Oxidation Reaction (at the Anode): $M \rightarrow M^{n+} + ne^-$

Reduction Reaction (at the Cathode): $A + ne^- \rightarrow A^{n-}$

Here, M represents the species undergoing oxidation, and A represents the species undergoing reduction. The n is the number of electrons transferred in the reaction.

Transition Metal Redox Reactions

Transition metals are commonly used in PEC detectors due to their multiple oxidation states, which facilitate redox reactions. For example, consider the redox reactions involving transition metal ions such as iron, cobalt, and nickel:

Iron (Fe) Redox Reactions:

$$Fe^{2+} \rightarrow Fe^{3+} + e^-(\text{Oxidation})$$

$$Fe^{3+} + e^- \rightarrow Fe^{2+}(\text{Reduction})$$

Cobalt (Co) Redox Reactions:

$$Co^{2+} \rightarrow Co^{3+} + e^-(\text{Oxidation})$$

$$Co^{3+} + e^- \rightarrow Co^{2+}(\text{Reduction})$$

Nickel (Ni) Redox Reactions:

$$Ni^{2+} + 2e^- \rightarrow Ni\ (\text{Reduction})$$

$$Ni \rightarrow Ni^{2+} + 2e^{-}(\text{Oxidation})$$

These redox reactions are integral to the functionality of PEC detectors, enabling the conversion of absorbed light into electrical signals through the transfer of electrons.

Electron Transfer Mechanisms

Electron transfer mechanisms describe how electrons move from the semiconductor photoelectrode to the redox species in the electrolyte. Two primary mechanisms govern this process: the inner-sphere and outer-sphere mechanisms [22].

Inner-Sphere Electron Transfer

In the inner-sphere mechanism, the electron transfer occurs through a direct bond between the oxidized and reduced species. This bond often involves a bridging ligand that connects the two redox centers, facilitating the electron transfer. For transition metals, inner-sphere mechanisms are common when the redox centers are close to each other or when a shared ligand can mediate the transfer.

For example, in the redox reaction between two transition metal complexes:

$$\left\{\left[M(H_2O)^6\right]^{3+} + \left[M'(CN)_6\right]^{4-} \rightarrow \left[M(H_2O)_6\right]^{2+} + \left[M'(CN)_6\right]^{3-}\right.$$

Here, M and M′ are different transition metals, and the bridging ligand (e.g., water or cyanide) facilitates the electron transfer [23].

Outer-Sphere Electron Transfer

In the outer-sphere mechanism, electron transfer occurs without any direct bond formation between the redox species. Instead, electrons tunnel through the solvent or medium that separates the redox centers. This mechanism is prevalent when the redox centers are not directly bonded and rely on the solvent's dielectric properties to facilitate electron transfer.

For example, the redox reaction between two metal ions in solution:

$$Fe^{2+} + Co^{3+} \rightarrow Fe^{3+} + Co^{2+}$$

In this reaction, the electron transfer happens through the medium, and the redox centers do not form a direct bond [24].

Equations and Kinetics

The kinetics of electron transfer reactions can be described by the Marcus theory, which provides insights into the rate of electron transfer based on the free energy change and reorganization energy of the system.

The rate constant (k_{ET}) for electron transfer can be expressed as:

$$k_{ET} = \mathrm{A} \exp\left(-\frac{\Delta G\ddagger}{RT}\right)$$

where

A is the pre-exponential factor, related to the frequency of attempts to transfer the electron.
ΔG‡ is the activation free energy barrier for the electron transfer.
R is the gas constant.
T is the temperature.

A schematical diagram for PEC sensor based on semiconductor is shown in Fig. 3.4. Redox reactions at the electrode–electrolyte interface and electron transfer mechanisms are essential for the functionality of PEC detectors. The ability to understand and optimize these processes, particularly through the use of transition metals and their multiple oxidation states, is crucial for enhancing the sensitivity and efficiency of these detectors. The interplay of inner-sphere and outer-sphere mechanisms, along with the kinetics of electron transfer, provides a comprehensive framework for designing advanced PEC systems for a variety of applications [25].

3.2.3 Signal Generation

The principles of photoelectrochemical (PEC) detection revolve around the generation of electrical signals through the interaction of light with semiconductor materials in an electrochemical cell. At the heart of PEC detection is the photoelectrode, a semiconductor material that absorbs incident light to initiate charge generation. When light photons with energy equal to or greater than the bandgap of the semiconductor strike the photoelectrode, they excite electrons from the valence band to the conduction band. This excitation creates electron–hole pairs, which are crucial for signal generation. The internal electric field of the semiconductor, established due to the difference in work function between the photoelectrode and the surrounding electrolyte, drives the separation of these charge carriers. Electrons migrate towards the photoelectrode surface, while holes move towards the electrolyte. Block diagram of the PEC sensing is shown in Fig. 3.5. At the interface between the photoelectrode and the electrolyte, these charge carriers participate in redox reactions. For instance, electrons can reduce species in the electrolyte, while holes can oxidize other species. The rate and extent of these electrochemical reactions directly influence the magnitude of the current generated. This current is proportional to the intensity of the absorbed light and the concentration of the electroactive species in the electrolyte. By measuring changes in current or potential, PEC detection translates optical signals into electrical signals. This conversion is critical for quantitative analysis, allowing precise measurement of analyte concentrations or detection of specific interactions. In essence, the PEC detection process hinges on the efficient generation, separation, and utilization of photoinduced charge carriers to produce a measurable electrical

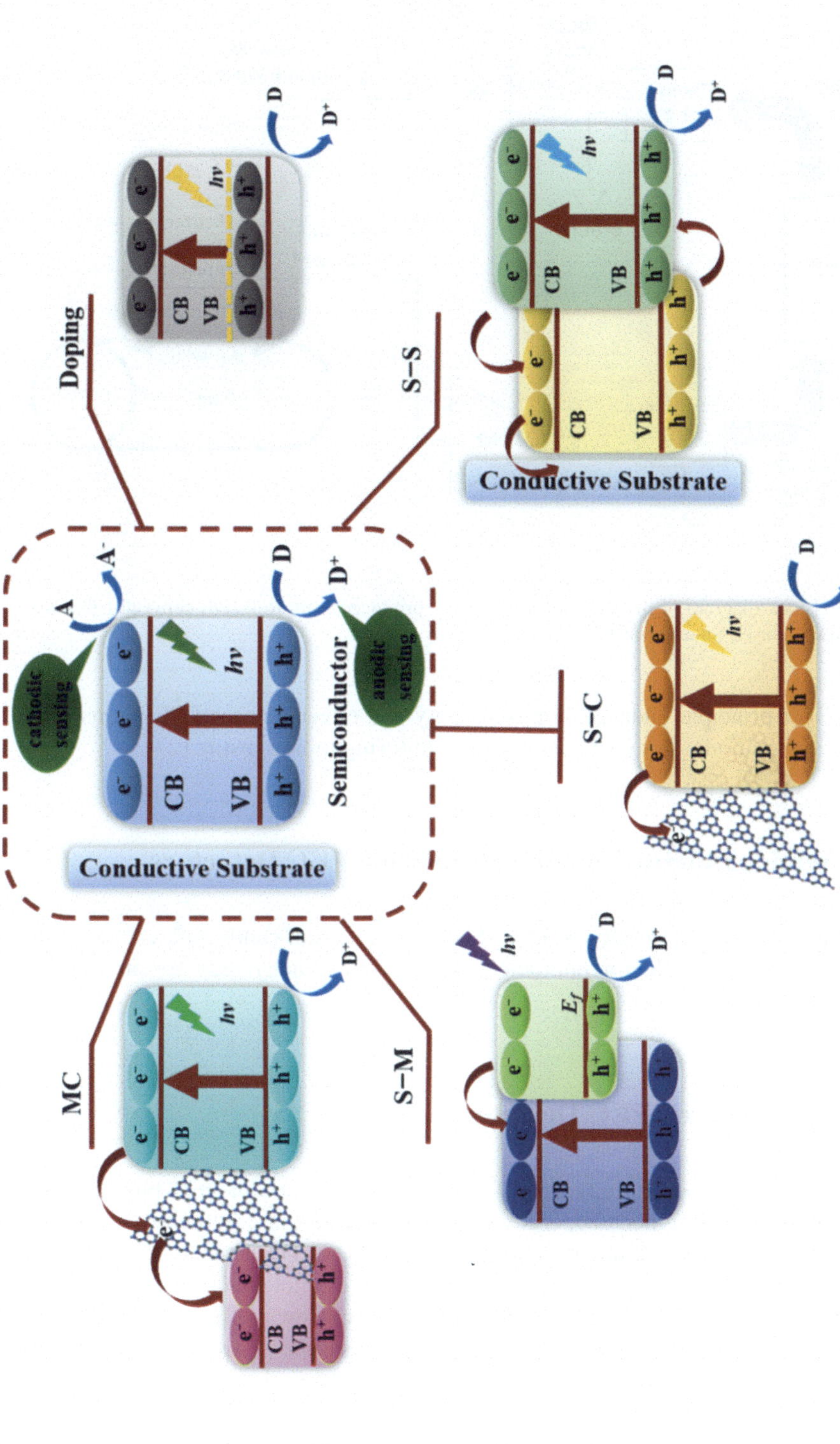

Fig. 3.4 Schematic diagram of a typical PEC sensor based on semiconductor photoactive materials reproduced from Ref. [26] copyright © 1996–2024 MDPI (Basel, Switzerland)

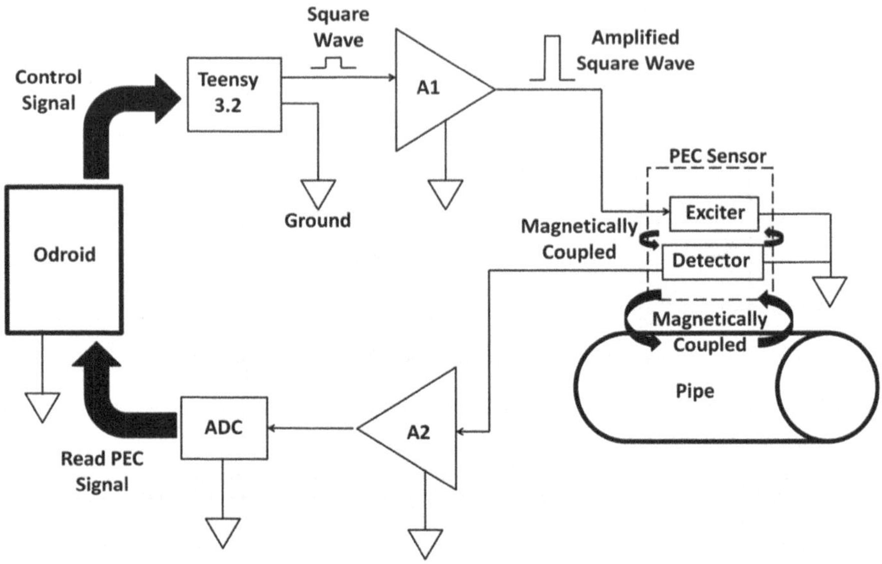

Fig. 3.5 Block diagram of the PEC sensing setup reproduced from Ref. [28] copyright © 1996–2024 MDPI (Basel, Switzerland)

signal. This principle underlies the sensitivity and effectiveness of PEC sensors in various applications, from environmental monitoring to biosensing [27].

3.3 Semiconductor Photoelectrodes for PEC Detectors

The efficiency and sensitivity of PEC detectors are significantly influenced by the properties of semiconductor photoelectrodes. Key characteristics are provided in Table 3.1.

3.4 Self-powered PEC Detectors

Self-powered PEC detectors represent an innovative advancement in photoelectrochemical sensing, designed to operate without the need for an external power source. These detectors harness ambient light to generate the electrical energy required for their operation. The core principle involves the use of photoelectrodes that convert light into electrical energy through the photoelectrochemical effect. When exposed to light, the semiconductor material in the photoelectrode generates electron–hole pairs, which drive electrochemical reactions at the electrode–electrolyte interface.

Table 3.1 Table depicting key characteristics and examples for PEC photodetector

Characteristic	Description	Examples
Bandgap energy	The bandgap of the semiconductor should match the wavelength of the incident light to maximize photon absorption. This ensures efficient excitation of electrons and optimal performance [29]	• TiO_2: ~ 3.0–3.2 eV • WO_3: ~ 2.6–2.8 eV • Fe_2O_3: ~ 2.0 eV
Surface properties	The surface area and morphology of the photoelectrode influence the rate of electrochemical reactions. Larger surface areas and favorable morphologies enhance the interaction between the photoelectrode and the electrolyte [30]	• Nanowires • Nanoparticles • Nanorods
Stability and durability	The photoelectrode material must be chemically stable and durable under operational conditions to ensure long-term reliability and consistent performance of the PEC detector [31]	• TiO_2: High chemical stability • WO_3: Good stability in various environments • Fe_2O_3: Moderate stability, often improved with coatings
Charge transport properties	Efficient charge transport within the semiconductor is essential to minimize recombination losses and enhance the photoresponse. High charge mobility contributes to better performance and sensitivity [32]	• TiO_2: Good electron mobility • WO_3: Moderate mobility • Fe_2O_3: Lower mobility, often enhanced with doping or composites

This process produces a current that powers the detector and provides the signal output [33].

3.4.1 Photoelectrode Material

The performance of self-powered photoelectrochemical (PEC) detectors hinges significantly on the choice of photoelectrode material. This material must efficiently convert light into electrical energy through the photoelectrochemical effect while maintaining stability and high activity under operational conditions. The selection of an appropriate photoelectrode material is critical because it influences the detector's sensitivity, energy conversion efficiency, and overall performance. The bandgap of the photoelectrode material should be well-suited to the wavelength of the incident light. Materials with bandgaps that align with the solar spectrum are preferable for maximizing photon absorption. Commonly used materials include Titanium Dioxide (TiO_2), which has a bandgap of about 3.0–3.2 eV, allowing it to absorb UV light effectively. Tungsten Trioxide (WO_3) and Iron Oxide (Fe_2O_3) are also popular due to their bandgaps of approximately 2.6–2.8 eV and 2.0 eV, respectively, which make them suitable for visible light absorption. Beyond bandgap energy, the photoelectrode material must exhibit high charge carrier mobility and minimal recombination losses. Materials such as Cadmium Sulfide (CdS) and Bismuth Vanadate ($BiVO_4$) are known for their high charge mobility and efficient photoelectrochemical conversion, making them ideal for self-powered PEC detectors. Additionally, the material should possess high chemical stability to ensure durability over time, particularly in various environmental conditions. To enhance performance, photoelectrode materials are often modified or combined with nanostructures such as nanoparticles, nanowires, or nanotubes. These modifications increase the surface area available for light absorption and electrochemical reactions, thereby improving the detector's sensitivity and efficiency. For instance, doping the photoelectrode material with other elements can also adjust the bandgap and enhance the material's performance under different light conditions. The choice of photoelectrode material is pivotal in the design of self-powered PEC detectors. It must balance factors such as bandgap energy, charge carrier mobility, chemical stability, and surface area to achieve high efficiency and stability in converting light into electrical energy [34].

3.4.2 Photoelectrochemical Cell Design

The design of a photoelectrochemical (PEC) cell is integral to the functionality of self-powered PEC detectors. This design encompasses several aspects, including the configuration of the photoelectrode, the electrolyte, and the overall cell architecture to optimize light absorption, charge separation, and energy conversion. The photoelectrode is typically positioned to maximize exposure to light while ensuring effective interaction with the electrolyte. Common configurations include flat electrodes and those with nanostructured surfaces. Nanostructured electrodes, such as those with nanotubes or nanowires, offer increased surface area and enhanced light absorption, which are crucial for improving the efficiency of the photoelectrochemical reactions. The choice of electrolyte is important for facilitating the redox reactions at the photoelectrode surface. The electrolyte should be compatible with the photoelectrode material and enhance the photoelectrochemical activity. Common electrolytes include aqueous solutions with salts or acids that support the desired electrochemical reactions. In some cases, solid-state electrolytes or gel-based electrolytes are used to improve stability and reduce leakage currents. The overall architecture of the PEC cell can influence its performance. Incorporating optical filters or reflectors within the cell design can enhance light absorption by increasing the effective path length of light within the photoelectrode material. Additionally, integrating a well-designed counter electrode that complements the photoelectrode material can improve the efficiency of the charge transfer process. For practical applications, the PEC cell must be integrated into a compact and durable package. This often involves encapsulating the photoelectrode and electrolyte in a way that protects them from environmental factors while allowing effective light entry and charge transfer. Advances in packaging technology also focus on ensuring that the cell remains efficient and stable over long periods of use. In essence, the design of a photoelectrochemical cell is a multi-faceted process that requires careful consideration of the photoelectrode configuration, electrolyte compatibility, overall architecture, and practical packaging. Each aspect plays a role in optimizing the performance and longevity of the self-powered PEC detector [35, 36].

3.4.3 Energy Harvesting

Energy harvesting in self-powered photoelectrochemical (PEC) detectors involves the efficient conversion of ambient light into electrical energy. This process is central to the operation of these detectors, as they rely on light as both the excitation source and the power supply. The core principle of energy harvesting in PEC detectors is the photoelectrochemical conversion of light. When the photoelectrode material absorbs photons, it generates electron–hole pairs, which drive electrochemical reactions at the electrode–electrolyte interface. This reaction produces a flow of current that powers the detector. The efficiency of this conversion depends on the material's bandgap, light absorption capabilities, and charge carrier dynamics. To enhance energy harvesting, various optimization techniques are employed. These include designing photoelectrodes with high light absorption efficiency, using advanced materials with better charge separation properties, and incorporating nanostructures to increase the effective surface area. Additionally, the use of anti-reflective coatings or light-trapping structures can improve the amount of light captured by the photoelectrode. In some designs, self-powered PEC detectors are integrated with energy storage components, such as supercapacitors or batteries. This integration allows the detector to store excess energy generated during periods of high light intensity, which can be used to power the device during low-light conditions or for extended periods. This approach enhances the reliability and functionality of the PEC detector, making it more versatile for different applications. One of the primary challenges in energy harvesting for self-powered PEC detectors is achieving high efficiency across a broad range of light intensities and conditions. Solutions to this challenge include developing photoelectrode materials with tunable bandgaps, optimizing cell designs for different light environments, and improving the durability of the materials to withstand various operating conditions. Overall, energy harvesting in self-powered PEC detectors is a complex process that requires careful material selection, design optimization, and integration with energy storage solutions. By addressing these factors, the efficiency and practicality of these detectors can be significantly enhanced [37].

3.4.4 Application Specifics

Self-powered photoelectrochemical (PEC) detectors are versatile devices with a range of applications across various fields, leveraging their ability to operate independently of external power sources. Each application exploits the unique advantages of self-powered PEC technology, including its portability, energy efficiency, and high sensitivity. In environmental monitoring, self-powered PEC detectors are used to detect pollutants and contaminants in air, water, and soil. Their ability to function autonomously makes them ideal for remote or inaccessible locations where external power sources are not feasible. These detectors can identify trace levels of harmful substances, such as heavy metals or organic pollutants, providing critical data for environmental protection and regulatory compliance. In biosensing applications, self-powered PEC detectors can identify biological molecules such as proteins, nucleic acids, and pathogens. By functionalizing the photoelectrode surface with specific ligands or antibodies, the PEC detector can selectively bind to target analytes, producing a detectable signal. This approach is valuable for medical diagnostics, allowing for rapid and sensitive detection of biomarkers and disease indicators without the need for external power. Self-powered PEC detectors are also employed in food safety to detect contaminants, toxins, and adulterants. These detectors can be integrated into portable devices for on-site testing, providing immediate results and enhancing food quality assurance. The ability to operate without external power is particularly advantageous for field testing and inspections. In medical diagnostics, self-powered PEC detectors are used to develop portable and low-cost diagnostic devices. These detectors can be employed in point-of-care testing for various diseases, offering rapid and reliable results. Their self-powered nature ensures that they can be used in resource-limited settings, making them suitable for global health applications. Beyond sensing, self-powered PEC detectors are explored for energy harvesting applications. By capturing ambient light and converting it into electrical energy, these detectors can power small electronic devices or sensors. This application is particularly useful in low-power devices and remote sensors where traditional power sources are impractical [38].

3.5 PEC Detector Configurations and Device Architectures

3.5.1 Flat-Band Electrodes

Flat-band electrodes are among the most traditional and straightforward configurations employed in photoelectrochemical (PEC) detectors. In this setup, the photoelectrode is characterized by a planar, uniform surface that is directly exposed to incident light. The primary operational principle of flat-band electrodes involves the direct illumination of the semiconductor material to generate photoelectrochemical signals. When light strikes the photoelectrode, it excites electrons from the valence band to the conduction band, leading to the generation of electron–hole pairs. These charge carriers facilitate electrochemical reactions at the semiconductor-electrolyte interface, producing measurable electrical signals. One of the main advantages of flat-band electrodes is their simplicity in design and fabrication. The uniform, flat surface allows for consistent and predictable light absorption, which simplifies the integration of the photoelectrode into various sensing systems. The straightforward nature of this configuration makes it relatively easy to manufacture and implement in practical applications. Additionally, the planar surface provides a large area for light exposure, which can be beneficial for certain applications where uniform light distribution is critical. Despite these advantages, flat-band electrodes come with notable limitations. A significant challenge is the relatively low surface area available for electrochemical reactions. The flat surface does not utilize the increased surface area that can enhance the rates of these reactions and improve the overall sensitivity of the PEC detector. In applications requiring high sensitivity or detection of low-concentration analytes, this limitation can result in reduced performance compared to more advanced configurations. To address these limitations, researchers and engineers have explored various modifications to enhance the performance of flat-band electrodes. One approach involves applying surface coatings or treatments to improve the photoelectrode's light absorption properties or to modify its electrochemical characteristics. For example, the addition of light-absorbing materials or catalysts can increase the efficiency of light-to-electrical energy conversion. These modifications aim to compensate for the inherent constraints of the flat-band design and improve the overall performance of the PEC detector. However, even with these enhancements, the fundamental design of flat-band electrodes remains constrained by their limited surface area and light absorption efficiency. The flat surface does not take full advantage of the potential benefits offered by advanced photoelectrode materials and structures. For instance, nanostructured semiconductors, such as nanowires or nanoparticles, provide significantly increased surface areas and better charge transport properties compared to flat surfaces. These nanostructures can enhance reaction rates and improve sensitivity, making them more effective for a range of applications. As a result, while flat-band electrodes serve as a foundational approach to PEC detection, they are often overshadowed by more advanced configurations in terms of sensitivity and efficiency. The flat-band design provides a basic framework for PEC sensors but may not fully exploit the capabilities of modern photoelectrode materials. For applications that demand higher sensitivity and performance, more sophisticated configurations, such as nanostructured electrodes or photoelectrochemical cells with reflectors, are frequently preferred. These advanced designs leverage the benefits of increased surface area and improved light absorption to achieve superior performance [39].

3.5.2 Nanostructured Electrodes

Nanostructured electrodes have significantly advanced photoelectrochemical (PEC) detector technology, surpassing the performance of traditional flat-band electrodes. This innovative configuration employs semiconductor materials engineered at the nanoscale—such as nanowires, nanoparticles, and nanorods—which markedly enhance PEC detectors' effectiveness. The key to these improvements lies in the enhanced surface area and unique properties of nanostructured materials. The primary advantage of nanostructured electrodes is their exceptionally high surface-to-volume ratio. Nanowires and nanoparticles offer a substantial increase in surface area compared to flat-band electrodes. This expanded surface area provides more active sites for electrochemical reactions, resulting in increased reaction rates and heightened sensitivity. The larger surface area facilitates more interactions between the photoelectrode and the electrolyte, leading to a more pronounced electrochemical response. In addition to increased surface area, nanostructured electrodes exhibit improved light absorption properties. Nanostructures, due to their unique geometries and sizes, are highly effective at trapping and scattering light. This ability to manipulate light enhances photon absorption within the semiconductor material. When light is absorbed more efficiently, it generates more photoinduced charge carriers—electrons and holes—at the photoelectrode. This enhanced generation of charge carriers translates to stronger and more reliable electrical signals, improving the overall sensitivity and accuracy of the PEC detector. Charge transport within nanostructured electrodes is also notably superior to that in flat-band electrodes. Nanostructures reduce the distance that charge carriers must travel, minimizing recombination losses where charge carriers might otherwise recombine before contributing to the electrochemical process. By shortening the path length for charge carriers, nanostructured electrodes enhance the efficiency of the photoelectrochemical reaction. This reduction in recombination losses and improved charge transport contribute to faster response times, allowing for quicker detection and analysis. Despite these advantages, nanostructured electrodes pose certain challenges, particularly in terms of fabrication and stability. The production of uniform and reproducible nanostructures requires precise control over synthesis methods. Techniques such as chemical vapor deposition, sol–gel processes, or electrospinning are often used to create nanostructures, each requiring careful optimization to achieve the desired size, shape, and distribution. Variability in these parameters can lead to inconsistent performance, making quality control a critical factor. Maintaining the stability of nanostructured electrodes under operational conditions also presents challenges. Nanostructures can be susceptible to environmental factors such as moisture, temperature, and chemical exposure, which may affect their performance and longevity. Protective coatings or stabilization strategies are sometimes employed to enhance durability, but these additional steps can complicate the fabrication process and increase costs. Moreover, integrating nanostructured electrodes into practical PEC devices often involves complex processing and assembly steps. The alignment and integration of nanostructures with other components in a PEC system require meticulous engineering and design. The increased complexity can pose technical challenges and may necessitate advanced fabrication techniques. Despite these challenges, the performance benefits of nanostructured electrodes make them a compelling choice for advanced PEC detection applications. Their ability to significantly enhance light absorption, improve charge transport, and increase sensitivity has positioned them as a leading technology in the field of photoelectrochemical sensing. As research and development continue, further refinements in fabrication techniques and stability strategies are likely to address current limitations, expanding the potential applications and impact of nanostructured electrodes in various sensing and analytical technologies [40]. The typical biosensing digram is shown in Fig. 3.6.

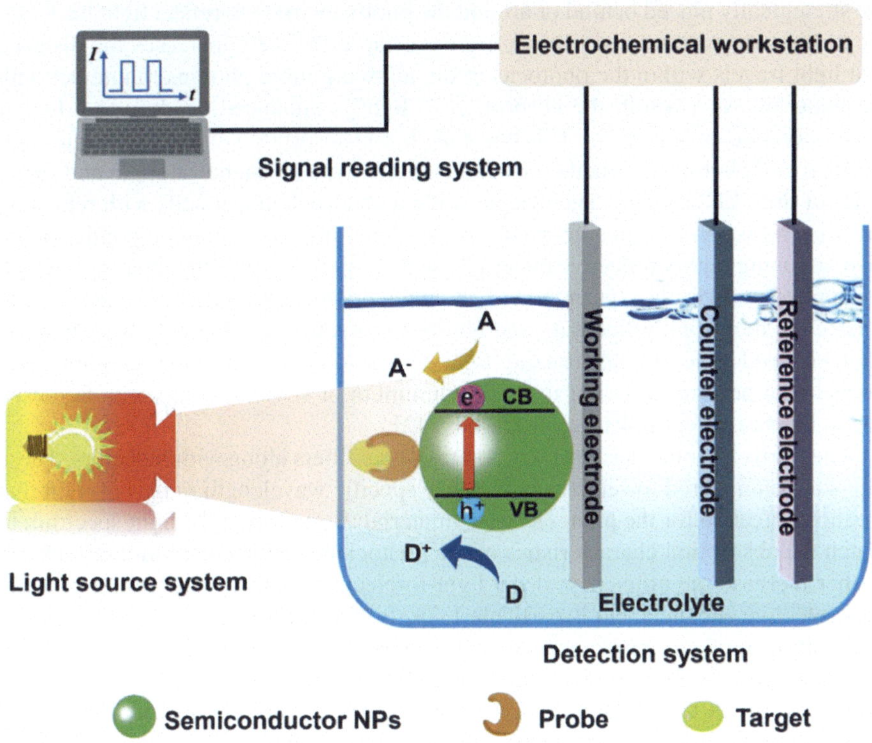

Fig. 3.6 Diagram of typical PEC biosensing system reproduced from Ref. [41] copyright © 1996–2024 MDPI (Basel, Switzerland)

3.5.3 Photoelectrochemical Cells with Reflectors

Photoelectrochemical (PEC) cells with reflectors represent a cutting-edge advancement in enhancing the performance of PEC detectors by optimizing light utilization. This innovative approach integrates optical reflectors or filters into the design to significantly improve photon absorption within the photoelectrode material. By extending the effective path length of light, these reflectors enhance the efficiency of the photoelectrochemical process, leading to more robust and sensitive detection capabilities. In a typical PEC cell, the photoelectrode is exposed to incident light, which is absorbed to generate photoinduced charge carriers—electrons and holes—that drive the electrochemical reactions necessary for detection. Reflectors are strategically placed behind or around the photoelectrode to reflect light back into the semiconductor material. This reflective setup effectively increases the distance that light travels within the photoelectrode, allowing more photons to interact with the material. As a result, the absorption of light is enhanced, leading to a higher generation of charge carriers. This boost in photon absorption directly translates into an improved photoresponse, thereby enhancing the overall signal strength and sensitivity of the PEC detector. The design of photoelectrochemical cells with reflectors can vary based on the specific needs and applications. One common configuration involves using highly reflective materials, such as metal or dielectric coatings, which are applied to surfaces adjacent to or behind the photoelectrode. These materials are chosen for their high reflectivity and ability to direct light efficiently back into the photoelectrode, thereby maximizing light absorption and improving the photoelectrochemical performance. For instance, aluminum or silver coatings are often used due to their excellent reflective properties [42].

Alternatively, some designs incorporate optical filters along with reflectors. These filters are engineered to selectively enhance specific wavelength ranges of light that are most effective for the photoelectrode material. By tailoring the light spectrum to match the absorption characteristics of the photoelectrode, these configurations can further increase the efficiency of the light-to-electrical energy conversion process. This selective enhancement is particularly useful for optimizing the performance of PEC cells in applications requiring precise wavelength sensitivity. Beyond improving light absorption, reflectors also play a crucial role in mitigating light scattering and loss within the PEC cell. In the absence of reflectors, light that is not absorbed by the photoelectrode can scatter or escape, reducing the overall effectiveness of the detection process. Reflectors help to minimize these losses by redirecting scattered light back towards the photoelectrode, thus increasing the likelihood of photon absorption and further enhancing the efficiency of the PEC detector. This configuration is especially advantageous in scenarios where maximizing light absorption is critical. For example, in low-light environments or for detecting analytes at low concentrations, the enhanced light utilization provided by reflectors can significantly improve the performance of PEC detectors. By ensuring that more of the incident light is absorbed and converted into electrical signals, reflectors enable more accurate and reliable detection even under challenging conditions [43].

However, the integration of reflectors into PEC cells introduces additional complexity in the design and fabrication processes. Achieving the desired performance improvements requires careful optimization of various factors, including the choice of reflector materials, their placement, and their alignment relative to the photoelectrode. The reflector's properties must be matched to the specific photoelectrode material and the operational conditions of the PEC cell to achieve the best results. This added complexity can make the design and production of PEC detectors with reflectors more challenging compared to simpler configurations. Despite these challenges, the benefits of incorporating reflectors into PEC cells are substantial. The ability to enhance light absorption and improve signal strength makes this configuration a valuable tool for advancing PEC sensing technologies. By leveraging reflectors to optimize light utilization, researchers and engineers can develop more sensitive, accurate, and efficient PEC detectors for a wide range of applications. Whether for environmental monitoring, biosensing, or other critical fields, photoelectrochemical cells with reflectors offer a promising pathway to achieving superior performance and enhanced detection capabilities [44].

3.5.4 Integrated Devices

Integrated devices in photoelectrochemical (PEC) detection represent a cutting-edge approach by seamlessly combining multiple functionalities into a single, compact system. These advanced configurations often merge PEC detectors with other critical components, such as energy-harvesting elements or additional sensing technologies, resulting in multifunctional devices that offer enhanced capabilities and operational efficiency. A prominent example of integrated PEC devices is the fusion of PEC sensors with photovoltaic cells. This integration allows the combined device to perform dual functions: detecting analytes through photoelectrochemical processes and simultaneously harvesting ambient light to power itself or other electronic components. By harnessing light not only for detection but also for energy production, these self-sustaining systems reduce dependence on external power sources. This integration is particularly advantageous in remote or field applications where access to power is limited. The ability to generate and utilize energy from the same light source used for detection enhances the device's versatility and operational longevity, making it an ideal solution for continuous and autonomous monitoring. Another significant approach in integrated PEC devices is the incorporation of microfluidic systems or lab-on-a-chip (LoC) technologies. Microfluidics involves the manipulation of small fluid volumes within microscale channels, enabling precise control over sample handling and reaction conditions. When integrated with PEC sensors, microfluidics allows for on-chip analysis of samples, facilitating real-time, rapid, and efficient detection of various analytes within complex matrices. This combination offers numerous advantages, including reduced reagent consumption, minimized sample volumes, and enhanced reaction kinetics. The ability to perform comprehensive analyses on a single chip not only streamlines the detection process but also improves accuracy and speed, making it particularly useful for applications in diagnostics, environmental monitoring, and chemical analysis [45].

Furthermore, integrated PEC devices often incorporate advanced signal processing and data acquisition systems to enhance their functionality and usability. The integration of microelectronics, such as embedded processors or analog-to-digital converters, allows for sophisticated data processing and real-time monitoring. Wireless communication components, such as Bluetooth or Wi-Fi modules, enable remote data transmission and access, making these devices suitable for field applications where immediate feedback is crucial. The ability to collect, analyze, and transmit data remotely transforms PEC detectors into powerful tools for continuous environmental monitoring, clinical diagnostics, and smart sensing applications. Despite the numerous advantages offered by integrated devices, there are inherent challenges associated with their design, fabrication, and integration. The combination of multiple technologies requires careful consideration of compatibility, performance trade-offs, and manufacturing processes. Ensuring that all components work harmoniously within a compact system can be complex, particularly when dealing with different materials and technologies that have distinct requirements and constraints. Design optimization is critical to balancing the trade-offs between functionality, size, and cost while maintaining high performance and reliability. Fabrication of integrated devices also presents challenges. The process of assembling multiple technologies into a single device often involves sophisticated techniques and precision engineering. Ensuring that all components are seamlessly integrated and function as intended requires advanced manufacturing capabilities and rigorous quality control. Additionally, the long-term stability and durability of integrated devices must be addressed to ensure reliable performance under various operational conditions. Nevertheless, the potential benefits of integrated PEC devices make them a promising direction for future developments in photoelectrochemical sensing. By combining multiple functionalities into a unified system, these devices offer enhanced capabilities, greater efficiency, and increased versatility. The ability to perform complex analyses, harvest energy, and communicate data remotely positions integrated PEC devices at the forefront of innovative sensing technologies, paving the way for new applications and advancements in environmental monitoring, diagnostics, and beyond. As technology continues to evolve, addressing the challenges associated with integrated devices will be crucial for realizing their full potential and achieving widespread adoption in various fields [46].

3.6 Applications of PEC Detectors

This Table 3.2 outlines the diverse applications of PEC detectors, highlighting their versatility and the benefits they offer in each field.

Table 3.2 Table outline different applications and the key feature with examples

Application	Description	Key features	Examples
Environmental monitoring	PEC detectors are employed to identify and quantify pollutants and contaminants in environmental samples, including air, water, and soil. They are particularly effective for detecting trace amounts of hazardous substances due to their high sensitivity and selectivity [47, 48]	• High sensitivity to low concentration pollutants • Ability to detect multiple contaminants • Real-time monitoring capabilities	• Detection of heavy metals in water (e.g., Pb, Cd) • Monitoring of air quality for pollutants (e.g., NO_2, SO_2) • Soil contamination analysis for pesticides and industrial chemicals
Biosensing	In biosensing applications, PEC detectors are utilized to identify and quantify biological molecules such as proteins, nucleic acids, and pathogens. Functionalized photoelectrodes can specifically interact with target biomolecules, enabling sensitive and selective detection [49, 50]	• Specific interactions with biological molecules • High sensitivity for low-abundance targets • Potential for multiplexed detection	• Detection of specific proteins or antibodies in clinical samples • Nucleic acid detection for genetic testing • Pathogen detection in food or clinical samples
Food safety	PEC detectors are used to ensure the quality and safety of food products by detecting contaminants, toxins, and adulterants. This application helps in consumer protection and compliance with regulatory standards [51, 52]	• Detection of foodborne pathogens • Monitoring for toxic substances and adulterants • Ensuring regulatory compliance	• Detection of pesticide residues on fruits and vegetables • Identification of food adulterants (e.g., melamine in milk) • Monitoring for bacterial contamination (e.g., *E. coli*, *Salmonella*)
Medical diagnostics	PEC detectors are integrated into diagnostic devices to detect biomarkers and disease indicators. Their rapid and sensitive detection capabilities are advantageous for early diagnosis and monitoring of various health conditions [53, 54]	• Rapid and sensitive detection of biomarkers • Integration with diagnostic devices for point-of-care testing • Potential for personalized medicine applications	• Detection of cancer biomarkers in blood samples • Monitoring of glucose levels in diabetes management • Detection of infectious disease markers (e.g., HIV, hepatitis)

(continued)

Table 3.2 (continued)

Application	Description	Key features	Examples
Energy harvesting	Self-powered PEC detectors are explored for energy harvesting, converting ambient light into electrical energy to power small electronic devices. This application leverages the photoelectric effect to create sustainable, self-sufficient systems [55, 56]	• Conversion of light into electrical energy • Self-sustaining operation without external power • Suitable for small, low-power devices	• Powering wireless sensors or IoT devices • Energy harvesting for remote environmental monitoring stations • Self-powered portable electronic devices (e.g., calculators, low-power gadgets)

3.7 Conclusion and Future Prospective

Photoelectrochemical (PEC) detectors represent a transformative advancement in sensing technology, combining the principles of photoelectrochemistry with innovative materials and device architectures. Throughout this chapter, we have explored the fundamental principles of PEC detection, including the mechanisms of signal generation and the role of semiconductor photoelectrodes. We also examined various configurations and device architectures, such as flat-band electrodes, nanostructured electrodes, photoelectrochemical cells with reflectors, and integrated devices, each offering unique advantages and addressing specific challenges in PEC sensing. PEC detectors have demonstrated their efficacy across a broad spectrum of applications, from environmental monitoring and biosensing to food safety, medical diagnostics, and energy harvesting. Their ability to provide sensitive, real-time, and accurate measurements makes them invaluable tools in these fields. The integration of PEC detectors with other technologies, such as photovoltaic cells and microfluidic systems, has further enhanced their capabilities, leading to the development of multifunctional and self-sustaining devices that push the boundaries of conventional sensing technologies [57, 58].

Looking ahead, the future of PEC detectors is poised for significant advancements driven by ongoing research and technological innovation. Several key areas hold promise for enhancing the performance and expanding the applications of PEC detectors:

Material Development: The discovery and development of new semiconductor materials with optimized bandgap energies, enhanced stability, and improved charge transport properties will be crucial. Advanced materials such as 2D materials, composites, and doped semiconductors could offer superior performance and enable new functionalities in PEC detection [58, 59].

Nanotechnology and Microfabrication: Continued progress in nanotechnology and microfabrication techniques will facilitate the creation of highly efficient nanostructured photoelectrodes and integrated devices. These advancements will enhance the sensitivity, response time, and overall performance of PEC detectors, enabling their use in more demanding applications [60–62].

Integration with Emerging Technologies: The integration of PEC detectors with emerging technologies, such as flexible electronics, wearable devices, and smart systems, will open new avenues for real-time monitoring and diagnostics. This integration could lead to the development of portable, user-friendly devices with enhanced capabilities for various applications [63, 64].

Sustainability and Energy Efficiency: The pursuit of self-powered and energy-efficient PEC detectors will be a key focus. Innovations in energy harvesting, coupled with improvements in device efficiency, could lead to sustainable solutions for remote and off-grid applications, reducing the reliance on external power sources [65, 66].

Data Management and Connectivity: The incorporation of advanced data management and connectivity features will enable seamless integration with digital platforms and cloud-based systems. This will facilitate real-time data analysis, remote monitoring, and the development of intelligent sensing networks for diverse applications [67, 68].

In conclusion, PEC detectors hold immense potential for revolutionizing various fields through their unique sensing capabilities and integration with other technologies. Continued research and development will drive further advancements, leading to more sophisticated, efficient, and versatile PEC sensing solutions. The future of PEC detectors promises to be dynamic and impactful, with the potential to address emerging challenges and contribute to advancements in science and technology.

References

1. Sivula, K., Van De Krol, R.: Semiconducting materials for photoelectrochemical energy conversion. Nat. Rev. Mater. **1**(2), 1–16 (2016). https://doi.org/10.1038/natrevmats.2015.10
2. Zang, Y., Fan, J., Ju, Y., Xue, H., Pang, H.: Current advances in semiconductor nanomaterial-based photoelectrochemical biosensing. Chem. Eur. J. **24**(53), 14010–14027 (2018). https://doi.org/10.1002/CHEM.201801358
3. Shi, L., Yin, Y., Zhang, L.C., Wang, S., Sillanpää, M., Sun, H.: Design and engineering heterojunctions for the photoelectrochemical monitoring of environmental pollutants: a review. Appl. Catal. B **248**, 405–422 (2019). https://doi.org/10.1016/J.APCATB.2019.02.044
4. Wang, Y., Rong, Y., Ma, T., Li, L., Li, X., Zhu, P., Zhou, S., Yu, J., Zhang, Y.: Photoelectrochemical sensors based on paper and their emerging applications in point-of-care testing. Biosens. Bioelectron. **236**, 115400 (2023). https://doi.org/10.1016/J.BIOS.2023.115400
5. Shu, J., Tang, D.: Recent advances in photoelectrochemical sensing: from engineered photoactive materials to sensing devices and detection modes. Anal. Chem. **92**(1), 363–377 (2020). https://doi.org/10.1021/ACS.ANALCHEM.9B04199/ASSET/ACS.ANALCHEM.9B04199.FP.PNG_V03

6. Zang, Y., Lei, J., Ju, H.: Principles and applications of photoelectrochemical sensing strategies based on biofunctionalized nanostructures. Biosens. Bioelectron. **96**, 8–16 (2017). https://doi.org/10.1016/J.BIOS.2017.04.030
7. Qureshi, A., Shaikh, T., Niazi, J.H.: Semiconductor quantum dots in photoelectrochemical sensors from fabrication to biosensing applications. Analyst **148**(8), 1633–1652 (2023). https://doi.org/10.1039/D2AN01690G
8. Kaur, M., Kumar, P., Ghotra, H.S.: A review on advances in photoelectrochemical (PEC-type) photodetectors: a trending thrust research area. Int. J. Hydrogen Energy **49**, 1095–1112 (2024). https://doi.org/10.1016/J.IJHYDENE.2023.11.018
9. Wang, Y., Liu, J., Lin, F.: A photoelectrochemical sensor for the sensitive detection of cysteine based on cadmium sulfide/tungsten disulfide nanocomposites. Nanomaterials **14**(5), 427 (2024). https://doi.org/10.3390/NANO14050427/S1
10. Yanagi, R., Zhao, T., Solanki, D., Pan, Z., Hu, S.: Charge separation in photocatalysts: mechanisms, physical parameters, and design principles. ACS Energy Lett. **7**(1), 432–452 (2022). https://doi.org/10.1021/ACSENERGYLETT.1C02516/ASSET/IMAGES/MEDIUM/NZ1C02516_0010.GIF
11. Bott Neto, J.L., Martins, T.S., Machado, A.S.S., Oliveira, O.N.: Enhanced photocatalysis on graphitic carbon nitride sensitized with gold nanoparticles for photoelectrochemical immunosensors. Appl. Surf. Sci. **606**, 154952 (2022). https://doi.org/10.1016/J.APSUSC.2022.154952
12. Zhao, Y., Hoivik, N., Wang, K.: Recent advance on engineering titanium dioxide nanotubes for photochemical and photoelectrochemical water splitting. Nano Energy **30**, 728–744 (2016). https://doi.org/10.1016/J.NANOEN.2016.09.027
13. Tan, R., Sivanantham, A., Jansi Rani, B., Jeong, Y.J., Cho, I.S.: Recent advances in surface regulation and engineering strategies of photoelectrodes toward enhanced photoelectrochemical water splitting. Coord. Chem. Rev. **494**, 215362 (2023). https://doi.org/10.1016/J.CCR.2023.215362
14. Zayed, M., Nasser, N., Shaban, M., Alshaikh, H., Hamdy, H., Ahmed, A.M.: Effect of morphology and plasmonic on Au/ZnO films for efficient photoelectrochemical water splitting. Nanomaterials **11**(9), 2338 (2021). https://doi.org/10.3390/NANO11092338
15. Chen, Y., Yan, C., Dong, J., Zhou, W., Rosei, F., Feng, Y., Wang, L.N.: Structure/Property control in photocatalytic organic semiconductor nanocrystals. Adv. Funct. Mater. **31**(36), 2104099 (2021). https://doi.org/10.1002/ADFM.202104099
16. Wang, Y., Huang, J., Chen, Y., Yang, H., Ye, K.H., Huang, Y.: Modulating built-in electric field via Bi-VO_4-Fe interfacial bridges to enhance charge separation for efficient photoelectrochemical water splitting. J. Colloid Interface Sci. **672**, 12–20 (2024). https://doi.org/10.1016/J.JCIS.2024.05.218
17. Qiu, Z., Tang, D.: Nanostructure-based photoelectrochemical sensing platforms for biomedical applications. J. Mater. Chem. B **8**(13), 2541–2561 (2020). https://doi.org/10.1039/C9TB02844G
18. Chang, A.M., Chen, Y.H., Lai, C.C., Pu, Y.C.: Synergistic effects of surface passivation and charge separation to improve photo-electrochemical performance of BiOI nanoflakes by au nanoparticle decoration. ACS Appl. Mater. Interfaces **13**(4), 5721–5730 (2021). https://doi.org/10.1021/ACSAMI.0C18430/SUPPL_FILE/AM0C18430_SI_001.PDF
19. Liao, L., Wu, B., Kovalska, E., Oliveira, F.M., Azadmanjiri, J., Mazánek, V., Valdman, L., Spejchalová, L., Xu, C., Levinský, P., Hejtmánek, J., Sofer, Z.: InSe:Ge-doped inse van Der Waals heterostructure to enhance photogenerated carrier separation for self-powered photoelectrochemical-type photodetectors. Nanoscale **14**(14), 5412–5424 (2022). https://doi.org/10.1039/D1NR07150E
20. Tan, C.F., Ong, W.L., Ho, G.W.: Self-biased hybrid piezoelectric-photoelectrochemical cell with photocatalytic functionalities. ACS Nano **9**(7), 7661–7670 (2015). https://doi.org/10.1021/ACSNANO.5B03075/SUPPL_FILE/NN5B03075_SI_001.PDF
21. Zhao, W.W., Xu, J.J., Chen, H.Y.: Photoelectrochemical enzymatic biosensors. Biosens. Bioelectron. **92**, 294–304 (2017). https://doi.org/10.1016/J.BIOS.2016.11.009

22. Tu, W., Wang, Z., Dai, Z.: Selective photoelectrochemical architectures for biosensing: design, mechanism and responsibility. TrAC Trends Anal. Chem. **105**, 470–483 (2018). https://doi.org/10.1016/J.TRAC.2018.06.007
23. Seo, D., Won, S., Kim, J.T., Chung, T.D.: Adopting back reduction current as an additional output signal for achieving photoelectrochemical differentiated detection. Anal. Chem. **94**(4), 2063–2071 (2022). https://doi.org/10.1021/ACS.ANALCHEM.1C04129/SUPPL_FILE/AC1C04129_SI_001.PDF
24. Ramakrishnan, V., Tsyganok, A., Davydova, E., Pavan, M.J., Rothschild, A., Visoly-Fisher, I.: Competitive photo-oxidation of water and hole scavengers on hematite photoanodes: photoelectrochemical and operando Raman spectroelectrochemistry study. ACS Catal. **13**(1), 540–549 (2023). https://doi.org/10.1021/ACSCATAL.2C02849/SUPPL_FILE/CS2C02849_SI_001.PDF
25. Govindaraju, G.V., Wheeler, G.P., Lee, D., Choi, K.S.: Methods for electrochemical synthesis and photoelectrochemical characterization for photoelectrodes. Chem. Mater. **29**(1), 355–370 (2017). https://doi.org/10.1021/ACS.CHEMMATER.6B03469/ASSET/IMAGES/MEDIUM/CM-2016-03469A_0014.GIF
26. Mao, Y., Liu, X., Bao, Y., Niu, L.: Recent advances in photoelectrochemical sensors for analysis of toxins and abused drugs in the environment. Chemosensors **11**(7), 412 (2023). https://doi.org/10.3390/CHEMOSENSORS11070412
27. Zhao, W.W., Xu, J.J., Chen, H.Y.: Photoelectrochemical bioanalysis: the state of the art. Chem. Soc. Rev. **44**(3), 729–741 (2015). https://doi.org/10.1039/C4CS00228H
28. Ulapane, N., Alempijevic, A., Vidal Calleja, T., Miro, J.V.: Pulsed eddy current sensing for critical pipe condition assessment. Sensors **17**(10), 2208 (2017). https://doi.org/10.3390/S17102208
29. Zappia, M.I., Bianca, G., Bellani, S., Serri, M., Najafi, L., Oropesa-Nuñez, R., Martín-García, B., Bouša, D., Sedmidubský, D., Pellegrini, V., Sofer, Z., Cupolillo, A., Bonaccorso, F.: Solution-processed GaSe nanoflake-based films for photoelectrochemical water splitting and photoelectrochemical-type photodetectors. Adv. Funct. Mater. **30**(10), 1909572 (2020). https://doi.org/10.1002/ADFM.201909572
30. Chawla, P., Tripathi, M.: Surface modification of semiconductor photoanode for photoelectrochemical water splitting. Int. J. Hydrogen Energy **41**(19), 7987–7992 (2016). https://doi.org/10.1016/J.IJHYDENE.2015.11.118
31. Yang, W., Prabhakar, R.R., Tan, J., Tilley, S.D., Moon, J.: Strategies for enhancing the photocurrent, photovoltage, and stability of photoelectrodes for photoelectrochemical water splitting. Chem. Soc. Rev. **48**(19), 4979–5015 (2019). https://doi.org/10.1039/C8CS00997J
32. Yang, L., Xiong, Y., Guo, W., Zhou, M., Song, K., Xiao, P., Cao, G.: Manipulation of charge transport in ferroelectric-semiconductor hybrid for photoelectrochemical applications. Nano Energy **44**, 63–72 (2018). https://doi.org/10.1016/J.NANOEN.2017.11.066
33. Yang, P., Hou, X., Gao, X., Peng, Y., Li, Q., Niu, Q., Liu, Q.: Recent trends in self-powered photoelectrochemical sensors: from the perspective of signal output. ACS Sens. **9**(2), 577–588 (2024). https://doi.org/10.1021/ACSSENSORS.3C02198/ASSET/IMAGES/MEDIUM/SE3C02198_0007.GIF
34. Zhu, J., Shao, D., Wen, W., Tian, Z., Zhang, X., Wang, S.: Self-powered electrochemical sensor based on photoelectrode: an up-to-date review. Coord. Chem. Rev. **518**, 216095 (2024). https://doi.org/10.1016/J.CCR.2024.216095
35. Zhang, Z., Peng, B., Ouyang, X., Zhu, X., Chen, L., Fan, X., Zhou, Z., Wang, J., Tang, L.: A self-powered photoelectrochemical aptasensor based on dual-photoelectrode photofuel cell for chloramphenicol detection. Sens. Actuators B Chem. **368**, 132144 (2022). https://doi.org/10.1016/J.SNB.2022.132144
36. Smith, W.A.: Photoelectrochemical cell design, efficiency, definitions, standards, and protocols. In: Photoelectrochemical Solar Fuel Production: From Basic Principles to Advanced Devices, pp. 163–197 (2016). https://doi.org/10.1007/978-3-319-29641-8_4
37. Zi, Y., Hu, Y., Pu, J., Wang, M., Huang, W.: Recent progress in interface engineering of nanostructures for photoelectrochemical energy harvesting applications. Small **19**(19), 2208274 (2023). https://doi.org/10.1002/SMLL.202208274

38. Hun, X., Meng, Y.: Electron acceptors co-regulated self-powered photoelectrochemical strategy and its application for circulating tumor nucleic acid detection coupled with recombinase polymerase amplification. Anal. Chem. **92**(17), 11771–11778 (2020). https://doi.org/10.1021/ACS.ANALCHEM.0C01893/SUPPL_FILE/AC0C01893_SI_001.PDF
39. Li, C., Luo, Z., Wang, T., Gong, J.: Surface, bulk, and interface: rational design of hematite architecture toward efficient photo-electrochemical water splitting. Adv. Mater. **30**(30), 1707502 (2018). https://doi.org/10.1002/ADMA.201707502
40. Zhao, H., Lei, Y.: 3D nanostructures for the next generation of high-performance nanodevices for electrochemical energy conversion and storage. Adv. Energy Mater. **10**(28), 2001460 (2020). https://doi.org/10.1002/AENM.202001460
41. Shi, X.-M., Mei, L.-P., Wang, Q., Zhao, W.-W., Xu, J.-J., Chen, H.-Y.: Recent advances of nanostructured materials for photoelectrochemical bioanalysis. Chemosensors **10**(1), 14 (2021). https://doi.org/10.3390/CHEMOSENSORS10010014
42. Wang, W., Qi, L.: Light management with patterned micro- and nanostructure arrays for photocatalysis, photovoltaics, and optoelectronic and optical devices. Adv. Funct. Mater. **29**(25), 1807275 (2019). https://doi.org/10.1002/ADFM.201807275
43. He, X., Tian, W., Yang, L., Bai, Z., Li, L.: Optical and electrical modulation strategies of photoelectrodes for photoelectrochemical water splitting. Small Methods **8**(2), 2300350 (2024). https://doi.org/10.1002/SMTD.202300350
44. Jian, J., Jiang, G., van de Krol, R., Wei, B., Wang, H.: Recent advances in rational engineering of multinary semiconductors for photoelectrochemical hydrogen generation. Nano Energy **51**, 457–480 (2018). https://doi.org/10.1016/J.NANOEN.2018.06.074
45. Panaritis, C., Couillard, M., Baranova, E.A., Sugawara, Y., Inoue, W., Yomogida, A., Bukola, S., Mayura Silva, L., Creager, S.E.: Editors' choice—a monolithic photoelectrochemical device evolving hydrogen in pure water. J. Electrochem. Soc. **166**(13), H656 (2019). https://doi.org/10.1149/2.1151913JES
46. Han, Q., Wang, H., Wang, J.: Multi-mode/signal biosensors: electrochemical integrated sensing techniques. Adv. Funct. Mater. 2403122 (2024). https://doi.org/10.1002/ADFM.202403122
47. Cao, S., Jiao, Y., Xiao, G., Wu, W., Xie, Z., Li, J., Liu, X., Zhao, E., Yue, Z.: Miniaturized photoelectrochemical sensing system for reusable detection of macromolecules and its applications for unattended environmental monitoring. Sens. Actuators B Chem. **421**, 136515 (2024). https://doi.org/10.1016/J.SNB.2024.136515
48. Wang, H., Xu, Y., Xu, D., Chen, L., Qiu, X., Zhu, Y.: Graphitic carbon nitride for photoelectrochemical detection of environmental pollutants. ACS ES&T Eng **2**(2), 140–157 (2022). https://doi.org/10.1021/ACSESTENGG.1C00337/ASSET/IMAGES/MEDIUM/EE1C00337_0016.GIF
49. Wang, B., Cao, J.T., Liu, Y.M.: Recent progress of heterostructure-based photoelectrodes in photoelectrochemical biosensing: a mini review. Analyst **145**(4), 1121–1128 (2020). https://doi.org/10.1039/C9AN02448D
50. Svitkova, V., Palchetti, I.: Functional polymers in photoelectrochemical biosensing. Bioelectrochemistry **136**, 107590 (2020). https://doi.org/10.1016/J.BIOELECHEM.2020.107590
51. Chen, Y., Gu, W., Zhu, C., Hu, L.: Recent advances in photoelectrochemical sensing for food safety. Anal. Chem. **96**(22), 8855–8867 (2024). https://doi.org/10.1021/ACS.ANALCHEM.4C01062/ASSET/ACS.ANALCHEM.4C01062.FP.PNG_V03
52. Ge, L., Liu, Q., Hao, N., Kun, W.: Recent developments of photoelectrochemical biosensors for food analysis. J. Mater. Chem. B **7**(46), 7283–7300 (2019). https://doi.org/10.1039/C9TB01644A
53. Cao, S., Xie, Z., Xiao, G., Sun, X., Diao, H., Zhou, X., Yue, Z.: Photoelectrochemical sensors based on heterogeneous nanostructures for in vitro diagnostics. Biosens. Bioelectron. X **11**, 100200 (2022). https://doi.org/10.1016/J.BIOSX.2022.100200
54. Xiao, G., Ge, H., Yang, Q., Zhang, Z., Cheng, L., Cao, S., Ji, J., Zhang, J., Yue, Z.: Light-addressable photoelectrochemical sensors for multichannel detections of GPC1, CEA and GSH and its applications in early diagnosis of pancreatic cancer. Sens. Actuators B Chem. **372**, 132663 (2022). https://doi.org/10.1016/J.SNB.2022.132663

55. Wang, Y., Tang, J., Peng, Z., Wang, Y., Jia, D., Kong, B., Elzatahry, A.A., Zhao, D., Zheng, G.: Fully solar-powered photoelectrochemical conversion for simultaneous energy storage and chemical sensing. Nano Lett. **14**(6), 3668–3673 (2014). https://doi.org/10.1021/NL5014579/SUPPL_FILE/NL5014579_SI_001.PDF
56. Alhalaili, B., Dryden, D.M., Vidu, R., Ghandiparsi, S., Cansizoglu, H., Gao, Y., Saif Islam, M.: High-aspect ratio micro- and nanostructures enabled by photo-electrochemical etching for sensing and energy harvesting applications. Appl. Nanosci. (Switzerland) **8**(5), 1171–1177 (2018). https://doi.org/10.1007/S13204-018-0737-5/METRICS
57. Feng, Z.Y., Jiang, J.C., Meng, L.Y.: Carbon-based photoelectrochemical sensors: recent developments and future prospects. Dalton Trans. **53**(27), 11192–11215 (2024). https://doi.org/10.1039/D4DT00534A
58. Shi, J., Chen, Z., Zhao, C., Shen, M., Li, H., Zhang, S., Zhang, Z.: Photoelectrochemical biosensing platforms for tumor marker detection. Coord. Chem. Rev. **469**, 214675 (2022). https://doi.org/10.1016/J.CCR.2022.214675
59. Li, L., Chen, J., Xiao, C., Luo, Y., Zhong, N., Xie, Q., Chang, H., Zhong, D., Xu, Y., Zhao, M., Liao, Q.: Recent advances in photoelectrochemical sensors for detection of ions in water. Chin. Chem. Lett. **34**(6), 107904 (2023). https://doi.org/10.1016/J.CCLET.2022.107904
60. Sultan, S., Nimal, R., Aftab, S., Kurbanoglu, S., Shah, A., Ozkan, S.A.: Photoelectrochemical nanosensors. In: New Developments in Nanosensors for Pharmaceutical Analysis, pp. 197–229 (2019). https://doi.org/10.1016/B978-0-12-816144-9.00007-9
61. Zhou, Y., Yin, H., Ai, S.: Applications of two-dimensional layered nanomaterials in photoelectrochemical sensors: a comprehensive review. Coord. Chem. Rev. **447**, 214156 (2021). https://doi.org/10.1016/J.CCR.2021.214156
62. Paras, Yadav, K., Kumar, P., Teja, D.R., Chakraborty, S., Chakraborty, M., Mohapatra, S.S., Sahoo, A., Chou, M.M.C., Liang, C.-Te., Hang, D.R.: A review on low-dimensional nanomaterials: nanofabrication, characterization and applications. Nanomaterials **13**(1), 160 (2023). https://doi.org/10.3390/NANO13010160
63. Zhou, J., Chen, L., Wang, Y., He, Y., Pan, X., Xie, E.: An overview on emerging photoelectrochemical self-powered ultraviolet photodetectors. Nanoscale **8**(1), 50–73 (2015). https://doi.org/10.1039/C5NR06167A
64. Un Nisa, M., Ajaz, M.N., Rehman, A., Wahad, F., Gulzar, S., Abid, Z.: Challenges and future of photoelectrochemical bioanalysis. In: Photoelectrochemical Bioanalysis: Fundamentals and Emerging Applications, pp. 139–170 (2023). https://doi.org/10.1016/B978-0-443-18955-5.00015-1
65. Dashtian, K., Shahbazi, S., Tayebi, M., Masoumi, Z.: A review on metal-organic frameworks photoelectrochemistry: a headlight for future applications. Coord. Chem. Rev. **445**, 214097 (2021). https://doi.org/10.1016/J.CCR.2021.214097
66. Sathre, R., Greenblatt, J.B., Walczak, K., Sharp, I.D., Stevens, J.C., Ager, J.W., Houle, F.A.: Opportunities to improve the net energy performance of photoelectrochemical water-splitting technology. Energy Environ. Sci. **9**(3), 803–819 (2016). https://doi.org/10.1039/C5EE03040D
67. Pilarczyk, K., Wlaźlak, E., Przyczyna, D., Blachecki, A., Podborska, A., Anathasiou, V., Konkoli, Z., Szaciłowski, K.: Molecules, semiconductors, light and information: towards future sensing and computing paradigms. Coord. Chem. Rev. **365**, 23–40 (2018). https://doi.org/10.1016/J.CCR.2018.03.018
68. Victorious, A., Saha, S., Pandey, R., Didar, T.F., Soleymani, L.: Affinity-based detection of biomolecules using photo-electrochemical readout. Front. Chem. **7**, 478309 (2019). https://doi.org/10.3389/FCHEM.2019.00617/BIBTEX

Chapter 4
Photoelectrochemical Water Splitting

4.1 Introduction to Photoelectrochemical (PEC) Water Splitting

The quest for sustainable and clean energy solutions has led to significant advancements in various technologies, with photoelectrochemical (PEC) water splitting emerging as a promising approach for hydrogen production as shown in Fig. 4.1. This process harnesses solar energy to drive the electrochemical splitting of water into hydrogen and oxygen gases, offering a potential pathway to a clean and renewable energy future [1]. PEC water splitting leverages the unique properties of semiconductor materials, known as photoelectrodes, which absorb sunlight and convert it into electrical energy. This energy is then used to initiate and sustain the water-splitting reactions, namely the hydrogen evolution reaction (HER) and the oxygen evolution reaction (OER) [2]. The development of PEC technology not only represents a significant stride in energy conversion but also holds the promise of providing a scalable and efficient method for producing hydrogen fuel, which is vital for addressing global energy demands and environmental concerns. This section provides an overview of PEC water splitting by exploring its fundamental principles, historical development, and significance in the context of current and future energy needs. By understanding the origins and advancements of this technology, we can better appreciate its role in the broader landscape of renewable energy solutions and its potential impact on sustainable development.

PEC water splitting represents a promising approach to sustainable hydrogen production, a crucial component of the transition towards a clean energy future. This technology leverages solar energy to drive the electrochemical splitting of water into hydrogen and oxygen gases [3]. Hydrogen, as a clean fuel, has the potential to revolutionize energy systems due to its high energy content per unit mass and its only by-product when used in fuel cells—water. The process of PEC water splitting involves the use of semiconductor materials, known as photoelectrodes, which absorb sunlight and convert it into electrical energy [4]. This electrical energy is then used to

P. Phogat et al., *Electrochemical Devices*, Engineering Materials,
https://doi.org/10.1007/978-981-96-0527-9_4

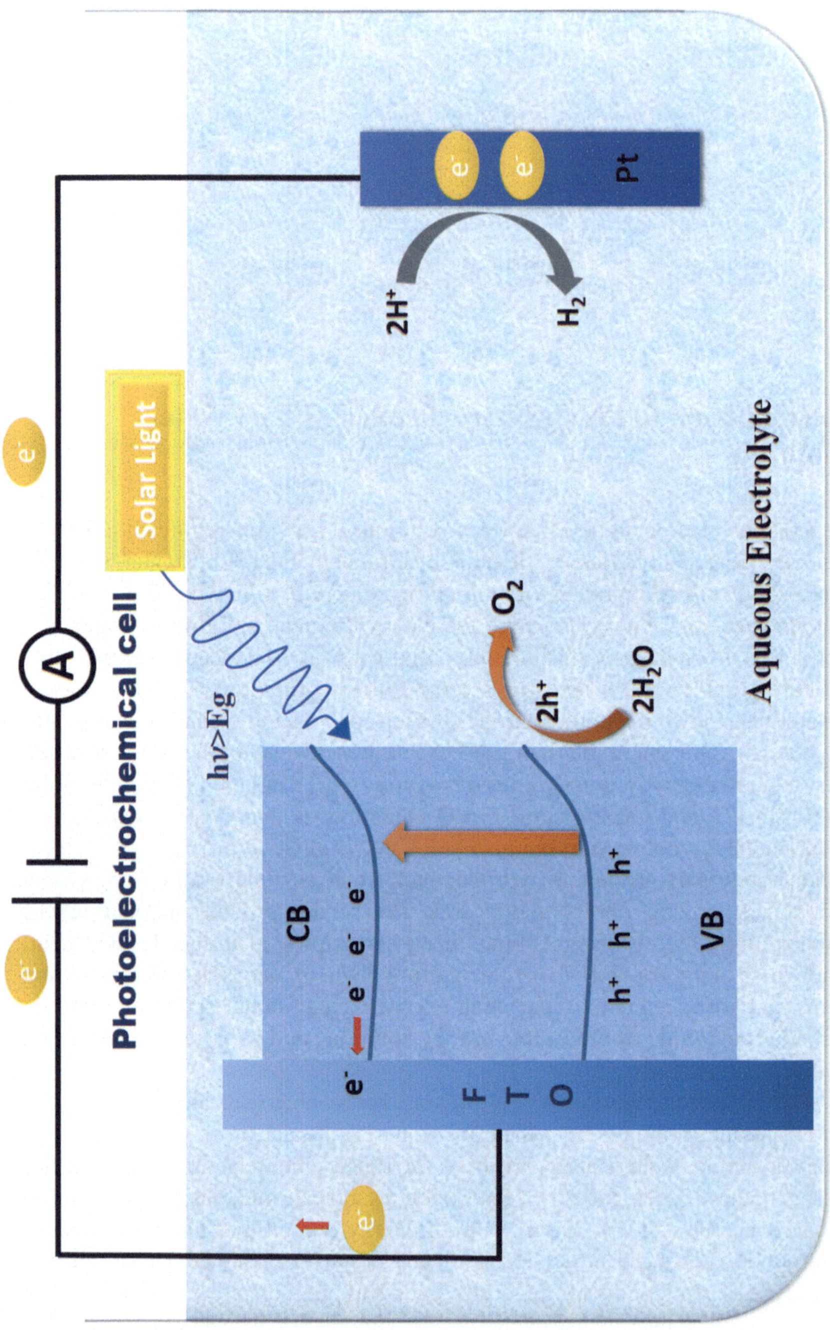

Fig. 4.1 Schematic diagram of PEC water splitting

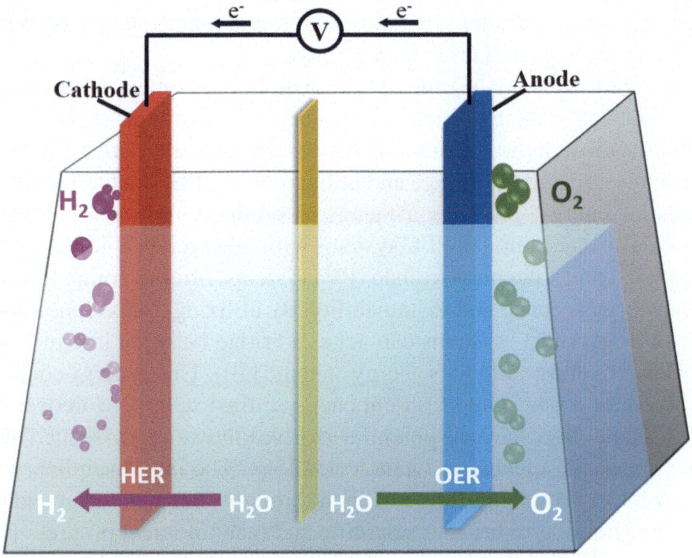

Fig. 4.2 Hydrogen evolution reaction (HER) and oxygen evolution reaction (OER)

drive the water-splitting reactions: hydrogen evolution reaction (HER) and oxygen evolution reaction (OER) as shown in Fig. 4.2. The primary advantage of PEC water splitting over traditional methods is its direct conversion of solar energy into chemical energy without requiring an external power source [5], thereby potentially reducing overall system costs and improving efficiency.

The concept of PEC water splitting dates back to the early 1970s, a period when researchers first demonstrated the feasibility of using semiconductors for water splitting. In 1972, Fujishima and Honda made a groundbreaking contribution by demonstrating the use of titanium dioxide (TiO_2) photoelectrodes for splitting water under illumination [6]. This seminal work laid the foundation for subsequent research in PEC technology. Following this initial breakthrough, the field witnessed significant advancements. In the 1980s, researchers improved the performance of PEC cells by developing new semiconductor materials and optimizing cell configurations. Notable progress included the discovery of materials with direct bandgaps and suitable conduction and valence band positions, which enhanced light absorption and charge separation efficiency [7]. The 1990s and early 2000s saw the advent of nanotechnology, which further revolutionized PEC water splitting. Nanostructured materials, such as nanoparticles, nanowires, and nanotubes, were introduced to enhance the surface area and improve charge transport properties. The development of co-catalysts and the integration of advanced materials enabled more efficient and stable PEC systems. In recent years, the focus has shifted towards the integration of novel materials such as perovskites, 2D materials, and hybrid systems to push the boundaries of efficiency and stability. Researchers are also exploring innovative PEC cell architectures and novel approaches to mitigate challenges such as photocorrosion and low solar-to-hydrogen (STH) efficiency [8].

PEC water splitting holds significant promise in the context of global efforts to transition to renewable energy sources. As traditional fossil fuels deplete and climate change concerns grow, hydrogen has emerged as a viable alternative due to its potential for zero-emission energy storage and utilization [9]. PEC technology, by directly harnessing solar energy, provides a means to produce hydrogen sustainably and economically. The integration of PEC systems with other renewable energy technologies, such as wind and solar photovoltaic (PV) systems, offers a compelling approach to achieving energy security and sustainability. By utilizing excess renewable energy to produce hydrogen, PEC systems can act as a bridge between intermittent renewable energy sources and continuous energy demand [10]. This synergy could facilitate the development of a robust and resilient energy infrastructure. Moreover, advancements in PEC technology have the potential to drive innovations in related fields, such as catalysis, materials science, and nanotechnology. The interdisciplinary nature of PEC research fosters collaborations and breakthroughs that extend beyond hydrogen production, contributing to broader scientific and technological progress [11].

Ongoing research and development (R&D) efforts are crucial to addressing the challenges and advancing PEC water splitting technology. Key areas of focus include improving the efficiency and stability of photoelectrodes, developing scalable manufacturing processes, and optimizing cell designs for practical applications. Collaboration between academic institutions, industry stakeholders, and government agencies is essential to accelerate progress and translate laboratory achievements into commercial solutions [12]. Research in PEC water splitting also benefits from advancements in related fields, such as computational modeling [13], which enables the prediction and optimization of material properties and cell performance. The integration of theoretical and experimental approaches ensures a comprehensive understanding of the underlying mechanisms and guides the development of next-generation PEC systems. In summary, photoelectrochemical water splitting is a transformative technology with the potential to significantly impact hydrogen production and the broader energy landscape [14]. The historical milestones and ongoing research highlight the continuous evolution and promise of PEC systems in addressing the global energy challenge. As we advance towards a sustainable future, PEC water splitting will play a pivotal role in harnessing solar energy for clean and efficient hydrogen production.

4.2 Fundamental Principles of PEC Water Splitting

Photoelectrochemical (PEC) water splitting is a sophisticated process that integrates principles from semiconductor physics, electrochemistry, and material science to convert solar energy into chemical energy through the splitting of water molecules into hydrogen and oxygen. Understanding the fundamental principles of PEC water splitting [15] is essential for optimizing the performance of PEC systems and advancing this technology. This section delves into the core aspects of PEC water splitting, including the thermodynamics and kinetics involved, energy band alignment and charge carrier dynamics [16], and the key reactions of hydrogen and oxygen evolution.

4.2.1 *Thermodynamics and Kinetics of Water Splitting*

Photoelectrochemical (PEC) water splitting is a complex process that converts solar energy into chemical energy by splitting water into hydrogen and oxygen. Understanding the thermodynamics and kinetics of water splitting is crucial for optimizing the efficiency of PEC systems [17]. This section delves into the fundamental thermodynamic principles and kinetic processes that govern the water-splitting reaction.

4.2.2 *Thermodynamics of Water Splitting*

Water splitting can be represented by the overall reaction:

$$H_2O \rightarrow H_2 + \frac{1}{1}O_2$$

This reaction involves the cleavage of O–H bonds in water molecules and the formation of H–H and O=O bonds in hydrogen and oxygen molecules, respectively. The process requires the input of energy, which is why sunlight, a renewable energy source, is employed in PEC systems. The Gibbs free energy change can be expressed as:

$$\Delta G = \Delta H - T\Delta S$$

where ΔH is the enthalpy change, T is the temperature, and ΔS is the entropy change. The Gibbs free energy change (ΔG) for the water-splitting reaction under standard conditions (25 °C, 1 atm) is + 237.13 kJ/mol. This positive ΔG indicates that the reaction is non-spontaneous, meaning that an external energy source (such as light) is needed to drive the reaction forward. In terms of electrical energy, this corresponds to a theoretical minimum voltage of 1.23 V that must be applied to overcome the Gibbs free energy barrier [18]. However, in practical PEC systems, the applied voltage needs to be higher than 1.23 V due to overpotentials associated with the various steps in the reaction. These overpotentials arise from inefficiencies in charge transfer, mass transport, and other kinetic barriers, leading to a required practical voltage often exceeding 1.5–2.0 V [19].

4.2.3 Kinetics of Water Splitting

The kinetics of water splitting involves multiple steps, each with its own energy barrier, influencing the overall reaction rate. The rate of the PEC water-splitting reaction is determined by two primary half-reactions: the Hydrogen Evolution Reaction (HER) and the Oxygen Evolution Reaction (OER). These reactions occur at the cathode and anode, respectively, and their rates are influenced by the kinetics of charge transfer at the electrode–electrolyte interface. The HER, which takes place at the cathode, is the reduction half-reaction where protons (H^+) in the electrolyte gain electrons to form hydrogen gas:

$$2H^+ + 2e^- \rightarrow H_2$$

This reaction, while seemingly simple, is highly dependent on the electrocatalyst used and the surface properties of the cathode. The kinetics of HER are influenced by factors such as the rate of proton transfer from the electrolyte to the electrode surface, the adsorption strength of hydrogen intermediates, and the rate at which hydrogen atoms combine to form H_2 molecules. The kinetics of HER can be improved by using electrocatalysts with high activity, such as platinum, or by optimizing the surface properties of the electrode to facilitate faster electron transfer [20]. On the other hand, the OER, which occurs at the anode, is the oxidation half-reaction where water molecules are oxidized to produce oxygen gas, electrons, and protons:

$$2H_2O \rightarrow O_2 + 4H^+ + 4e^-$$

The OER is more kinetically challenging than the HER due to the involvement of a four-electron transfer process and the formation of oxygen–oxygen bonds. The kinetics of OER are often the rate-limiting step in PEC water splitting [21], leading to significant overpotentials. This necessitates the use of highly active oxygen evolution catalysts, such as transition metal oxides (e.g., RuO_2, IrO_2), to lower the energy barrier and enhance the reaction rate [22]. The efficiency of the OER is also affected by the structure and surface properties of the anode, the pH of the electrolyte, and the presence of intermediate species that may adsorb on the electrode surface and influence the reaction pathway.

4.2.4 Overpotentials and Efficiency Losses

The overpotentials associated with HER and OER contribute to the efficiency losses in PEC water splitting. Overpotentials arise from various factors, including the activation energy required to initiate the reactions, the resistance to electron flow (ohmic losses) [23], and the mass transport limitations in the electrolyte. The sum of these overpotentials determines the additional voltage required beyond the theoretical

1.23 V to sustain the water-splitting reaction at a practical rate. To minimize overpotentials, it is crucial to design and select materials with high catalytic activity and low resistance to electron and ion transport [24]. For instance, the use of nanostructured electrodes with high surface area can reduce the overpotentials by providing more active sites for the reactions and enhancing charge carrier mobility. Additionally, optimizing the electrolyte composition and concentration can help mitigate mass transport limitations and improve the overall kinetics of the reactions.

4.2.5 Thermodynamic Considerations in Real-World Applications

While the theoretical thermodynamics of water splitting provide a foundation for understanding the energy requirements, real-world PEC systems must contend with various practical considerations that influence the overall efficiency and feasibility of hydrogen production [25]. Factors such as the stability of the photoelectrode materials under prolonged illumination, the efficiency of light absorption and charge transfer, and the integration of PEC systems with existing hydrogen production infrastructure all play a role in determining the practical thermodynamics and kinetics of the process. For instance, photocorrosion is a significant challenge in PEC water splitting, where the photoelectrode materials degrade over time due to the oxidative or reductive environment. This degradation not only affects the longevity of the PEC system but also introduces additional energy barriers that can increase the overpotentials and reduce the overall efficiency. Developing photoelectrode materials with enhanced stability [17] and resistance to photocorrosion is therefore a critical area of research in PEC water splitting. Moreover, the scalability of PEC systems is influenced by the ability to integrate the technology with existing hydrogen production methods, such as steam methane reforming or electrolysis. The thermodynamic and kinetic principles of water splitting must be carefully considered in the design of PEC systems to ensure that they can operate efficiently at large scales and compete with conventional hydrogen production technologies in terms of cost and efficiency.

4.3 Energy Band Alignment and Charge Carrier Dynamics

The efficiency of photoelectrochemical (PEC) water splitting is heavily dependent on the proper alignment of energy bands and the dynamics of charge carriers within the photoelectrode materials. This section explores the principles of energy band alignment, the generation and separation of charge carriers, and their transport to the electrode surfaces, which are critical for driving the hydrogen and oxygen evolution reactions.

4.3.1 Energy Band Alignment

In PEC water splitting, the energy band structure of the semiconductor material used as the photoelectrode is crucial. The band structure consists of the valence band (VB), which is filled with electrons, and the conduction band (CB), which is typically empty but can accept excited electrons as illustrated in Fig. 4.3 [26]. The energy difference between these two bands is known as the bandgap, and it dictates the wavelength of light the material can absorb.

For efficient water splitting, the band edges of the semiconductor must be appropriately aligned with the redox potentials of the water-splitting reactions. The conduction band edge of the semiconductor should be positioned above the reduction potential of water (0 V vs. NHE) to enable the transfer of photogenerated electrons to the electrolyte, where they reduce protons to form hydrogen. Simultaneously, the valence band edge should be positioned below the oxidation potential of water (1.23 V vs. NHE) to allow photogenerated holes to oxidize water molecules, producing oxygen. This alignment ensures that the photogenerated electrons and holes have sufficient energy to drive the respective half-reactions. However, achieving ideal band alignment is challenging and often requires careful material selection, doping, or the creation of heterojunctions. For instance, in titanium dioxide (TiO_2), a widely used photoanode material, the conduction band is well-aligned for reducing water, but its wide bandgap (~ 3.2 eV) restricts absorption to the UV region, which comprises only a small fraction of the solar spectrum [27]. Figure 4.4 shows titanium dioxide (TiO_2) photoelectrodes for splitting water under illumination. Bandgap engineering through doping with elements like nitrogen or carbon can introduce mid-gap states, narrowing

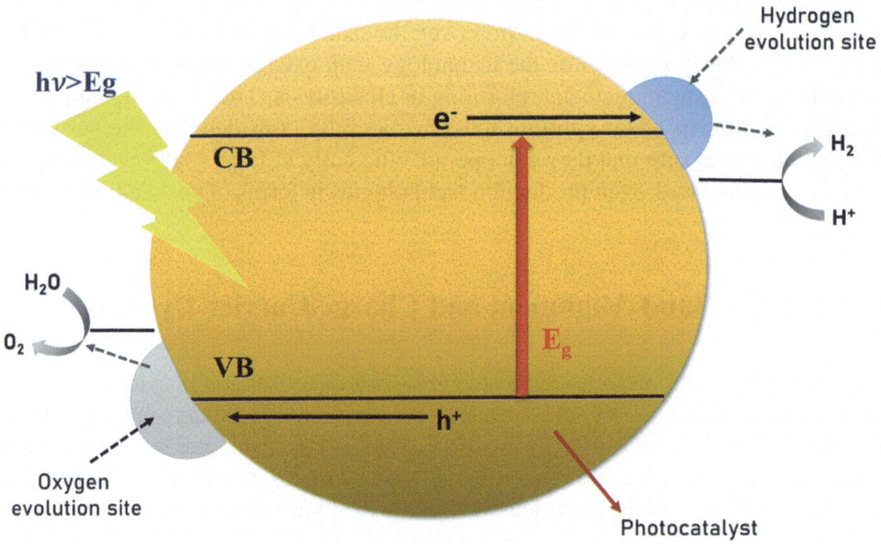

Fig. 4.3 Energy band diagram for HER and OER

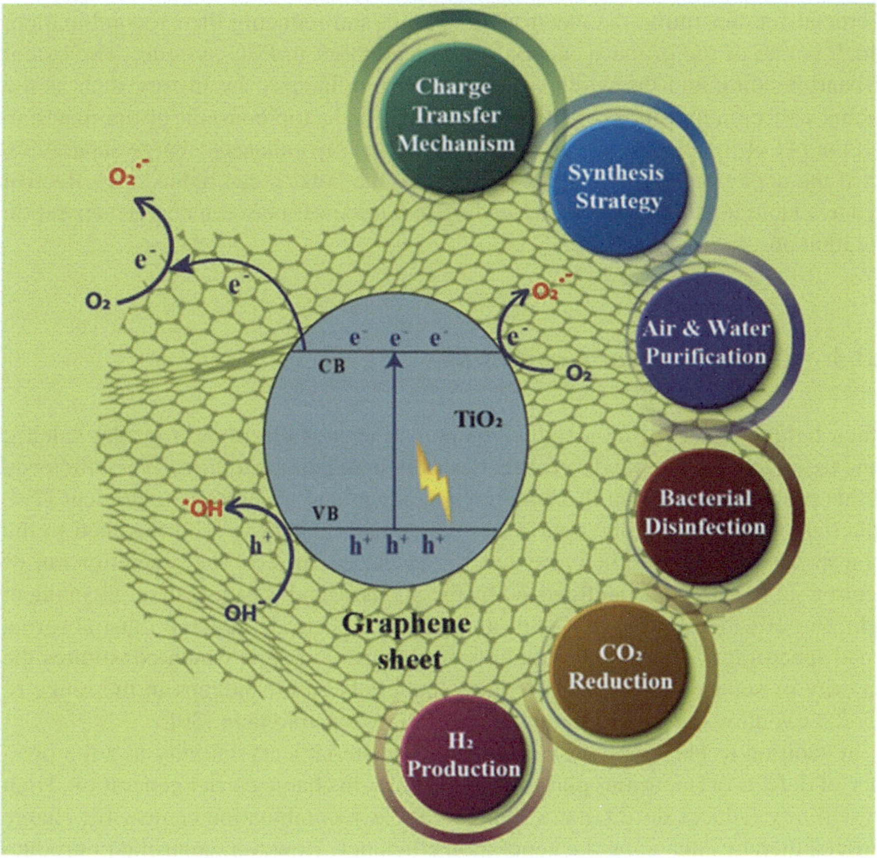

Fig. 4.4 Titanium dioxide (TiO_2) photoelectrodes for splitting water under illumination from Ref. [28]. Copyright © 2021 The Authors. Published by Elsevier Ltd.

the bandgap and extending absorption into the visible range, thereby improving the overall efficiency of the PEC system.

4.3.2 *Band Bending and Space Charge Regions*

When a semiconductor comes into contact with an electrolyte, a space charge region (SCR) or depletion layer forms near the surface due to the difference in chemical potentials (Fermi levels) between the two phases. This leads to band bending, where the energy bands curve near the semiconductor-electrolyte interface. In n-type semiconductors, which are commonly used in PEC water splitting, the conduction band bends downward near the surface, creating an electric field that drives photogenerated electrons towards the bulk and holes towards the surface. This built-in electric field is crucial for separating the electron–hole pairs and reducing their recombination, which is one of the primary causes of efficiency loss in PEC systems. The extent of band bending and the width of the SCR are influenced by factors such as the doping concentration in the semiconductor, the dielectric constant of the material, and the pH of the electrolyte. A wider SCR generally enhances charge separation, but if the depletion region extends too far into the bulk, it can reduce the effective area available for light absorption, leading to a trade-off between charge separation and light absorption efficiency.

4.3.3 Charge Carrier Generation

When light is absorbed by a semiconductor, it excites electrons from the valence band to the conduction band, creating electron–hole pairs (excitons). The efficiency of this process depends on the material's bandgap and absorption coefficient [29]. Materials with a higher absorption coefficient and a bandgap well-matched to the solar spectrum are more efficient at generating charge carriers. The generation rate of electron–hole pairs is also influenced by the intensity and wavelength of the incident light. For effective PEC water splitting, the photoelectrode material must absorb a broad spectrum of sunlight, particularly in the visible range, which constitutes the majority of solar radiation. This requires materials with bandgaps in the range of 1.5–2.4 eV, allowing them to harness both UV and visible light [30].

In addition to bandgap considerations, the material's crystallinity and the presence of defects or impurities play significant roles in charge carrier generation. High crystallinity reduces defect states that can act as recombination centers for charge carriers, thereby enhancing the generation efficiency. However, controlled introduction of defects or doping can also create mid-gap states that facilitate light absorption in sub-bandgap energies, further boosting charge generation.

4.3.4 Charge Carrier Separation and Transport

Once electron–hole pairs are generated, they must be efficiently separated and transported to the respective reaction sites to participate in the water-splitting reactions. Charge separation is driven by the internal electric field created by band bending, as well as by the material's intrinsic properties, such as mobility and diffusion length [31]. The diffusion length, which is the average distance a charge carrier travels before recombining, is a critical parameter in PEC water splitting. If the diffusion length is shorter than the distance to the reaction site (e.g., the surface of the photoelectrode), many carriers will recombine before contributing to the water-splitting reaction, reducing the overall efficiency. Thus, materials with long diffusion lengths, such as certain metal oxides or perovskites, are preferred for PEC applications. Nanostructuring of the photoelectrode can enhance charge separation by reducing the distance that charge carriers need to travel to reach the electrode surface. For example, the use of nanorods, nanowires, or nanoporous structures increases the surface area available for light absorption and provides shorter pathways for charge carriers to reach the reaction sites, thereby minimizing recombination losses [32]. Additionally, the interface between the semiconductor and the electrolyte plays a crucial role in charge transport. The transfer of electrons from the semiconductor to the electrolyte is influenced by the energy barrier at the interface, the nature of the surface states, and the presence of surface catalysts. Surface passivation and the use of co-catalysts can reduce the recombination of charge carriers at the interface and improve the kinetics of the charge transfer process.

4.3.5 Recombination Mechanisms and Mitigation Strategies

Recombination of electron–hole pairs is a major loss mechanism in PEC water splitting. Recombination can occur through various pathways, including radiative recombination (where an electron recombines with a hole, emitting a photon), non-radiative recombination [32] (where the energy is dissipated as heat), and surface recombination (where charge carriers recombine at surface defects or traps). Surface recombination is particularly detrimental in PEC systems because it not only reduces the number of charge carriers available for the water-splitting reaction but also contributes to photocorrosion of the photoelectrode material [33]. Strategies to mitigate recombination include the use of surface passivation layers, which can reduce surface defects, and the incorporation of co-catalysts that facilitate faster charge transfer to the electrolyte, thus reducing the likelihood of recombination. Heterojunctions, where two materials with different bandgaps or work functions are combined, can also enhance charge separation by creating an additional electric field at the interface. This electric field can drive electrons and holes in opposite directions, further reducing recombination rates and improving overall efficiency.

4.3.6 Impact of Surface Catalysts on Charge Carrier Dynamics

The dynamics of charge carriers are significantly influenced by the presence of surface catalysts, which lower the activation energy required for the water-splitting reactions and improve the kinetics of charge transfer. Catalysts such as platinum for HER and iridium oxide for OER provide active sites where the charge carriers can efficiently participate in the redox reactions, thereby reducing the overpotentials and enhancing the overall efficiency of the PEC system [17]. The interaction between the charge carriers and the surface catalysts is a critical aspect of PEC water splitting. The catalyst must have good electrical conductivity to facilitate fast electron transfer and must be well-integrated with the semiconductor material to ensure efficient charge carrier transport. Additionally, the surface catalyst should be stable under the operating conditions of the PEC cell to prevent degradation and maintain long-term performance.

4.4 Key PEC Reactions: Hydrogen Evolution Reaction (HER) and Oxygen Evolution Reaction (OER)

Photoelectrochemical (PEC) water splitting is fundamentally driven by two critical half-reactions: the Hydrogen Evolution Reaction (HER) and the Oxygen Evolution Reaction (OER). These reactions occur at the cathode and anode of the PEC cell, respectively, and are responsible for the generation of hydrogen and oxygen gases from water. Understanding the mechanisms, kinetics, and challenges associated with these reactions is essential for optimizing PEC systems and improving their efficiency.

4.4.1 Hydrogen Evolution Reaction (HER)

The Hydrogen Evolution Reaction (HER) is the cathodic half-reaction in PEC water splitting, where protons (H^+) in the electrolyte are reduced to produce hydrogen gas (H_2). The overall reaction can be represented as:

$$2H^+ + 2e^- \rightarrow H_2$$

HER involves a series of steps that depend on the nature of the electrocatalyst and the operating conditions. The reaction mechanism typically follows either the Volmer–Heyrovsky or Volmer–Tafel pathways [34]. The initial step in both mechanisms is the Volmer reaction, where a proton is adsorbed onto the catalyst surface, forming a hydrogen atom (H^*), which is often referred to as the adsorbed intermediate:

$$H^+ + e^- \rightarrow H^*$$

Following the Volmer step, the reaction can proceed via the Heyrovsky step, where a second proton combines with the adsorbed hydrogen atom and an electron to release hydrogen gas:

$$H^* + H^+ + e^- \rightarrow H_2$$

Alternatively, in the Tafel step, two adsorbed hydrogen atoms combine to form hydrogen gas:

$$2H^* \rightarrow H_2$$

The efficiency of the HER process is largely determined by the binding strength of the adsorbed hydrogen intermediate on the catalyst surface. If the binding is too weak, the Volmer step becomes rate-limiting, as protons are not efficiently adsorbed [35]. Conversely, if the binding is too strong, the release of hydrogen gas becomes sluggish, hampering the overall reaction kinetics. Platinum (Pt) is considered the benchmark catalyst for HER due to its near-optimal hydrogen binding energy, resulting in very low overpotentials and high catalytic activity. However, the high cost and scarcity of platinum have driven research into alternative materials such as transition metal sulfides (e.g., MoS_2), phosphides, and carbides, which show promising HER activity, though often with slightly higher overpotentials [36]. The kinetics of HER are also influenced by factors such as the pH of the electrolyte, the applied potential, and the nature of the semiconductor material used in the PEC cell. In acidic electrolytes, HER proceeds more readily due to the high concentration of protons, whereas in alkaline conditions, the reaction rate can be slower because water molecules must first dissociate to provide the necessary protons.

4.4.2 Oxygen Evolution Reaction (OER)

The Oxygen Evolution Reaction (OER) is the anodic half-reaction in PEC water splitting, where water molecules are oxidized to produce oxygen gas (O_2). The overall reaction can be represented as:

$$2H_2O \rightarrow O_2 + 4H^+ + 4e^-$$

OER is a more complex reaction than HER, involving the transfer of four electrons and the formation of an O–O bond. This multi-electron process makes OER kinetically sluggish, requiring higher overpotentials and more effective catalysts to proceed efficiently. The reaction mechanism typically involves the following steps [37]:

(a) **Adsorption of Hydroxide**: Water molecules adsorb onto the catalyst surface, forming hydroxide species (OH*):

$$H_2O \rightarrow OH^* + H^+ + e^-$$

(b) **Formation of OOH Intermediate**: The adsorbed hydroxide species undergo further oxidation to form an oxy-hydroxy (OOH) intermediate:

$$OH^* \rightarrow O^* + H^+ + e^- O^* + H_2O \rightarrow OOH^* + H^+ + e^-$$

(c) **Release of Oxygen Gas**: The OOH intermediate releases oxygen gas, completing the reaction:

$$OOH^* \rightarrow O_2 + H^+ + e^-$$

The OER process is highly dependent on the catalyst material, with oxides of transition metals such as iridium (IrO_2) and ruthenium (RuO_2) being among the most effective due to their ability to facilitate the formation and desorption of the OOH intermediate. However, these materials are expensive and scarce, leading to significant research interest in alternative catalysts such as cobalt oxides (Co_3O_4), nickel–iron oxides ($NiFeO_x$), and perovskite oxides, which offer a balance between cost and performance.

4.4.3 Overpotentials and Efficiency Considerations

Both HER and OER require an additional driving force beyond the theoretical potential to overcome kinetic barriers, known as the overpotential. The overpotential is a measure of the extra energy needed to drive the reaction at a significant rate and is influenced by the catalytic activity, the surface area of the electrodes, and the reaction environment (e.g., electrolyte composition, pH). In PEC water splitting, the overpotentials for HER and OER are additive, meaning that minimizing overpotentials for both reactions is crucial for achieving high overall efficiency. The total voltage required to drive water splitting is the sum of the thermodynamic potential (1.23 V) [13] and the overpotentials for HER and OER. Strategies to reduce overpotentials include the development of more active catalysts, optimizing the surface structure and morphology of the electrodes to increase active sites, and engineering the semiconductor-catalyst interface to improve charge transfer kinetics.

4.4.4 Interplay Between HER and OER

The interplay between HER and OER in a PEC system is complex and requires careful balancing to optimize overall performance. For instance, the current generated by the photoelectrode during HER must be matched by the current sustained by the OER catalyst to maintain a stable reaction environment and prevent the buildup of charge carriers [33], which can lead to recombination losses. Moreover, the chemical environment in the electrolyte, including the pH and ionic strength, can affect the relative rates of HER and OER. In some cases, the choice of electrolyte can favor one reaction over the other, necessitating careful tuning of electrolyte composition to achieve optimal performance.

4.5 Photoelectrode Materials: Design and Optimization

The performance of photoelectrochemical (PEC) water splitting systems is significantly influenced by the choice and design of photoelectrode materials. These materials must efficiently absorb solar energy, facilitate charge separation, and drive the electrochemical reactions for water splitting. This section explores the various types of photoelectrode materials, their design considerations, and optimization strategies to enhance their performance [38].

4.5.1 Semiconductor Materials: Metal Oxides, Sulfides, and Nitrides

In photoelectrochemical (PEC) water splitting, semiconductor materials play a crucial role in determining the overall efficiency of the process. These materials are responsible for absorbing sunlight, generating charge carriers (electrons and holes), and driving the electrochemical reactions that split water into hydrogen and oxygen. The choice of semiconductor material is therefore critical to optimizing the performance of PEC cells. Among the various semiconductor materials explored for PEC applications, metal oxides, sulfides, and nitrides have garnered significant attention due to their suitable electronic properties, stability, and availability as depicted in Fig. 4.5 [39].

4.5.2 Metal Oxides

Metal oxides are among the most widely studied semiconductor materials for PEC water splitting. Their popularity stems from several factors, including their stability in aqueous environments, abundance, and favorable bandgap energies for visible light absorption. Titanium dioxide (TiO_2), iron oxide (Fe_2O_3), and bismuth vanadate ($BiVO_4$) are some of the most commonly used metal oxides in PEC systems [40].

(a) **Titanium Dioxide (TiO_2)**: TiO_2 is often regarded as the benchmark material for PEC applications due to its excellent chemical stability, non-toxicity, and strong photocatalytic activity. However, TiO_2 has a wide bandgap (~ 3.2 eV for the anatase phase), which limits its absorption to the ultraviolet (UV) region of the solar spectrum, capturing only a small fraction of the available sunlight. To enhance its solar absorption, various strategies have been employed, such as doping with elements like nitrogen or carbon to narrow the bandgap, and coupling with narrow-bandgap semiconductors or plasmonic materials [41] to extend its photoresponse into the visible range.

(b) **Iron Oxide (Fe_2O_3)**: Hematite (α-Fe_2O_3) is another promising metal oxide for PEC water splitting due to its narrow bandgap (~ 2.1 eV), which allows it to

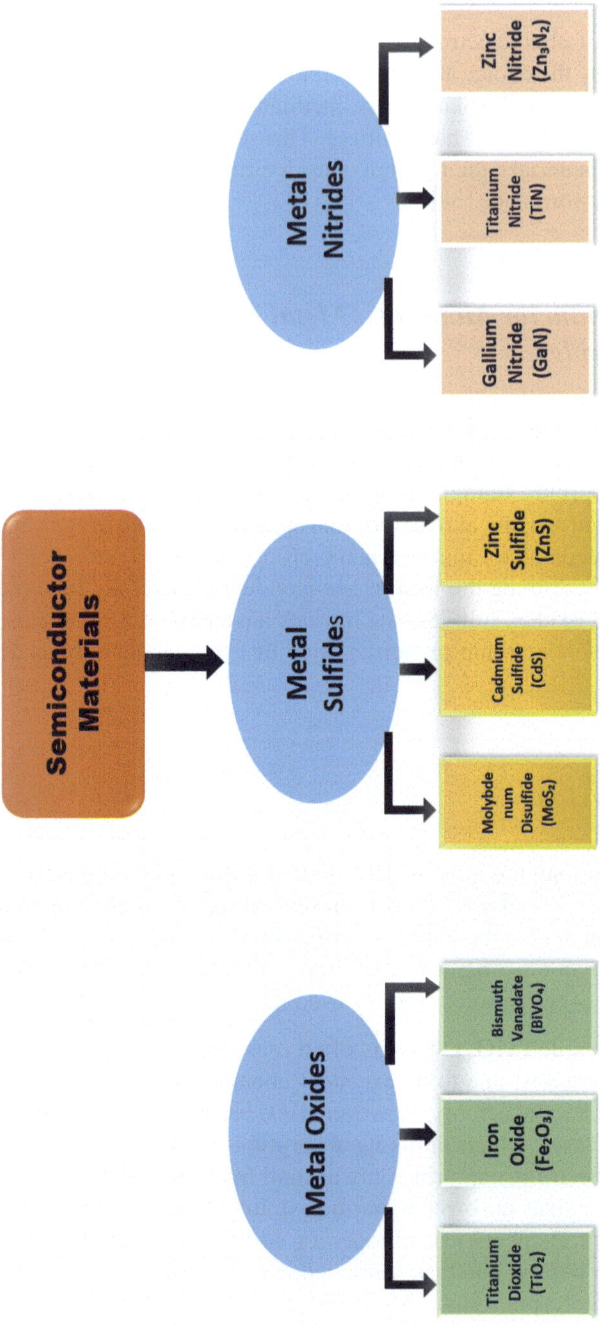

Fig. 4.5 Various semiconductor materials used for PEC water splitting

absorb a significant portion of the visible light spectrum. Additionally, Fe_2O_3 is abundant, inexpensive, and stable under aqueous conditions. However, its short hole diffusion length (~ 2–4 nm) and poor conductivity limit its PEC performance. Strategies such as nanostructuring, doping with elements like tin (Sn) or titanium (Ti), and the use of surface passivation layers [42] have been explored to overcome these limitations and improve charge separation and transport.

(c) **Bismuth Vanadate ($BiVO_4$)**: $BiVO_4$ has emerged as a leading material for visible-light-driven PEC water splitting due to its relatively narrow bandgap (~ 2.4 eV), which allows for efficient sunlight absorption, and its favorable band alignment for oxygen evolution. Despite these advantages, $BiVO_4$ suffers from slow charge transport and recombination losses. To address these challenges, researchers have developed strategies such as doping with elements like molybdenum (Mo) or tungsten (W), creating composite structures with conductive materials, and employing surface modification techniques to enhance its catalytic activity [40] and stability.

4.5.3 Metal Sulfides

Metal sulfides are another class of semiconductor materials that have been explored for PEC water splitting. These materials typically have narrower bandgaps than metal oxides, allowing for better absorption of visible light. However, they often suffer from poor stability in aqueous environments, particularly under oxidative conditions, which can lead to photocorrosion.

a. **Molybdenum Disulfide (MoS_2)**: MoS_2 is a layered transition metal dichalcogenide (TMD) that has gained attention for its catalytic activity in the hydrogen evolution reaction (HER). Its narrow bandgap (~ 1.2 eV) makes it a good candidate for visible light absorption, and its unique two-dimensional (2D) structure offers high surface area and active edge sites for catalysis. However, MoS_2 is prone to photocorrosion in aqueous environments, which limits its long-term stability [43]. Strategies to enhance its stability include the use of protective coatings, combining it with more stable semiconductors, or developing hybrid materials that integrate MoS_2 with other catalytic or conductive materials.
b. **Cadmium Sulfide (CdS)**: CdS is a well-known metal sulfide with a direct bandgap of~ 2.4 eV, making it suitable for visible light absorption. CdS is often used as a photoanode material in PEC cells due to its good charge transport properties and ability to generate a significant photocurrent under illumination. However, like many metal sulfides, CdS suffers from photocorrosion, particularly under oxidative conditions. To address this issue, CdS is often used in combination with other materials, such as protective oxide layers or co-catalysts [44], to improve its stability and overall PEC performance.
c. **Zinc Sulfide (ZnS)**: ZnS is another metal sulfide with a wide bandgap (~ 3.6 eV), making it primarily active in the UV region. While its photocatalytic activity in the visible range is limited, ZnS is often used in composite structures with other

semiconductors to enhance charge separation and transfer. For instance, ZnS can be combined with narrow-bandgap materials to form heterojunctions [45] that improve the overall efficiency of PEC water splitting by utilizing a broader range of the solar spectrum.

4.5.4 Metal Nitrides

Metal nitrides represent a relatively newer class of semiconductor materials for PEC water splitting, characterized by their high chemical stability, tunable electronic properties, and narrow bandgaps suitable for visible light absorption.

a. **Gallium Nitride (GaN)**: GaN is a wide-bandgap semiconductor (~ 3.4 eV) with excellent chemical stability and high electron mobility. While its bandgap limits its absorption to the UV region, GaN can be alloyed with other elements, such as indium, to form indium gallium nitride (InGaN), which has a tunable bandgap that can be adjusted to absorb visible light. InGaN alloys have shown promise in PEC applications, particularly in tandem cell architectures where they can serve as the top absorber layer in combination with other semiconductors.
b. **Titanium Nitride (TiN)**: TiN is a metallic nitride with unique properties, including high electrical conductivity and good catalytic activity, particularly for the oxygen evolution reaction (OER). While TiN itself is not a typical semiconductor used for light absorption, it is often employed as a conductive support or protective layer in PEC cells [46]. Its stability under harsh conditions and ability to form strong bonds with other materials make it a valuable component in composite photoelectrodes.
c. **Zinc Nitride (Zn_3N_2)**: Zn_3N_2 is a relatively less explored nitride material for PEC applications, but it has shown potential due to its narrow bandgap (~ 1.0–1.5 eV) and good absorption in the visible light range. However, Zn_3N_2 is sensitive to air and moisture, which can lead to degradation and reduced performance. Research is ongoing to develop strategies to improve its stability, such as through surface passivation or by creating protective coatings

4.6 Bandgap Engineering for Enhanced Solar Absorption

In photoelectrochemical (PEC) water splitting, the ability of semiconductor materials to efficiently absorb sunlight and convert it into chemical energy is paramount. The bandgap of a semiconductor, defined as the energy difference between the valence band and the conduction band, dictates the portion of the solar spectrum that can be absorbed and utilized for photogenerated charge carriers. To maximize the solar-to-hydrogen (STH) [47] conversion efficiency, it is crucial to engineer the bandgap of photoelectrode materials so that they can absorb a broader range of the solar spectrum, particularly in the visible region where the majority of solar energy lies.

Bandgap engineering thus plays a central role in the design and optimization of photoelectrode materials for PEC water splitting.

4.6.1 The Importance of Optimal Bandgap

An optimal bandgap for a PEC photoelectrode should be narrow enough to absorb a significant portion of the solar spectrum, particularly in the visible range, yet wide enough to generate sufficient photovoltage to drive the water-splitting reactions. The theoretical ideal bandgap for a single-junction PEC system is around 1.23 eV, which corresponds to the energy required to split water into hydrogen and oxygen. However, considering overpotentials and other system losses, a practical bandgap closer to 1.6–2.2 eV is often targeted for visible light absorption and efficient water splitting.

Materials with bandgaps that are too wide, such as titanium dioxide (TiO_2), primarily absorb ultraviolet (UV) light, which constitutes only a small fraction of the solar spectrum, leading to lower overall efficiency. Conversely, materials with bandgaps that are too narrow may absorb more of the solar spectrum but lack sufficient photovoltage to efficiently drive the electrochemical reactions [48]. Therefore, engineering the bandgap to strike the right balance between light absorption and photovoltage generation is crucial for the performance of PEC systems.

4.6.2 Approaches to Bandgap Engineering

Various strategies have been developed to engineer the bandgap of semiconductor materials to enhance their solar absorption capabilities. These approaches include doping with foreign elements, alloying, creating heterojunctions, and utilizing quantum confinement effects in nanostructures.

a. **Doping and Alloying**

Doping involves introducing foreign atoms into the crystal lattice of a semiconductor material to modify its electronic properties, including the bandgap. For example, doping titanium dioxide (TiO_2) with non-metal elements such as nitrogen (N), carbon (C), or sulfur (S) can reduce its bandgap from around 3.2 eV to a value more favorable for visible light absorption. materials with bandgaps that are too narrow may absorb more of the solar spectrum but lack sufficient photovoltage to efficiently drive the electrochemical reactions [49].

Similarly, alloying involves combining two or more semiconductors to form a material with a tunable bandgap. For instance, by alloying indium phosphide (InP) with gallium arsenide (GaAs), it is possible to create a material with a bandgap that can be adjusted to absorb a broader spectrum of sunlight. This approach is particularly useful in designing tandem cells, where multiple semiconductors with different bandgaps are stacked to absorb different parts of the solar spectrum.

b. **Quantum Confinement in Nanostructures**
Nanostructuring of semiconductor materials can lead to quantum confinement effects, which can significantly alter their bandgap. When the dimensions of a semiconductor are reduced to the nanoscale (typically below 10 nm), the motion of charge carriers becomes confined in one or more dimensions, leading to discrete energy levels rather than continuous bands. This quantum confinement effect can result in an increase in the bandgap of the material, allowing for the fine-tuning of the absorption properties.
For example, quantum dots (QDs) of cadmium selenide (CdSe) exhibit size-dependent bandgaps, where smaller QDs have larger bandgaps and absorb higher-energy (shorter-wavelength) light, while larger QDs have smaller bandgaps and absorb lower-energy (longer-wavelength) light. By carefully controlling the size of the QDs, it is possible to engineer their bandgap to match the desired absorption characteristics for PEC water splitting.

c. **Formation of Heterojunctions**
Heterojunctions are interfaces between two different semiconductor materials with different bandgaps and electronic properties. The formation of a heterojunction can lead to improved charge separation and reduced recombination of photogenerated carriers, as well as enhanced light absorption through the combination of the bandgaps of the constituent materials.
For example, the combination of titanium dioxide (TiO_2) with cadmium sulfide (CdS) forms a TiO_2/CdS heterojunction, where TiO_2 serves as the electron acceptor and CdS as the light absorber. The CdS layer absorbs visible light and generates electron–hole pairs, while the TiO_2 layer helps to separate the charges and prevent recombination. This heterojunction not only extends the absorption of TiO_2 into the visible region but also improves the overall efficiency of the PEC cell by facilitating charge transfer [50].

d. **Bandgap Grading**
Bandgap grading is another strategy used in PEC cells, where the bandgap of the photoelectrode material gradually changes across its thickness. This can be achieved through compositional grading or by varying the doping concentration. Bandgap grading allows for better light absorption and charge separation, as different regions of the material can absorb different wavelengths of light and generate charge carriers with varying energies. This approach is particularly useful in tandem and multi-junction PEC cells, where different layers of the photoelectrode absorb different parts of the solar spectrum.

4.6.3 *Applications, Challenges and Future Directions*

Bandgap engineering has been successfully applied to various semiconductor materials to enhance their performance in PEC water splitting. For example, the development of nitrogen-doped TiO_2, alloyed InGaN, and heterojunctions like TiO_2/CdS has led to significant improvements in solar absorption and overall PEC efficiency. However, challenges remain in achieving the optimal balance between bandgap tuning, material stability, and charge carrier dynamics [51]. One of the key challenges is maintaining the stability of the engineered bandgap materials under the harsh conditions of PEC water splitting. Doping and alloying can introduce defects and trap states that may lead to increased recombination and reduced efficiency. Moreover, nanostructured materials with quantum confinement effects may suffer from surface recombination and stability issues. Future research in bandgap engineering for PEC water splitting will likely focus on developing new materials and methods to achieve better control over bandgap tuning while addressing stability concerns. Advanced techniques such as high-throughput computational screening, machine learning, and combinatorial synthesis could accelerate the discovery of novel materials with optimized bandgaps [52]. Additionally, integrating these materials into innovative cell architectures, such as tandem or multi-junction PEC cells, could further enhance solar absorption and drive the efficiency of PEC water splitting towards practical and sustainable hydrogen production.

4.7 Nanostructuring Techniques for Improved Charge Separation and Transport

Nanostructuring has emerged as a powerful approach to enhance the performance of photoelectrochemical (PEC) water-splitting systems by improving charge separation and transport within photoelectrode materials. The inherently high surface area, short carrier diffusion paths, and unique quantum properties of nanostructured materials make them particularly suitable for addressing the challenges of charge carrier recombination and transport, which are critical factors in determining the overall efficiency of PEC devices [53]. This section explores the various nanostructuring techniques used to optimize charge separation and transport, focusing on the underlying principles and their impact on PEC performance.

4.7.1 *The Importance of Charge Separation and Transport*

In PEC water splitting, the efficient separation and transport of photogenerated charge carriers—electrons and holes—are crucial for maximizing the solar-to-hydrogen (STH) conversion efficiency [54]. Upon light absorption, electron–hole pairs are generated within the photoelectrode material. For these carriers to drive the water-splitting reactions, they must be rapidly separated and transported to the appropriate reaction sites: electrons to the hydrogen evolution reaction (HER) site and holes to the oxygen evolution reaction (OER) site. However, several challenges, such as bulk and surface recombination, long carrier diffusion distances, and poor charge transport properties, can significantly impede this process, leading to efficiency losses. Nanostructuring techniques are designed to mitigate these challenges by modifying the material's morphology and structure at the nanoscale.

(a) **High Surface Area and Shortened Diffusion Paths**
Nanostructuring photoelectrode materials into one-dimensional (1D) nanostructures such as nanowires, nanotubes, and nanorods can greatly enhance charge separation and transport. These 1D nanostructures provide a direct pathway for charge carriers, reducing the distance electrons and holes must travel to reach the surface reaction sites. This short diffusion path minimizes the likelihood of bulk recombination, where electrons and holes recombine before reaching the surface, thus enhancing the photocurrent and overall PEC efficiency. For example, titanium dioxide (TiO_2) nanowires and nanotubes have been widely studied as photoanode materials for PEC water splitting. The high aspect ratio of these nanostructures allows for efficient charge transport along the length of the nanowire or nanotube while also providing a large surface area for light absorption and reaction sites. This configuration has been shown to improve the charge separation efficiency and reduce recombination, leading to higher photocurrents compared to bulk or planar TiO_2.

(b) **Enhanced Light Absorption and Scattering**
Nanostructuring can also enhance light absorption through mechanisms such as light scattering and trapping, which are particularly important for materials with poor intrinsic light absorption properties. By designing nanostructures with specific dimensions and morphologies, it is possible to create optical effects that increase the amount of light absorbed within the photoelectrode material, thereby generating more charge carriers. For instance, plasmonic nanostructures, such as metal nanoparticles embedded in or deposited on semiconductor surfaces, can induce localized surface plasmon resonances (LSPRs). These LSPRs can enhance the local electromagnetic field near the metal-semiconductor interface, leading to increased light absorption in the semiconductor. Additionally, the scattering of light by these plasmonic nanostructures can increase the optical path length within the material, further enhancing absorption. This approach has been used to improve the performance of PEC photoelectrodes such as gold (Au) or silver (Ag) nanoparticles incorporated into TiO_2 or other metal oxide photoanodes.

(c) **Improved Electron and Hole Transport**
Nanostructuring techniques can also be employed to improve the transport of electrons and holes within the photoelectrode material by reducing defects and optimizing crystallinity. Defects such as grain boundaries, dislocations, and impurities can act as recombination centers, trapping charge carriers and

reducing their mobility. By carefully controlling the synthesis and fabrication processes, it is possible to create nanostructured materials with fewer defects and better crystallinity, leading to improved carrier mobility and reduced recombination [55]. One example of this approach is the synthesis of highly crystalline hematite (α-Fe_2O_3) nanostructures for use as photoanodes. Hematite is a promising photoanode material due to its suitable bandgap and chemical stability; however, it suffers from poor charge transport properties due to low carrier mobility and high recombination rates. By nanostructuring hematite into well-defined nanorods or nanoplates with controlled crystallinity, researchers have been able to significantly improve its charge transport properties, resulting in enhanced PEC performance.

(d) **Quantum Confinement and Charge Carrier Dynamics**
Quantum confinement effects in nanostructured materials can also play a role in improving charge separation and transport. When the size of a semiconductor material is reduced to the nanoscale, quantum confinement can lead to the formation of discrete energy levels and an increase in the material's bandgap. This can alter the charge carrier dynamics, potentially leading to more efficient charge separation and reduced recombination. For example, quantum dots (QDs) of cadmium sulfide (CdS) or cadmium selenide (CdSe) have been explored as light absorbers in PEC systems. The quantum confinement effect in these QDs allows for the tuning of their bandgap and absorption properties, as well as the creation of energy level alignments that favor efficient charge separation [56]. Moreover, the small size of QDs leads to short diffusion distances for charge carriers, reducing the probability of recombination and enhancing the overall photocurrent.

(e) **Hierarchical Nanostructures for Synergistic Effects**
In recent years, the development of hierarchical nanostructures has gained attention as a way to combine the benefits of different nanostructuring techniques. Hierarchical nanostructures consist of multiple levels of nanostructuring, such as combining nanowires with nanodots or incorporating nanorods into porous scaffolds. These structures can offer synergistic effects, such as enhanced light absorption, improved charge separation, and efficient transport pathways. For example, a hierarchical photoanode composed of TiO_2 nanowires decorated with CdS QDs has been demonstrated to enhance PEC performance by leveraging the light absorption and quantum confinement effects of CdS QDs along with the efficient charge transport properties of TiO_2 nanowires [57]. Figure 4.6 shows a hierarchical photoanode composed of TiO_2 nanowires decorated with CdS QDs. This combination results in a material that can absorb more sunlight, generate more charge carriers, and transport them efficiently to the reaction sites, leading to higher overall efficiency.

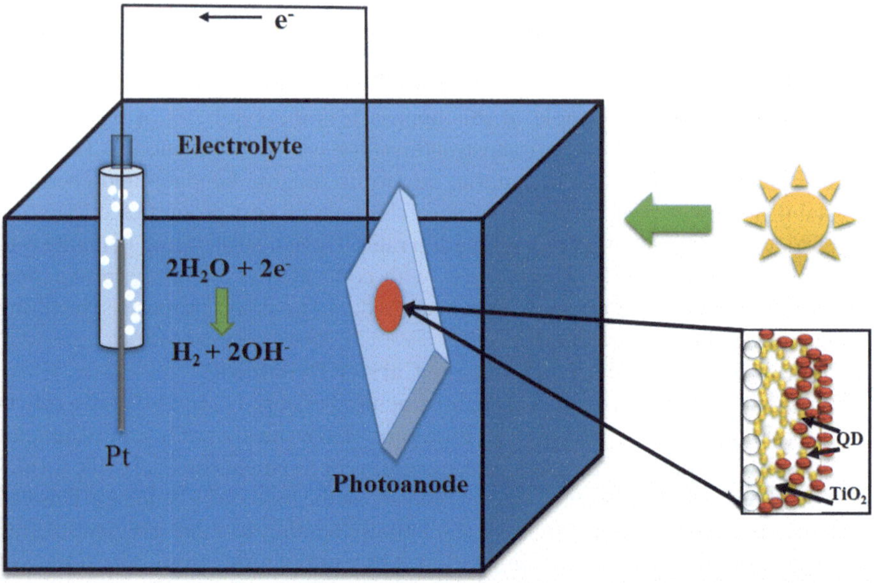

Fig. 4.6 Hierarchical photoanode composed of TiO_2 nanowires decorated with CdS QDs

4.8 Surface Modifications and Catalysts

Surface modifications and the incorporation of catalysts play a pivotal role in optimizing the performance of photoelectrochemical (PEC) water-splitting systems. The surface of photoelectrode materials directly interfaces with the electrolyte, where the critical hydrogen evolution reaction (HER) and oxygen evolution reaction (OER) occur [58]. Surface modifications can enhance the reaction kinetics, reduce recombination losses, and improve overall efficiency by addressing surface-related challenges. This section delves into the various surface modification techniques and the role of catalysts in improving PEC performance, highlighting the underlying mechanisms and their impact on water-splitting efficiency.

4.8.1 The Role of Surface Properties in PEC Performance

The efficiency of PEC water splitting is heavily influenced by the surface properties of the photoelectrode materials. The surface must not only facilitate the absorption of light and the generation of charge carriers but also efficiently transfer these carriers to the electrolyte for the water-splitting reactions. However, the surface of many photoelectrode materials, particularly metal oxides, often suffers from issues such as slow reaction kinetics, high overpotentials, and surface recombination of charge carriers. These challenges can significantly reduce the overall photocurrent and hydrogen production efficiency [59]. Surface modifications aim to address these challenges by tailoring the chemical, physical, and electronic properties of the photoelectrode surface. By optimizing these properties, it is possible to enhance charge carrier separation, reduce recombination, and improve the catalytic activity for HER and OER.

(a) **Surface Passivation and Defect Engineering**
One common approach to surface modification is surface passivation, which involves the creation of a protective layer on the surface of the photoelectrode to reduce surface recombination and improve stability. Surface recombination occurs when photogenerated electrons and holes recombine at surface defects or trap states before participating in the water-splitting reactions. Passivation layers can mitigate this by covering surface defects, reducing the number of recombination sites, and creating an energy barrier that promotes charge separation. For example, the passivation of hematite (α-Fe_2O_3) photoanodes with ultrathin layers of materials such as alumina (Al_2O_3) or titanium dioxide (TiO_2) has been shown to reduce surface recombination and enhance photocurrent. These passivation layers can be deposited using techniques like atomic layer deposition (ALD), which allows for precise control over thickness and uniformity. Additionally, passivation layers can improve the stability of the photoelectrode by protecting it from photocorrosion [60], which is a common issue with many semiconductor materials under PEC operating conditions. Defect engineering is another surface modification strategy that involves deliberately introducing or controlling defects to enhance PEC performance. By creating specific defect sites, it is possible to tailor the electronic properties of the surface, such as band bending, which can improve charge separation and transfer. For instance, oxygen vacancies in TiO_2 have been found to enhance its photocatalytic activity by creating mid-gap states that facilitate charge transfer to the electrolyte.

(b) **Surface Catalysts for HER and OER**
The efficiency of PEC water splitting is also highly dependent on the catalytic activity of the photoelectrode surface for the HER and OER. These reactions require the transfer of multiple electrons and protons, which can be kinetically slow and require significant overpotentials to proceed. Surface catalysts are essential for lowering these overpotentials and accelerating the reaction kinetics [33], thereby increasing the overall efficiency of the PEC system. For the HER, which involves the reduction of protons to hydrogen gas (H_2), platinum (Pt) is considered the benchmark catalyst due to its exceptional catalytic activity and low overpotential. However, Pt is expensive and scarce, driving the search for alternative, cost-effective catalysts. Transition metal dichalcogenides (TMDs), such as molybdenum disulfide (MoS_2) and tungsten disulfide (WS_2), have emerged as promising HER catalysts. These materials possess active edge sites that are highly efficient for proton reduction, and their catalytic activity can be further enhanced through nanostructuring or doping. For the OER, which involves the oxidation of water to oxygen gas (O_2), iridium oxide (IrO_2) and ruthenium oxide (RuO_2) are the most widely used catalysts due to their high

activity and stability. However, similar to Pt, these materials are expensive and scarce. As a result, alternative catalysts based on earth-abundant transition metals, such as nickel (Ni), cobalt (Co), and iron (Fe), have been extensively studied. Nickel–iron layered double hydroxides (NiFe LDHs) and cobalt oxide (Co_3O_4) are examples of such catalysts that have shown promising OER activity in PEC systems [61].

(c) **Hybrid and Composite Catalysts**
To further enhance the catalytic performance and stability of photoelectrodes, hybrid and composite catalysts have been developed. These catalysts combine the strengths of different materials to achieve synergistic effects, such as enhanced catalytic activity, improved stability, and better charge transfer properties. For example, composite catalysts that combine TMDs with noble metals or metal oxides have been shown to exhibit superior HER and OER performance. A hybrid catalyst composed of MoS_2 nanosheets decorated with Pt nanoparticles, for instance, can achieve higher HER activity than either MoS_2 or Pt alone. The Pt nanoparticles provide high catalytic activity, while the MoS_2 nanosheets offer a large surface area and additional active sites, resulting in a highly efficient and stable HER catalyst. Similarly, hybrid catalysts for OER, such as NiFe LDHs combined with carbon-based materials like graphene or carbon nanotubes (CNTs), have been developed to improve charge transfer and stability. The carbon materials enhance electrical conductivity and provide a robust support structure, while the NiFe LDHs [62] offer high catalytic activity. This combination leads to improved PEC performance and durability under operating conditions.

(d) **Surface Functionalization and Molecular Catalysts**
Surface functionalization involves the modification of the photoelectrode surface with organic or inorganic molecules to enhance its catalytic activity and stability. Molecular catalysts, such as metal–organic complexes or porphyrins, can be attached to the surface to facilitate specific reactions, such as proton reduction or water oxidation. For example, cobalt-based molecular catalysts, such as cobalt porphyrins [63], have been grafted onto semiconductor surfaces to improve OER activity. These molecular catalysts can provide high catalytic efficiency and selectivity, and their activity can be tuned through molecular design. Additionally, molecular catalysts can be integrated into hybrid systems with inorganic catalysts to achieve complementary effects and enhance overall PEC performance.

(e) **Advanced Surface Modification Techniques**
Recent advances in surface modification techniques, such as plasma treatment [64], electrochemical deposition, and photochemical grafting, have opened new avenues for optimizing the surface properties of photoelectrodes. Plasma treatment, for instance, can be used to introduce functional groups or create surface defects that enhance catalytic activity. Electrochemical deposition allows for the precise control of catalyst loading and distribution, enabling the formation of uniform and well-adhered catalyst layers. Photochemical grafting involves

the use of light to initiate chemical reactions on the surface, enabling the selective attachment of molecules or catalysts. These advanced techniques offer new possibilities for tailoring the surface properties of photoelectrodes to achieve higher efficiency, better stability, and longer lifetimes in PEC water-splitting systems.

4.9 PEC Cell Configurations and Integration Strategies

This section focuses on the various design and integration approaches that are crucial for optimizing the performance of photoelectrochemical (PEC) cells in water-splitting applications. The efficiency of a PEC system is not solely dependent on the materials used for photoelectrodes; the architecture of the cell and the strategies employed to integrate various components also play a significant role. This section explores the different PEC cell configurations, discusses the integration of photoelectrodes with co-catalysts, and examines how the composition of electrolytes influences overall PEC performance. In the context of PEC water splitting, the architecture of the cell is central to determining its efficiency and practicality. PEC cells can be designed in various configurations, each with its unique set of advantages and challenges. These configurations include single-junction, tandem, and multi-junction cells, each of which utilizes different strategies for light absorption and charge separation to optimize hydrogen production. The choice of architecture often depends on the specific materials used for photoelectrodes, the desired level of efficiency, and the operational stability required for the application.

In addition to the cell architecture, the integration of co-catalysts with photoelectrodes is another critical aspect that significantly affects the performance of PEC cells. Co-catalysts are employed to enhance the kinetics of the hydrogen evolution reaction (HER) and oxygen evolution reaction (OER), reducing the overpotentials required for these processes and thereby improving overall efficiency. The successful integration of co-catalysts involves ensuring strong adhesion to the photoelectrode surface, maintaining the stability of the catalyst under operational conditions, and achieving a balanced catalytic activity [44] for both HER and OER. The composition of the electrolyte used in PEC cells is also a vital factor in determining the efficiency and stability of the water-splitting process. The electrolyte serves as the medium through which ions are transported between the anode and cathode, and its composition can influence the reaction kinetics, charge carrier dynamics, and the overall stability of the photoelectrodes. Different electrolytes, ranging from acidic to alkaline solutions, offer varying benefits and challenges, and their selection must be carefully considered in the design of PEC systems. This section will delve into the intricacies of these factors, providing a comprehensive understanding of how PEC cell configurations and integration strategies can be optimized to achieve high-efficiency solar-driven water splitting. The subsequent subsections will explore the specific details of PEC cell architectures, the role of co-catalysts, and the influence of electrolyte composition [65], highlighting the interplay between these elements in the quest for efficient and sustainable hydrogen production.

4.9.1 PEC Cell Architectures

The architecture of a photoelectrochemical (PEC) cell plays a crucial role in determining its performance, efficiency, and practicality for water-splitting applications. PEC cells are designed to harness solar energy for the electrochemical splitting of water into hydrogen and oxygen. The design of these cells involves optimizing various factors, including light absorption, charge carrier separation, and the interface between photoelectrodes and electrolytes [66]. Understanding the different PEC cell architectures and their respective advantages and challenges is essential for advancing this technology.

4.9.2 Single-Junction PEC Cells

Single-junction PEC cells as depicted in Fig. 4.7a are the simplest and most straightforward design, consisting of a single layer of photoelectrode material. In this configuration, a semiconductor photoelectrode is exposed to sunlight and operates as both the photoanode and photocathode in a two-electrode system. The primary function of the single-junction cell is to absorb sunlight and generate electron–hole pairs, which are then utilized to drive HER and OER. The primary advantage of single-junction PEC cells is their simplicity and ease of fabrication. They are typically less complex and more cost-effective to produce compared to more advanced configurations. However, their efficiency is often limited by the photoelectrode material's ability to absorb a broad spectrum of sunlight and the intrinsic limitations in charge carrier separation and transport. The performance of single-junction cells can be constrained by factors such as the bandgap of the semiconductor, which affects the range of the solar spectrum that can be absorbed.

4.9.3 Tandem PEC Cells

Tandem PEC cells as depicted in Fig. 4.7b are designed to overcome some of the limitations associated with single-junction cells by stacking two or more photoelectrode layers, each with different bandgaps. The idea behind the tandem configuration is to use a series of photoelectrodes that can absorb different parts of the solar spectrum, thereby enhancing overall light absorption and improving efficiency [67]. In a typical tandem cell, one layer is optimized for absorbing high-energy photons, while another layer is tailored for low-energy photons. The tandem configuration can significantly

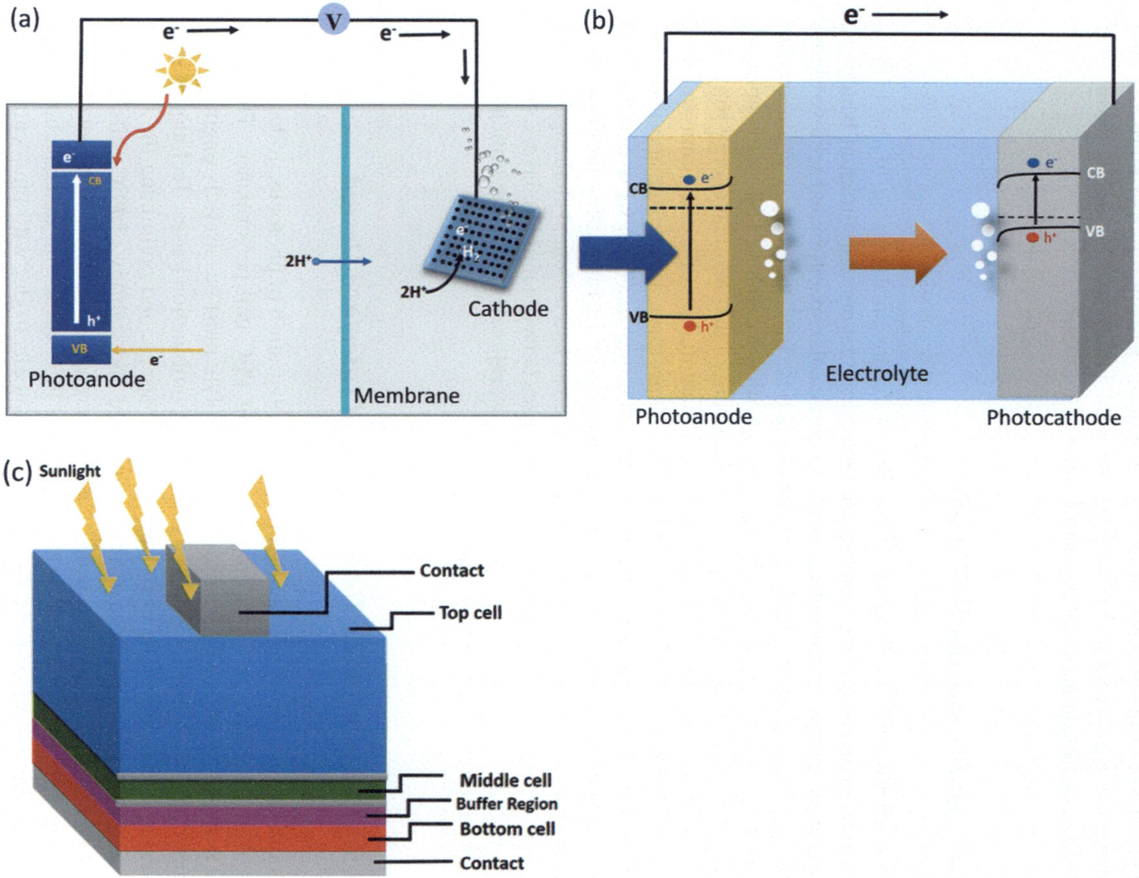

Fig. 4.7 **a** Single-junction PEC cell **b** tandem PEC cell **c** multi-junction PEC cell

improve the solar-to-hydrogen (STH) efficiency by utilizing a broader range of the solar spectrum and reducing energy losses due to the bandgap mismatch. This design also allows for better charge carrier separation, as each layer can be optimized for specific photogenerated charge carriers. However, tandem cells introduce additional complexity in terms of fabrication and integration. Ensuring proper alignment and contact between layers and managing potential issues such as light filtering and parasitic absorption are challenges that must be addressed.

4.9.4 Multi-junction PEC Cells

Multi-junction PEC cells as shown in Fig. 4.7c are an extension of the tandem design, incorporating three or more photoelectrode layers with different bandgaps. These cells aim to maximize the absorption of the solar spectrum and achieve higher efficiency by stacking multiple junctions in series. Each junction is engineered to absorb a specific range of the solar spectrum, with the combined effect of capturing a wider spectrum and converting more solar energy into chemical energy [68]. The multi-junction configuration offers the highest potential for efficiency improvements due to its ability to utilize nearly the entire solar spectrum. This design approach also facilitates better charge carrier management and minimizes energy losses associated with bandgap mismatches. However, the complexity of fabricating multi-junction cells is significantly higher, and the associated costs and technical challenges can be substantial. Issues such as managing the interfaces between multiple layers, ensuring uniform light absorption, and optimizing the overall cell design are critical for the success of multi-junction PEC cells.

4.9.5 Photoelectrode Integration and Interface Design

In addition to the basic cell architectures, the integration of photoelectrodes and the design of interfaces are vital considerations in PEC cell performance. The interface between the photoelectrode and the electrolyte must be carefully engineered to ensure efficient charge transfer and minimal recombination losses. Techniques such as surface passivation, the use of protective coatings, and optimizing the interface chemistry can enhance the overall performance and stability of PEC cells. Recent developments have also introduced hybrid and composite architectures that combine different types of photoelectrodes or integrate PEC cells with other technologies. For example, hybrid systems that integrate photoelectrodes with photocatalysts or other energy conversion devices can offer enhanced performance and versatility. Composite architectures that combine multiple photoelectrode materials in a single system can also improve light absorption, charge separation, and stability.

4.9.6 *Challenges and Future Directions*

While various PEC cell architectures offer promising advancements, several challenges remain. Fabrication complexity, material stability, and cost-effectiveness are ongoing concerns. The development of scalable and economically viable PEC cell designs is crucial for the practical implementation of water-splitting technologies. Future research is likely to focus on optimizing cell architectures, improving material performance, and addressing the integration challenges associated with advanced PEC systems. PEC cell architectures play a fundamental role in determining the efficiency and practicality of solar-driven water-splitting technologies. By understanding the strengths and limitations of different designs, researchers and engineers can work towards developing more efficient, stable, and cost-effective PEC cells for sustainable hydrogen production.

4.10 Integration of Photoelectrodes with Co-catalysts

The integration of photoelectrodes with co-catalysts is a pivotal aspect in optimizing photoelectrochemical (PEC) cells for efficient water splitting. Co-catalysts enhance the kinetics of the hydrogen evolution reaction (HER) and oxygen evolution reaction (OER), which are crucial for the effective conversion of solar energy into chemical energy. This section explores the methods and strategies for incorporating co-catalysts into PEC systems, highlighting their role in improving performance and efficiency.

4.10.1 *Role of Co-catalysts in PEC Systems*

In PEC cells, photoelectrodes absorb sunlight and generate electron–hole pairs that drive the HER and OER. However, the intrinsic kinetics of these reactions often pose limitations, with high overpotentials required to achieve significant rates [69]. Co-catalysts are employed to address these limitations by lowering the activation energies needed for the reactions and enhancing the overall efficiency of the PEC system. For the HER, co-catalysts such as platinum (Pt), palladium (Pd), and other noble metals are well-established due to their high catalytic activity. They facilitate the reduction of protons to hydrogen gas by providing active sites where the HER can proceed more efficiently. Despite their high performance, the use of noble metals is often limited by cost and scarcity. Consequently, there is a growing interest in alternative materials like transition metal dichalcogenides (TMDs) and transition metal phosphides (TMPs) [70], which offer competitive performance at a lower cost. These materials operate through mechanisms that involve creating active sites on their surfaces, enhancing the HER activity compared to the bare photoelectrode. In the context of OER, co-catalysts such as iridium oxide (IrO_2) and ruthenium oxide (RuO_2) are known for their high catalytic performance. They play a critical role in facilitating the oxidation of water to produce oxygen gas. However, due to the high cost and limited availability of these materials, research is shifting towards more affordable and abundant alternatives. Cobalt-based oxides and nickel–iron layered double hydroxides (NiFe LDHs) have emerged as promising options, showing significant improvements in OER performance. These co-catalysts help by providing active sites and reducing the energy barriers for the reaction, thereby addressing the slow kinetics often associated with OER.

4.10.2 Methods of Co-catalyst Integration

Several methods are employed to integrate co-catalysts into PEC cells, each offering different advantages and suited to various types of photoelectrode materials. These methods include deposition techniques, hybrid material formation, and interface engineering.

(a) **Deposition Techniques**
Deposition techniques are commonly used to apply co-catalysts onto photoelectrode surfaces. Chemical vapor deposition (CVD) and atomic layer deposition (ALD) are precise methods that allow for the controlled deposition of thin layers of co-catalysts. CVD involves the chemical reaction of gaseous precursors to form a solid coating on the photoelectrode surface, while ALD provides atomic-level control over the thickness and uniformity of the deposited layer [71]. These techniques are advantageous for creating well-adhered and uniform catalyst layers, which are essential for consistent performance. Electrochemical deposition is another technique where metal ions are reduced from a solution onto the photoelectrode. This method is particularly effective for producing high-loadings of co-catalysts and can be tailored to achieve specific catalyst morphologies. It is a versatile technique that allows for the deposition of a wide range of co-catalyst materials and can be adapted to different photoelectrode types.

(b) **Hybrid Material Formation**
Hybrid materials combine photoelectrodes with co-catalysts to create composite systems that leverage the strengths of both components. For example, integrating metal nanoparticles or metal–organic frameworks (MOFs) with photoelectrodes can enhance light absorption, charge transfer, and catalytic activity [72]. Hybrid systems are designed to optimize the interaction between the photoelectrode and co-catalyst, leading to improved performance compared to individual components. The combination of different materials in a hybrid system can result in synergistic effects that enhance the overall efficiency of the PEC cell. For instance, combining semiconductors with noble metal nanoparticles

can improve light absorption and charge separation, while also providing high catalytic activity for both HER and OER.

(c) **Interface Engineering**
Interface engineering involves optimizing the contact between the photoelectrode and co-catalyst to ensure efficient charge transfer and minimize recombination losses. Techniques such as surface passivation, the use of adhesion layers, and the design of protective coatings are employed to enhance the interaction between the photoelectrode and co-catalyst. Surface passivation techniques are used to reduce surface recombination and enhance the stability of the photoelectrode. Adhesion layers, often composed of conductive or adhesive materials, help in achieving strong and stable contact between the photoelectrode and co-catalyst. Additionally, protective coatings can prevent the degradation of the photoelectrode material and maintain the integrity of the co-catalyst under operational conditions.

4.10.3 Optimization and Performance Improvement

The integration of co-catalysts into PEC cells requires careful optimization to balance factors such as catalyst loading, distribution, and interaction with the photoelectrode. Properly integrated co-catalysts can significantly enhance the efficiency of water splitting by improving reaction kinetics and reducing overpotentials. The choice of co-catalyst materials, deposition techniques, and interface engineering strategies must be tailored to the specific requirements of the PEC system and the characteristics of the photoelectrode. The integration of photoelectrodes with co-catalysts is a crucial component of optimizing PEC cells for efficient water splitting. Co-catalysts play a significant role in improving the performance of PEC systems by enhancing the kinetics of HER and OER. Through various methods of integration and careful optimization, the performance of PEC cells can be significantly improved, contributing to more effective and practical solar-driven water-splitting technologies [73].

4.11 Influence of Electrolyte Composition on PEC Performance

The composition of the electrolyte in photoelectrochemical (PEC) cells is a fundamental factor influencing the performance of water splitting reactions. The electrolyte serves as the medium through which ions and charge carriers move between the photoelectrode and the counter-electrode, and its properties can significantly impact the efficiency, stability, and overall effectiveness of the PEC system. This section explores how electrolyte composition affects PEC performance, including aspects such as ionic conductivity, pH, and the role of various electrolytes in optimizing the water-splitting process.

4.11.1 Ionic Conductivity and Electrolyte Performance

One of the primary roles of the electrolyte in PEC cells is to facilitate the movement of ions between the photoelectrode and the counter-electrode. The ionic conductivity of the electrolyte is crucial for maintaining efficient charge transfer and reducing resistance within the cell. High ionic conductivity ensures that ions can move freely through the electrolyte, thereby minimizing ohmic losses and improving overall cell performance [74]. Common electrolytes used in PEC systems include aqueous solutions of acids, bases, and salts. Acidic electrolytes, such as sulfuric acid (H_2SO_4) and hydrochloric acid (HCl), are often employed because they provide high ionic conductivity and promote efficient hydrogen evolution. However, acidic environments can lead to the degradation of some photoelectrode materials over time. Basic electrolytes, such as sodium hydroxide (NaOH) or potassium hydroxide (KOH), are used for their favorable conditions for oxygen evolution and their compatibility with a broader range of photoelectrode materials. Basic solutions also offer high ionic conductivity and can be less corrosive to certain photoelectrodes. Saline electrolytes, which contain dissolved salts like sodium chloride (NaCl) or potassium chloride (KCl), are used in some PEC systems to enhance ionic conductivity. These electrolytes can be advantageous in certain configurations but may require careful management to avoid issues such as salt precipitation or electrode fouling [75].

4.11.2 pH and Reaction Kinetics

The pH of the electrolyte plays a significant role in the kinetics of the hydrogen evolution reaction (HER) and oxygen evolution reaction (OER). In acidic solutions, the HER typically proceeds more efficiently due to the availability of protons, which are readily reduced to hydrogen gas. Conversely, in alkaline solutions, the OER tends to be more favorable because hydroxide ions are readily available for oxidation. The choice of pH also affects the stability and performance of photoelectrode materials. Some photoelectrodes are more stable in acidic environments, while others perform better in alkaline conditions. Therefore, selecting an appropriate electrolyte pH is essential for optimizing the performance and longevity of the PEC cell. The pH also influences the solubility and activity of co-catalysts, which can impact their effectiveness in promoting the HER and OER.

4.11.3 Role of Electrolyte Additives

In addition to the primary electrolyte components, various additives can be introduced to enhance PEC performance. Additives such as buffering agents, surfactants, and redox mediators can modify the properties of the electrolyte to improve stability, reduce overpotentials, and enhance reaction kinetics. Buffering agents are used to maintain a stable pH within the electrolyte, which can be critical for ensuring consistent performance and preventing degradation of photoelectrode materials. Surfactants can alter the surface properties of photoelectrodes and co-catalysts, affecting their interaction with the electrolyte and improving catalytic activity. Redox mediators, which are compounds that participate in redox reactions without being consumed, can enhance charge transfer and reduce overpotentials, leading to more efficient water splitting.

4.11.4 Electrolyte Compatibility and Stability

The compatibility of the electrolyte with the photoelectrode and co-catalyst materials is another important consideration. Some electrolytes may react with or corrode the photoelectrode materials, leading to decreased performance and shorter lifetimes. For instance, highly acidic or basic conditions can cause dissolution or degradation of certain semiconductors and catalysts [76]. Therefore, selecting an electrolyte that is chemically compatible with all components of the PEC cell is crucial for maintaining long-term stability and performance.

4.11.5 Electrolyte Engineering for Enhanced Performance

Advancements in electrolyte engineering are focused on developing new formulations and strategies to optimize PEC performance. Researchers are exploring novel electrolyte compositions, including hybrid and non-aqueous electrolytes, to address the limitations of traditional aqueous solutions. Hybrid electrolytes, which combine elements of different electrolyte types, aim to achieve a balance of high ionic conductivity and stability. Non-aqueous electrolytes, such as ionic liquids or organic solvents [77], offer potential advantages in terms of stability and performance but require further investigation for practical application in PEC systems.

4.12 Challenges in PEC Water Splitting

Photoelectrochemical (PEC) water splitting is a promising approach for sustainable hydrogen production, leveraging solar energy to drive the water-splitting reaction. Despite its potential, several significant challenges hinder the widespread adoption and efficiency of PEC technology. Addressing these challenges is crucial for advancing the field and making PEC water splitting a viable option for large-scale hydrogen production.

4.12.1 Photocorrosion and Stability Issues

Photocorrosion represents a significant challenge in the field of photoelectrochemical (PEC) water splitting, impacting both the longevity and efficiency of photoelectrode materials. This phenomenon involves the degradation of photoelectrodes when exposed to light and electrochemical conditions, leading to a gradual decline in their performance. The primary cause of photocorrosion is the interaction between the photoelectrode material and the generated electron–hole pairs during the PEC process. These electron–hole pairs can react with the photoelectrode or electrolyte, causing oxidative damage and dissolution of the material. The mechanisms behind photocorrosion can vary depending on the type of photoelectrode material used. For instance, in oxide-based photoelectrodes, such as titanium dioxide (TiO_2), photocorrosion often involves the oxidation of the semiconductor surface. This oxidation can lead to the formation of unwanted by-products or phases that degrade the material's structural and electronic properties. In contrast, sulfide-based and phosphide-based photoelectrodes may experience dissolution or transformation of active phases under light and electrochemical stress. Understanding these specific mechanisms is essential for developing strategies to mitigate photocorrosion and enhance the stability of photoelectrodes.

Material selection is a critical factor in addressing photocorrosion. Choosing photoelectrode materials with inherent stability under PEC conditions is one approach. Materials such as TiO_2 are known for their strong chemical stability and resistance to oxidative environments, making them suitable candidates for PEC applications. However, even these materials can suffer from photocorrosion under certain conditions. Therefore, the development of new materials or modification of existing ones to enhance their stability is an active area of research. One effective strategy for mitigating photocorrosion is the application of protective coatings or layers on the photoelectrode surface. These coatings act as barriers, preventing direct interaction between the photoelectrode material and the electrolyte or light. For example, depositing a thin layer of a stable oxide or nitride on the photoelectrode can shield the underlying material from corrosive effects while still allowing effective light absorption and charge transfer [60]. This approach helps to maintain the integrity of the photoelectrode and prolong its operational lifespan. Surface modifications also play a significant role in improving photoelectrode stability. Techniques such as doping, alloying, and surface passivation can alter the electronic and chemical properties of the photoelectrode material to make it more resistant to photocorrosion. Doping with elements that stabilize the semiconductor structure or modify the band structure can reduce susceptibility to degradation. Alloying with other materials to form solid solutions can also enhance the material's resistance to photocorrosion.

Surface passivation involves applying a thin layer of a stable material that prevents or reduces corrosive reactions at the photoelectrode surface. This layer can act as a protective shield, minimizing direct exposure of the photoelectrode material to the reactive environment. For instance, applying a passivation layer of a material that is highly stable in the electrolyte environment can help maintain the integrity of the photoelectrode during operation. The composition of the electrolyte and operating conditions of the PEC system also influence photocorrosion. Highly acidic or basic electrolytes can accelerate the degradation of photoelectrode materials. Therefore, optimizing the electrolyte composition to balance ionic conductivity with stability considerations is crucial for minimizing photocorrosion. Additionally, controlling operating parameters such as light intensity, temperature, and applied potentials can help reduce the rate of degradation [78]. Advanced characterization techniques are essential for understanding and addressing photocorrosion. Techniques such as scanning electron microscopy (SEM), transmission electron microscopy (TEM), and X-ray photoelectron spectroscopy (XPS) provide detailed insights into the structural and chemical changes occurring during photocorrosion. These techniques allow researchers to identify degradation mechanisms, analyze surface changes, and evaluate the effectiveness of stability-enhancing strategies. The photocorrosion and stability issues are significant challenges in PEC water splitting that affect the performance and longevity of photoelectrodes. Addressing these issues involves selecting stable materials, applying protective coatings, modifying surfaces, optimizing electrolytes, and using advanced characterization techniques. Continued research and development in these areas are essential for advancing PEC technology and achieving practical, long-term solutions for sustainable hydrogen production.

4.12.2 Efficiency Losses: Recombination, Overpotentials, and Charge Transport Barriers

Efficiency losses in photoelectrochemical (PEC) water splitting are critical factors that hinder the overall performance of PEC systems. These losses can be attributed to several factors, including recombination of charge carriers, high overpotentials required for the hydrogen evolution reaction (HER) and oxygen evolution reaction (OER), and charge transport barriers within the photoelectrode materials. Understanding and addressing these inefficiencies are essential for enhancing the effectiveness and practical applicability of PEC technology. One of the primary causes of efficiency losses in PEC systems is the recombination of electron–hole pairs. In a PEC cell, light absorption generates electron–hole pairs in the photoelectrode material. For effective water splitting, these charge carriers must be separated and utilized in the electrochemical reactions. However, in many materials, a significant proportion of these charge carriers recombine before participating in the reactions. This recombination reduces the number of available charge carriers and diminishes the overall efficiency of the PEC system. Several strategies can be employed to minimize recombination losses. One approach involves optimizing the material properties to enhance charge separation and transfer. For instance, engineering the band structure or creating suitable heterojunctions within the photoelectrode material can improve charge carrier dynamics and reduce recombination rates. Additionally, incorporating materials with high charge carrier mobilities or adding passivation layers can further reduce recombination losses and enhance efficiency [79]. High overpotentials required for the hydrogen evolution reaction (HER) and oxygen evolution reaction (OER) are another significant source of efficiency loss in PEC systems. Overpotential refers to the additional potential required beyond the thermodynamic equilibrium potential to drive the electrochemical reactions. Large overpotentials increase the energy input needed for the reactions, thereby reducing the overall energy conversion efficiency of the PEC cell.

Reducing overpotentials involves improving the catalytic activity of the photoelectrode material. This can be achieved through various means, such as optimizing the composition and structure of the photoelectrode or incorporating efficient co-catalysts. For example, integrating metal or metal oxide co-catalysts with high intrinsic catalytic activity can lower the overpotentials required for HER and OER [80]. Additionally, surface modifications that enhance the active sites for catalysis can also contribute to reducing overpotentials and improving efficiency. Charge transport barriers within the photoelectrode material can also contribute to efficiency losses. These barriers can arise from poor charge carrier mobility, insufficient electronic conductivity, or mismatches between different material components. Efficient charge transport is essential for delivering charge carriers to the reaction sites and facilitating the PEC process. Any impedance in charge transport can lead to losses in the overall efficiency of the system.

To address charge transport barriers, various approaches can be utilized. Improving the intrinsic electronic conductivity of the photoelectrode material through doping or alloying can enhance charge carrier mobility. Additionally, designing optimal interfaces and junctions between different material components can reduce impedance and facilitate efficient charge transfer. Employing nanostructuring techniques, such as creating hierarchical or porous structures, can also enhance charge transport and improve overall performance [81]. The inherent properties of the photoelectrode material, including its bandgap, electronic structure, and surface characteristics, play a crucial role in determining the efficiency of PEC water splitting. Materials with suitable bandgap values and efficient charge carrier dynamics are more likely to exhibit high performance. Tailoring material properties through advanced synthesis techniques and design strategies can help optimize the photoelectrode's efficiency and reduce energy losses.

4.12.3 Scalability and Integration with Existing Hydrogen Production Technologies

The scalability of photoelectrochemical (PEC) water splitting technology and its integration with existing hydrogen production systems are critical factors that determine its practical viability and commercial adoption. For PEC water splitting to transition from laboratory research to large-scale applications, several challenges must be addressed, including scaling up the technology, optimizing system designs, and integrating with established hydrogen production methods. Scaling up PEC systems from laboratory to industrial scale involves several technical and economic challenges. One of the primary challenges is the production of large-area photoelectrodes that maintain high performance and stability. The synthesis and fabrication processes used for small-scale photoelectrodes may not be directly applicable to larger scales, necessitating the development of scalable methods that can produce high-quality materials consistently [82]. Additionally, maintaining uniform performance across large-area photoelectrodes is essential to ensure efficient water splitting over extended surfaces. Another challenge in scaling up PEC systems is managing the increased complexity and cost associated with larger-scale installations. Large PEC systems require substantial amounts of photoelectrode materials, electrolytes, and supporting infrastructure, which can significantly impact the overall cost. Developing cost-effective materials and fabrication techniques, as well as optimizing system designs to reduce material usage and operational expenses, are crucial for making large-scale PEC systems economically viable.

Integrating PEC water splitting technology with existing hydrogen production methods is essential for achieving practical and efficient hydrogen production on a commercial scale. Currently, most hydrogen production is dominated by steam methane reforming (SMR) and electrolysis, both of which have established infrastructures and technologies. Integrating PEC systems with these methods can offer synergistic benefits [83], such as combining the advantages of PEC water splitting with the existing hydrogen production infrastructure. One approach to integration is hybrid systems that combine PEC water splitting with conventional electrolysis. In such systems, PEC cells can be used to produce hydrogen directly from sunlight, while conventional electrolysis can be employed to further purify and enhance the hydrogen production process. This hybrid approach leverages the strengths of both technologies, potentially improving overall efficiency and reducing the reliance on external energy sources. Economic considerations play a significant role in the integration of PEC technology with existing hydrogen production methods. The cost of PEC systems, including materials, fabrication, and operation, must be competitive with conventional hydrogen production methods. Additionally, technical factors such as the compatibility of PEC systems with existing infrastructure, the ability to operate under varying conditions, and the overall system efficiency must be addressed to ensure successful integration [84]. To facilitate integration, research and development efforts should focus on optimizing PEC system designs for compatibility with existing hydrogen production technologies. This includes developing efficient interfaces between PEC cells and conventional systems, optimizing operational parameters for combined processes, and addressing any technical challenges that arise from integrating different technologies. Advances in system design and integration techniques are crucial for the successful deployment of PEC technology on a larger scale. Innovations in materials science, such as the development of new photoelectrode materials with improved stability and efficiency, can enhance the performance and scalability of PEC systems. Additionally, advancements in system engineering, including the design of efficient reactor systems and integration strategies, will contribute to overcoming scalability challenges and achieving commercial viability.

4.13 Recent Advances and Breakthroughs in PEC Research

The field of photoelectrochemical (PEC) water splitting has witnessed significant advancements and breakthroughs in recent years. These developments span various aspects of PEC technology, including materials innovation, surface and interface engineering, and characterization techniques. Recent research has led to the discovery of novel materials, improved understanding of reaction mechanisms, and enhanced system designs, all of which contribute to advancing the efficiency and practicality of PEC systems.

Recent developments in materials science have brought forward several promising candidates for enhancing the performance of PEC water splitting systems. Among these emerging materials, perovskites, two-dimensional (2D) materials, and hybrid structures represent significant advancements with the potential to revolutionize PEC technology as illustrated in Fig. 4.8. Each of these material classes offers unique properties and advantages that contribute to improved efficiency and functionality in PEC applications.

4.13.1 Perovskite Materials

Perovskite materials [85], characterized by their distinctive ABX_3 crystal structure, have gained considerable attention for their exceptional light absorption properties and tunable bandgap. This family of materials includes both organic–inorganic hybrids and inorganic perovskites, each offering different benefits for PEC water splitting. For instance, lead halide perovskites, such as $MAPbI_3$ (methylammonium lead iodide), have demonstrated remarkable photocatalytic activity and stability under illumination, making them suitable for PEC applications. One of the key advantages of perovskites is their ability to absorb a broad spectrum of light, including visible and near-infrared regions, which enhances their photocatalytic efficiency. Additionally, perovskites can be synthesized through low-cost and scalable methods, such as solution processing, which facilitates large-area fabrication. Recent research has also focused on improving the stability and durability of perovskites under operational conditions, addressing previous concerns about their long-term performance in PEC systems. Incorporating perovskite materials into PEC systems can lead to significant improvements in overall efficiency. For example, perovskite-based photoelectrodes have been shown to achieve high photocurrent densities and low onset potentials for water splitting reactions. By optimizing the composition and structure of perovskites, researchers are developing materials with enhanced

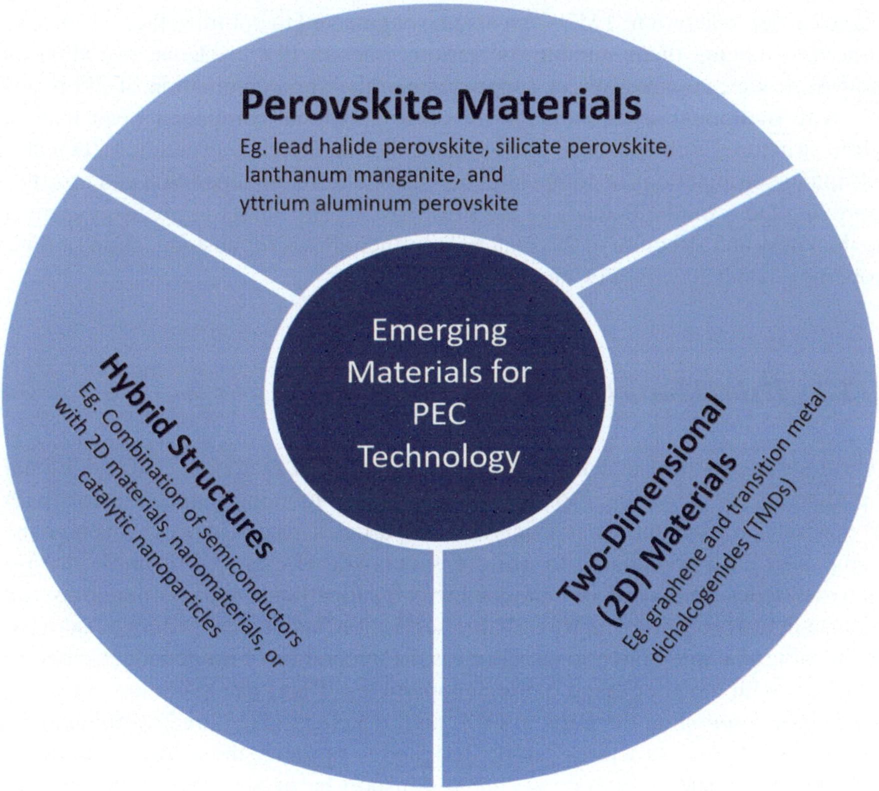

Fig. 4.8 Emerging materials for PEC water splitting

performance characteristics, such as improved charge carrier dynamics and reduced recombination losses.

4.13.2 Two-Dimensional (2D) Materials

Two-dimensional (2D) materials [86], such as graphene and transition metal dichalcogenides (TMDs), have emerged as versatile candidates for PEC water splitting due to their unique electronic and optical properties. Graphene, with its high electrical conductivity and large surface area, provides an excellent platform for enhancing charge transport and reducing recombination losses. When used as a conductive support or combined with other semiconductors, graphene can significantly improve the performance of PEC systems. Transition metal dichalcogenides (TMDs), such as MoS_2, WS_2, and TiS_2, exhibit strong light absorption, high surface area, and catalytic activity that are advantageous for PEC applications. These materials often possess direct bandgaps, which allow for efficient light absorption and charge carrier generation. TMDs can also be engineered to optimize their electronic properties, making them suitable for various roles in PEC systems, including as photoelectrodes, co-catalysts, or support materials. The combination of 2D materials with traditional semiconductors or other functional components can lead to hybrid structures with enhanced performance. For instance, incorporating 2D materials into heterojunctions or composites can facilitate charge separation and transfer, improving the overall efficiency of the PEC system. The ability to precisely control the thickness and structure of 2D materials further allows for customization of their properties to suit specific PEC applications.

4.13.3 Hybrid Structures

Hybrid structures [87] that integrate different types of materials offer additional opportunities for enhancing PEC performance. By combining semiconductors with 2D materials, nanomaterials, or catalytic nanoparticles, researchers can leverage the strengths of each component to achieve synergistic effects. For example, hybrid photoelectrodes composed of semiconductor nanoparticles and 2D materials can enhance light absorption, improve charge separation, and increase catalytic activity. One notable example of hybrid structures is the integration of semiconductor photoelectrodes with metal or metal oxide nanoparticles. These nanoparticles can act as co-catalysts, promoting the HER or OER and reducing the required overpotentials. Additionally, hybrid structures can be designed to optimize the interfaces between different components, minimizing charge transport barriers and enhancing overall efficiency. Recent advancements in the synthesis and fabrication of hybrid structures have enabled the development of novel PEC systems with superior performance. Techniques such as layer-by-layer assembly, self-assembly, and chemical vapor deposition have been employed to create well-defined hybrid architectures with controlled properties. These advances contribute to the creation of efficient and scalable PEC systems that can meet the demands of practical hydrogen production.

4.14 Innovative Surface and Interface Engineering Approaches

Surface and interface engineering are crucial aspects of enhancing the performance of photoelectrochemical (PEC) water splitting systems. By optimizing these interfaces, researchers can significantly improve the efficiency and stability of photoelectrodes. Recent advancements in surface modification techniques and interface design have led to substantial progress in PEC technology, addressing key challenges such as charge recombination, catalyst activity, and material stability.

4.14.1 Surface Modification Techniques

Surface modification plays a pivotal role in optimizing the performance of photoelectrodes by tailoring their chemical and physical properties. Recent innovations in surface modification techniques have enabled precise control over the photoelectrode surface, enhancing its interaction with light, charge carriers, and electrolytes. Techniques such as atomic layer deposition (ALD), chemical vapor deposition (CVD) [88], and plasma treatment have been employed to create thin, conformal coatings and functional layers on photoelectrode surfaces. Atomic layer deposition (ALD) [89] is a powerful technique that allows for the deposition of thin, uniform films with atomic precision. This method is particularly useful for creating passivation layers that protect photoelectrodes from degradation while maintaining high electrical conductivity. ALD coatings can also be tailored to introduce specific surface functionalities, such as increased catalytic activity or enhanced stability under operational conditions. Chemical vapor deposition (CVD) is another technique used to deposit thin films and create complex surface structures. CVD allows for the growth of high-quality, homogeneous layers that can improve the optical and electronic properties of photoelectrodes. Recent advancements in CVD techniques have led to the development of novel materials and coatings that enhance light absorption, charge separation, and catalytic performance. Plasma treatment is employed to modify the surface chemistry of photoelectrodes, improving their wettability, adhesion, and catalytic activity. Plasma processes can introduce functional groups, such as hydroxyl or carboxyl groups, that enhance the interaction between the photoelectrode and the electrolyte. This modification can lead to improved charge transfer and reduced overpotentials for the photoelectrochemical reactions [90–93].

4.14.2 Interface Engineering

Effective interface engineering is essential for optimizing charge transfer and minimizing recombination losses in PEC systems. Creating well-defined and stable interfaces between different material components can significantly enhance the overall efficiency of photoelectrodes. Innovations in interface engineering have focused on developing efficient heterojunctions, optimizing contact layers, and improving the integration of co-catalysts. Heterojunctions, where two or more semiconductor materials are combined, can improve charge separation and reduce recombination losses. Designing heterojunctions with appropriate band alignments ensures that photogenerated charge carriers are efficiently separated and transported to the catalytic sites. Recent research has focused on optimizing the composition and structure of heterojunctions to maximize their performance in PEC systems.

Contact layers, which serve as the interface between the photoelectrode and external electrical contacts, play a crucial role in charge transfer. Developing high-quality contact layers with low resistance and strong adhesion is essential for efficient electron transfer. Advances in contact layer materials and deposition techniques have contributed to improved performance and stability in PEC systems. The integration of co-catalysts is another critical aspect of interface engineering. Co-catalysts are often used to enhance the catalytic activity of photoelectrodes and reduce the required overpotentials for the hydrogen evolution reaction (HER) and oxygen evolution reaction (OER). Recent innovations in co-catalyst design, including the use of nanoparticles, molecular catalysts, and conductive supports, have led to significant improvements in PEC performance.

4.14.3 Recent Developments

Recent developments in surface and interface engineering have led to notable advancements in PEC technology. For example, the application of protective coatings and self-healing materials has improved the durability and longevity of photoelectrodes under harsh operational conditions. These coatings not only protect the photoelectrode from degradation but also maintain its performance over extended periods. Another significant development is the use of advanced characterization techniques to study and optimize surface and interface properties. Techniques such as in situ spectroscopy, scanning probe microscopy [94], and transmission electron microscopy [95] provide valuable insights into the behavior of photoelectrode surfaces and interfaces during PEC reactions. This information guides the design and optimization of materials and processes for improved performance.

4.15 Future Prospects

The future of photoelectrochemical (PEC) water splitting technology is vibrant with possibilities, driven by continuous advancements in material science, engineering, and system integration. As researchers and engineers strive to enhance the efficiency, scalability, and sustainability of PEC systems, several key areas are emerging as focal points. These include achieving higher solar-to-hydrogen (STH) efficiency, integrating PEC systems with renewable energy sources, and exploring novel applications in sustainable energy systems.

Attaining high solar-to-hydrogen (STH) [47] efficiency is a principal goal for PEC water splitting technology. Researchers are concentrating on optimizing the performance of photoelectrodes through various innovative approaches. One major focus is the development of advanced materials that offer improved light absorption, charge transport, and stability. Materials science has introduced new semiconductor materials, such as wide-bandgap oxides, perovskites, and 2D materials, which have the potential to significantly enhance PEC efficiency. For example, perovskite-based photoelectrodes are being explored for their tunable bandgaps and exceptional light-harvesting capabilities, while 2D materials like graphene and transition metal dichalcogenides are being investigated for their superior charge transport properties. Optimizing PEC cell architectures is another crucial area. Researchers are designing novel configurations, such as tandem cells, multi-junction systems, and advanced photonic structures, to maximize light absorption and facilitate efficient charge separation. Tandem cells, for instance, stack multiple photoelectrodes with different bandgaps to capture a broader spectrum of sunlight, potentially increasing overall efficiency. Additionally, integrating photonic structures, such as photonic crystals or plasmonic nanoparticles, can enhance light trapping and absorption, further boosting the efficiency of PEC systems. Improving the efficiency of the hydrogen evolution reaction (HER) and oxygen evolution reaction (OER) through better catalysts and co-catalysts is essential. Researchers are developing advanced catalytic materials that are both highly active and stable [96]. Innovations include the use of earth-abundant materials that offer cost-effective alternatives to precious metals, as well as the design of novel catalyst structures that enhance reaction kinetics. By addressing these aspects, researchers aim to reduce overpotentials and increase the overall performance of PEC systems.

Integrating PEC systems with renewable energy sources represents a significant opportunity to enhance their practicality and sustainability. Hybrid systems that combine PEC technology with other renewable energy sources, such as solar photovoltaics or wind turbines, can provide a continuous and reliable source of power for water splitting. This integration can help balance energy supply and demand, improving the overall efficiency and reliability of hydrogen production. For example, coupling PEC systems with solar panels can allow for continuous operation even when solar irradiance fluctuates, while wind turbines can provide supplemental energy during periods of low sunlight. Energy storage solutions also play a crucial role in integrating PEC systems with renewable energy sources. By incorporating energy storage technologies, such as batteries or supercapacitors, it is possible to manage intermittent energy sources and ensure a steady supply of electricity for water splitting. Advances in energy storage materials and technologies, including high-capacity batteries and fast-charging supercapacitors, will be essential for optimizing the performance and reliability of PEC systems. The integration of PEC systems into smart grids offers another avenue for enhancing their efficiency. Smart grids can manage energy flows from various sources, including PEC systems, and adjust to fluctuations in energy demand and supply. This dynamic management can optimize the distribution and utilization of energy, ensuring that PEC systems operate effectively and contribute to a more sustainable energy infrastructure.

The successful development and commercialization of PEC water splitting technology have the potential to revolutionize various aspects of sustainable energy systems. One of the most significant applications is the production of green hydrogen, a clean and versatile energy carrier. Hydrogen produced through PEC water splitting can be used in fuel cells, industrial processes, and transportation, providing a low-carbon alternative to fossil fuels. This application aligns with global efforts to reduce greenhouse gas emissions and transition to a more sustainable energy economy. Beyond hydrogen, PEC technology has the potential to be adapted for the production of other renewable fuels. Research is exploring ways to extend PEC systems to generate hydrocarbons or alcohols, which could offer additional energy storage and utilization options. This extension of PEC technology could expand its applications and contribute to a broader range of sustainable energy solutions. PEC systems also have potential applications in environmental remediation. By harnessing solar energy for chemical transformations, PEC technology can address environmental challenges such as pollutant removal and wastewater treatment. This capability aligns with efforts to develop sustainable technologies that not only produce clean energy but also contribute to environmental protection and resource management.

4.16 Conclusion

Photoelectrochemical (PEC) water splitting represents a transformative technology with the potential to address critical challenges in sustainable energy production. By harnessing solar energy to directly drive the splitting of water into hydrogen and oxygen, PEC systems offer a clean, renewable pathway for producing hydrogen fuel—a key component in the transition to a low-carbon economy. As explored throughout this chapter, PEC water splitting involves intricate principles and advanced materials, making it a highly dynamic field with significant ongoing research and development. The fundamental principles of PEC water splitting, encompassing thermodynamics, kinetics, and the intricacies of energy band alignment, lay the groundwork for understanding and optimizing this technology. By advancing our knowledge of these principles, researchers can enhance the efficiency of photoelectrodes and improve the overall performance of PEC systems. The exploration of key PEC reactions, including the hydrogen evolution reaction (HER) and the oxygen evolution reaction (OER), reveals the critical role of catalysis and reaction dynamics in achieving high efficiency and stability in water splitting processes. In the realm of photoelectrode materials, significant strides have been made in optimizing design and performance. The development of advanced semiconductor materials, effective bandgap engineering, and innovative nanostructuring techniques have all contributed to enhanced light absorption and improved charge carrier dynamics. Surface modifications and catalysts have also played a crucial role in boosting the efficiency of PEC systems by improving reaction kinetics and stability.

The integration of PEC systems into practical applications involves addressing several key factors, including cell architectures, the role of co-catalysts, and the influence of electrolyte composition. Designing efficient PEC cell configurations and optimizing integration strategies are essential for maximizing performance and ensuring scalability. The influence of electrolyte composition on PEC performance underscores the importance of material compatibility and reaction environment in achieving optimal results. As PEC technology progresses, addressing challenges such as photocorrosion, efficiency losses, and scalability will be critical. Future research will focus on achieving higher solar-to-hydrogen (STH) efficiency, integrating PEC systems with renewable energy sources, and exploring novel applications in sustainable energy systems. The advancement of materials, optimization of system components, and exploration of new applications will drive the continued evolution of PEC technology. PEC water splitting stands at the forefront of sustainable energy innovation, offering a promising solution for clean hydrogen production. The ongoing advancements in materials science, system design, and characterization techniques will play a pivotal role in realizing the full potential of PEC technology. By addressing current challenges and exploring future opportunities, the field of PEC water splitting is poised to make a significant impact on the global energy landscape, contributing to a more sustainable and resilient energy future.

References

1. Juneja, S., Bhattacharya, J.: Nanomaterial assisted photoelectrochemical water splitting. In: Materials Horizons: From Nature to Nanomaterials, pp. 249–273 (2022). https://doi.org/10.1007/978-981-16-7285-9_9
2. Liu, Y., et al.: Research progress of oxygen evolution reaction catalysts for electrochemical water splitting. ChemSusChem **14**(24), 5359–5383 (2021). https://doi.org/10.1002/CSSC.202101898
3. Imran, S., Hussain, M.: Emerging trends in water splitting innovations for solar hydrogen production: analysis, comparison, and economical insights. Int. J. Hydrogen Energy (2024). Accessed: 02 Sept 2024. [Online]. Available: https://www.sciencedirect.com/science/article/pii/S036031992402490X
4. Jiang, C., Moniz, S.J.A., Wang, A., Zhang, T., Tang, J.: Photoelectrochemical devices for solar water splitting—materials and challenges. Chem. Soc. Rev. **46**, 4645 (2017). pubs.rsc.org. https://doi.org/10.1039/c6cs00306k
5. Xiao, F.X., Liu, B.: Plasmon-dictated photo-electrochemical water splitting for solar-to-chemical energy conversion: current status and future perspectives. Adv. Mater. Interfaces **5**(6) (2018). https://doi.org/10.1002/ADMI.201701098
6. Ma, Y., Wang, X., Jia, Y., Chen, X., Han, H., Li, C.: Titanium dioxide-based nanomaterials for photocatalytic fuel generations. Chem. Rev. **114**(19), 9987–10043 (2014). https://doi.org/10.1021/CR500008U
7. Marschall, R.: Semiconductor composites: strategies for enhancing charge carrier separation to improve photocatalytic activity. Adv. Funct. Mater. **24**(17), 2421–2440 (2014). https://doi.org/10.1002/ADFM.201303214
8. Moss, B., Babacan, O., Kafizas, A., Hankin, A.: A review of inorganic photoelectrode developments and reactor scale-up challenges for solar hydrogen production. Adv. Energy Mater. **11**(13) (2021). https://doi.org/10.1002/AENM.202003286
9. Chakraborty, S., et al.: Hydrogen energy as future of sustainable mobility. Front. Energy Res. **10** (2022). https://doi.org/10.3389/FENRG.2022.893475/FULL
10. Armaroli, N., Balzani, V.: Solar electricity and solar fuels: status and perspectives in the context of the energy transition. Chem. Eur. J. **22**(1), 32–57 (2016). https://doi.org/10.1002/CHEM.201503580
11. Mckone, J.R., Crans, D.C., Martin, C., Turner, J., Duggal, A.R., Gray, H.B.: Translational science for energy and beyond. ACS Publ. **55**(18), 9131–9143 (2016). https://doi.org/10.1021/acs.inorgchem.6b01097
12. Awasthy, R., Flint, S., Sankarnarayana, R., Jones, R.L.: A framework to improve university—industry collaboration. J. Ind. Univ. Collab. **2**(1), 49–62 (2020). https://doi.org/10.1108/JIUC-09-2019-0016/FULL/PDF

13. Rajan, A.G., Martirez, J.M.P., Carter E.A.: Why do we use the materials and operating conditions we use for heterogeneous (photo) electrochemical water splitting? ACS Catal. **10**(19) (2020). https://doi.org/10.1021/ACSCATAL.0C01862
14. Toe, C.Y., Pan, J., Scott, J., Amal, R.: Identifying key design criteria for large-scale photocatalytic hydrogen generation from engineering and economic perspectives. ACS ES T Eng. **2**(6), 1130–1143 (2022). https://doi.org/10.1021/ACSESTENGG.2C00030
15. McKone, J.R., Crans, D.C., Martin, C., Turner, J., Duggal, A.R., Gray, H.B.: Translational science for energy and beyond. Inorg. Chem. **55**(18), 9131–9143 (2016). https://doi.org/10.1021/ACS.INORGCHEM.6B01097
16. Wu, H., et al.: Unveiling carrier dynamics in periodic porous $BiVO_4$ photocatalyst for enhanced solar water splitting. ACS Energy Lett. **6**(10), 3400–3407 (2021). https://doi.org/10.1021/ACSENERGYLETT.1C01454
17. Yao, B., Zhang, J., Fan, X., He, J., Li, Y.: Surface engineering of nanomaterials for photoelectrochemical water splitting. Small **15**(1) (2019). https://doi.org/10.1002/SMLL.201803746
18. Roduner, E., Radhakrishnan, S.G.: In command of non-equilibrium. Chem. Soc. Rev. (2016). pubs.rsc.org, Accessed: 03 Sept 2024. [Online]. Available: https://pubs.rsc.org/en/content/articlehtml/2016/cs/c6cs00115g
19. Yang, H., Driess, M., Menezes, P.W.: Self-supported electrocatalysts for practical water electrolysis. Adv. Energy Mater. **11**(39) (2021). https://doi.org/10.1002/AENM.202102074
20. Jin, M., et al.: Strategies for designing high-performance hydrogen evolution reaction electrocatalysts at large current densities above 1000 mA cm^{-2}. ACS Nano **16**(8), 11577–11597 (2022). https://doi.org/10.1021/ACSNANO.2C02820
21. Zhang, X., Bieberle-Hütter, A.: Modeling and simulations in photoelectrochemical water oxidation: from single level to multiscale modeling. ChemSusChem **9**(11), 1223–1242 (2016). https://doi.org/10.1002/CSSC.201600214
22. Sharma, K.K.: Unravelling the secrets: how catalyst reconstruction redefines superior oxygen-evolving chemistry. J. Surv. Fish. Sci. **10**(3) (2023). sifisheriessciences.com, Accessed: 03 Sept 2024. [Online]. Available: http://www.sifisheriessciences.com/index.php/journal/article/view/2784
23. Thompson, E.L., Jorne, J., Gasteiger, H.: Oxygen reduction reaction kinetics in subfreezing PEM fuel cells. In: ACS National Meeting Book of Abstracts (2007). https://doi.org/10.1149/1.2742305/META
24. Sun, H., Xu, X., Song, Y., Zhou, W., Shao, Z.: Designing high-valence metal sites for electrochemical water splitting. Adv. Funct. Mater. **31**(16) (2021). https://doi.org/10.1002/ADFM.202009779
25. Ohta, T.: Thermodynamics of water-splitting. Solar Hyd. Energy Syst. (1979). ui.adsabs.harvard.edu. Accessed: 03 Sept 2024. [Online]. Available: https://ui.adsabs.harvard.edu/abs/1979shes.book...25O/abstract
26. Ley, L., Pollak, R.A., McFeely, F.R., Kowalczyk, S.P., Shirley, D.A.: Total valence-band densities of states of III-V and II-VI compounds from x-ray photoemission spectroscopy. Phys. Rev. B **9**(2), 600–621 (1974). https://doi.org/10.1103/PHYSREVB.9.600
27. Ge, M., et al.: One-dimensional TiO_2 nanotube photocatalysts for solar water splitting. Adv. Sci. **4**(1) (2017). https://doi.org/10.1002/ADVS.201600152
28. Padmanabhan, N.T., et al.: Graphene coupled TiO_2 photocatalysts for environmental applications: a review. Chemosphere **271**, 129506 (2021). https://doi.org/10.1016/j.chemosphere.2020.129506
29. McKenna, B., Evans, R.C.: Towards efficient spectral converters through materials design for luminescent solar devices. Adv. Mater. **29**(28) (2017) https://doi.org/10.1002/ADMA.201606491
30. Thangamuthu, M., et al.: Polymer photoelectrodes for solar fuel production: progress and challenges. Chem. Rev. **122**(13), 11778–11829 (2022). https://doi.org/10.1021/ACS.CHEMREV.1C00971
31. Yue, X., Fan, J., Xiang, Q.: Internal electric field on steering charge migration: modulations, determinations and energy-related applications. Adv. Funct. Mater. **32**(12) (2022). https://doi.org/10.1002/ADFM.202110258

32. Niu, J., Albero, J., Atienzar, P., García, H.: Porous single-crystal-based inorganic semiconductor photocatalysts for energy production and environmental remediation: preparation, modification, and applications. Adv. Funct. Mater. **30**(15) (2020) https://doi.org/10.1002/ADFM.201908984
33. Marwat, M.A., et al.: Advanced catalysts for photoelectrochemical water splitting. ACS Appl. Energy Mater. **4**(11), 12007–12031 (2021). https://doi.org/10.1021/ACSAEM.1C02548
34. Mahmood, N., Yao, Y., Zhang, J.W., Pan, L., Zhang, X., Zou, J.J.: Electrocatalysts for hydrogen evolution in alkaline electrolytes: mechanisms, challenges, and prospective solutions. Adv. Sci. **5**(2) (2018). https://doi.org/10.1002/ADVS.201700464
35. Rebollar, L., et al.: 'Beyond adsorption' descriptors in hydrogen electrocatalysis. ACS Catal. **10**(24), 14747–14762 (2020). https://doi.org/10.1021/ACSCATAL.0C03801
36. Zhuang, Z., Huang, J., Li, Y., Zhou, L., Mai, L.: The holy grail in platinum-free electrocatalytic hydrogen evolution: molybdenum-based catalysts and recent advances. ChemElectroChem **6**(14), 3570–3589 (2019). https://doi.org/10.1002/CELC.201900143
37. Ul Haq, T., Haik, Y.: Electrocatalysis for the water splitting: recent strategies for improving the performance of electrocatalyst. In: Advances in Sustainable Energy: Policy, Materials and Devices, pp. 315–339 (2021). https://doi.org/10.1007/978-3-030-74406-9_11
38. Han, J., Liu, Z.: Optimization and modulation strategies of zinc oxide-based photoanodes for highly efficient photoelectrochemical water splitting. ACS Appl. Energy Mater. **4**(1), 1004–1013 (2021). https://doi.org/10.1021/ACSAEM.0C02985
39. Hassan, N.S., et al.: Photoelectrochemical water splitting using post-transition metal oxides for hydrogen production: a review. Environ. Chem. Lett. **20**(1), 311–333 (2022). https://doi.org/10.1007/S10311-021-01357-X
40. Zhang, Y., Liu, B., Xu, L., Ding, Z., Yang, R., Wang, S.: Failure mechanism analysis and emerging strategies for enhancing the photoelectrochemical stability of photoanodes. ChemSusChem (2024). https://doi.org/10.1002/CSSC.202401420
41. Jiang, L., et al.: Near-infrared light responsive TiO_2 for efficient solar energy utilization. Adv. Funct. Mater. **32**(12) (2022). https://doi.org/10.1002/ADFM.202108977
42. Wang, H., Rogach, A.L.: Hierarchical SnO_2 nanostructures: recent advances in design, synthesis, and applications. Chem. Mater. **26**(1), 123–133 (2014). https://doi.org/10.1021/CM4018248
43. Raizada, P., Nguyen, T., Patial, S., Singh, P., Bajpai, A.: Toward practical solar-driven photocatalytic water splitting on two-dimensional MoS_2 based solid-state Z-scheme and S-scheme heterostructure. Fuel (2021). Accessed: 03 Sept 2024. [Online]. Available: https://www.sciencedirect.com/science/article/pii/S0016236121011819
44. Hu, S.: Solid-solid interfaces in photoelectrochemistry: co-catalysts, surface passivation, and corrosion protection. In: Springer Handbooks, pp. 879–921 (2022). https://doi.org/10.1007/978-3-030-63713-2_30
45. Yao, T., An, X., Han, H., Chen, J.Q., Li, C.: Photoelectrocatalytic materials for solar water splitting. Adv. Energy Mater. **8**(21) (2018). https://doi.org/10.1002/AENM.201800210
46. Kawase, Y., Higashi, T., Domen, K., Takanabe, K.: Recent developments in visible-light-absorbing semitransparent photoanodes for tandem cells driving solar water splitting.Adv. Energy Sustain. Res. **2**(7) (2021). https://doi.org/10.1002/aesr.202100023
47. Li, Z., Fang, S., Sun, H., Chung, R.J., Fang, X., He, J.H.: Solar hydrogen. Adv. Energy Mater. **13**(8) (2023). https://doi.org/10.1002/AENM.202203019
48. Tilley, S.D.: Recent advances and emerging trends in photo-electrochemical solar energy conversion. Adv. Energy Mater. **9**(2) (2019). https://doi.org/10.1002/aenm.201802877
49. Kautek, W., Gobrecht, J., Gerischer, H.: Applicability of semiconducting layered materials for electrochemical solar energy conversion. Berichte der Bunsengesellschaft/Phys. Chem. Chem. Phys. **84**(10), 1034–1040 (1980). https://doi.org/10.1002/BBPC.19800841021.
50. Yu, Z., Liu, H., Zhu, M., Li, Y., Li, W.: Interfacial charge transport in 1D TiO_2 based photoelectrodes for photoelectrochemical water splitting. Small **17**(9) (2021). https://doi.org/10.1002/SMLL.201903378

51. Nawaz, A., et al.: Review of hybrid 1D/2D photocatalysts for light-harvesting applications. ACS Appl. Nano Mater. **4**(11), 11323–11352 (2021). https://doi.org/10.1021/ACSANM.1C01014
52. Luo, S., Li, T., Wang, X., Faizan, M., Zhang, L.: High-throughput computational materials screening and discovery of optoelectronic semiconductors. Wiley Interdiscip. Rev. Comput. Mol. Sci. **11**(1) (2021). https://doi.org/10.1002/WCMS.1489
53. Zhang, W., Liu, M.: Modulating carrier transport via defect engineering in solar water splitting devices. ACS Energy Lett. **4**(4), 834–843 (2019). https://doi.org/10.1021/ACSENERGYLETT.9B00276
54. Zhao, Y., Niu, Z., Zhao, J., Xue, L., Fu, X., Long, J.: Recent advancements in photoelectrochemical water splitting for hydrogen production. Electrochem. Energy Rev. **6**(1) (2023). https://doi.org/10.1007/S41918-022-00153-7
55. Hu, X., Li, G., Yu, J.C.: Design, fabrication, and modification of nanostructured semiconductor materials for environmental and energy applications. Langmuir **26**(5), 3031–3039 (2010). https://doi.org/10.1021/LA902142B
56. Smith, M.: Femtosecond-laser irradiation as a platform for tailoring the optoelectronic properties of silicon (2012). Accessed: 03 Sept 2024. [Online]. Available: https://dspace.mit.edu/handle/1721.1/75849
57. Wang, K., et al.: Unconventional strategies to break through the efficiency of light-driven water splitting: a review. Electron **1**(1) (2023). https://doi.org/10.1002/ELT2.4
58. Nellist, M.R., Laskowski, F.A.L., Lin, F., Mills, T.J., Boettcher, S.W.: Semiconductor-electrocatalyst interfaces: theory, experiment, and applications in photoelectrochemical water splitting. Acc. Chem. Res. **49**(4), 733–740 (2016). https://doi.org/10.1021/ACS.ACCOUNTS.6B00001
59. Li, W., Wang, K.: Photocatalytic oxygen evolution. In: Photo- and Electro-Catalytic Processes: Water Splitting, N_2 Fixing, CO_2 Reduction, pp. 485–519 (2022). https://doi.org/10.1002/9783527830084.CH14
60. Scheuermann, A.G., McIntyre, P.C.: Atomic layer deposited corrosion protection: a path to stable and efficient photoelectrochemical cells. J. Phys. Chem. Lett. **7**(14), 2867–2878 (2016). https://doi.org/10.1021/ACS.JPCLETT.6B00631
61. Sreenivasulu, M., Hadrihalli, A., Alshehri, M.A., Shetti, N.P.: Rational designing of nickel-iron containing layered double hydroxide [NiFe@LDH] electrocatalysts for effective water splitting. Energy Fuels **38**(14), 12888–12899 (2024). https://doi.org/10.1021/ACS.ENERGYFUELS.4C01899
62. Lin, Y., et al.: Co-induced electronic optimization of hierarchical NiFe LDH for oxygen evolution. Small **16**(38) (2020). https://doi.org/10.1002/SMLL.202002426
63. Artero, V., Chavarot-Kerlidou, M., Fontecave, M.: Splitting water with cobalt. Angew. Chem. Int. Ed. **50**(32), 7238–7266 (2011). https://doi.org/10.1002/ANIE.201007987
64. Liston, E.M.: Plasma treatment for improved bonding: a review. J. Adhes. **30**(1–4), 199–218 (1989). https://doi.org/10.1080/00218468908048206
65. Luan, P., Zhang, J.: Stepping towards solar water splitting: recent progress in bismuth vanadate photoanodes. ChemElectroChem **6**(13), 3227–3243 (2019). https://doi.org/10.1002/CELC.201900398
66. Sheng, X., Xu, T., Feng, X.: Rational design of photoelectrodes with rapid charge transport for photoelectrochemical applications. Adv. Mater. **31**(11) (2019). https://doi.org/10.1002/ADMA.201805132
67. Prévot, M.S., Sivula, K.: Photoelectrochemical tandem cells for solar water splitting. J. Phys. Chem. C **117**(35), 17879–17893 (2013). https://doi.org/10.1021/JP405291G
68. Ginley, D., Green, M.A., Collins, R.: Solar energy conversion toward 1 terawatt. MRS Bull. **33**(4), 355–364 (2008). https://doi.org/10.1557/MRS2008.71
69. Wang, S., Liu, G., Wang, L.: Crystal facet engineering of photoelectrodes for photoelectrochemical water splitting. Chem. Rev. **119**(8), 5192–5247 (2019). https://doi.org/10.1021/ACS.CHEMREV.8B00584

70. Weng, C.C., Ren, J.T., Yuan, Z.Y.: Transition metal phosphide-based materials for efficient electrochemical hydrogen evolution: a critical review. ChemSusChem **13**(13), 3357–3375 (2020). https://doi.org/10.1002/CSSC.202000416
71. Yanagi, R., Zhao, T., Solanki, D., Pan, Z., Hu, S.: Charge separation in photocatalysts: mechanisms, physical parameters, and design principles. ACS Energy Lett. **7**(1), 432–452 (2022). https://doi.org/10.1021/ACSENERGYLETT.1C02516
72. Yang, H., et al.: Metal–organic framework coated titanium dioxide nanorod array p–n heterojunction photoanode for solar water-splitting. Nano Res. **12**(3), 643–650 (2019). https://doi.org/10.1007/S12274-019-2272-4
73. Thalluri, S.M., Bai, L., Lv, C., Huang, Z., Hu, X., Liu, L.: Strategies for semiconductor/electrocatalyst coupling toward solar-driven water splitting. Adv. Sci. **7**(6) (2020). https://doi.org/10.1002/ADVS.201902102
74. Zhang, J., Ricote, S., Hendriksen, P.V., Chen, Y.: Advanced materials for thin-film solid oxide fuel cells: recent progress and challenges in boosting the device performance at low temperatures. Adv. Funct. Mater. **32**(22) (2022). https://doi.org/10.1002/ADFM.202111205
75. Sassenburg, M., Kelly, M., Subramanian, S., Smith, W.A., Burdyny, T.: Zero-gap electrochemical CO_2 reduction cells: challenges and operational strategies for prevention of salt precipitation. ACS Energy Lett. **8**(1), 321–331 (2023). https://doi.org/10.1021/ACSENERGYLETT.2C01885
76. Weng, B., Qi, M.Y., Han, C., Tang, Z.R., Xu, Y.J.: Photocorrosion inhibition of semiconductor-based photocatalysts: basic principle, current development, and future perspective. ACS Catal. **9**(5), 4642–4687 (2019). https://doi.org/10.1021/ACSCATAL.9B00313
77. Xiong, S., Regula, M., Wang, D., Song, J.: Toward better lithium-sulfur batteries: functional non-aqueous liquid electrolytes. Electrochem. Energy Rev. **1**(3), 388–402 (2018). https://doi.org/10.1007/S41918-018-0015-Y
78. Samu, G.F., Janáky, C.: Photocorrosion at irradiated perovskite/electrolyte interfaces. J. Am. Chem. Soc. **142**(52), 21595–21614 (2020). https://doi.org/10.1021/JACS.0C10348
79. Cao, Y., et al.: Defects passivation strategy for efficient and stable perovskite solar cells. Adv. Mater. Interfaces **9**(21) (2022). https://doi.org/10.1002/ADMI.202200179
80. Rosman, N.N., et al.: An overview of co-catalysts on metal oxides for photocatalytic water splitting. Int. J. Energy Res. **46**(9), 11596–11619 (2022). https://doi.org/10.1002/ER.8001
81. Cong, L., Xie, H., Li, J.: Hierarchical structures based on two-dimensional nanomaterials for rechargeable lithium batteries. Adv. Energy Mater. **7**(12) (2017). https://doi.org/10.1002/AENM.201601906
82. Li, J., Fleetwood, J., Hawley, W.B., Kays, W.: From materials to cell: state-of-the-art and prospective technologies for lithium-ion battery electrode processing. Chem. Rev. **122**(1), 903–956 (2022). https://doi.org/10.1021/ACS.CHEMREV.1C00565
83. Nnabuife, S.G., Hamzat, A.K.,Whidborne, J., Kuang, B., Jenkins, K.W.: Integration of renewable energy sources in tandem with electrolysis: a technology review for green hydrogen production. Int. J. Hydrogen Energy (2024). Accessed: 03 Sept 2024. [Online]. Available: https://www.sciencedirect.com/science/article/pii/S0360319924025825
84. Meegahapola, L., Mancarella, P., Flynn, D., Moreno, R.: Power system stability in the transition to a low carbon grid: a techno-economic perspective on challenges and opportunities. Wiley Interdiscip. Rev. Energy Environ. **10**(5) (2021). https://doi.org/10.1002/WENE.399
85. Jung, H.S., Park, N.G.: Perovskite solar cells: from materials to devices. Small **11**(1), 10–25 (2015). https://doi.org/10.1002/SMLL.201402767
86. Zhang, H.: Introduction: 2D materials chemistry. Chem. Rev. **118**(13), 6089–6090 (2018). https://doi.org/10.1021/ACS.CHEMREV.8B00278
87. Chen, X., et al.: Graphene hybrid structures for integrated and flexible optoelectronics. Adv. Mater. **32**(27) (2020). https://doi.org/10.1002/ADMA.201902039
88. Spear, K.E.: Principles and applications of chemical vapor deposition (CVD). Pure Appl. Chem. **54**(7), 1297–1311 (1982). https://doi.org/10.1351/PAC198254071297/HTML
89. Leskelä, M., Ritala, M.: Atomic layer deposition chemistry: recent developments and future challenges. Angew. Chem. Int. Ed. **42**(45), 5548–5554 (2003). https://doi.org/10.1002/ANIE.200301652

90. Stiles, P.L., Dieringer, J.A., Shah, N.C.,Van Duyne, R.P.: Surface-enhanced Raman spectroscopy. Annu. Rev. Ana. Chem. **1**(1), 601–626 (2008). annualreviews.org. https://doi.org/10.1146/annurev.anchem.1.031207.112814
91. Hansma, P.K., Elings, V.B., Marti, O., Bracker, C.E.: Scanning tunneling microscopy and atomic force microscopy: application to biology and technology. Science **242**(4876), 209–216 (1988). https://doi.org/10.1126/SCIENCE.3051380
92. Liao, X., et al.: Density functional theory for electrocatalysis. Energy Environ. Mater. **5**(1), 157–185 (2022). https://doi.org/10.1002/EEM2.12204
93. Andrade, J.D.: X-ray photoelectron spectroscopy (XPS). In: Surface and Interfacial Aspects of Biomedical Polymers, pp. 105–195 (1985). https://doi.org/10.1007/978-1-4684-8610-0_5
94. Poggi, M.A., Gadsby, E.D., Bottomley, L.A., King, W.P., Oroudjev, E., Hansma, H.: Scanning probe microscopy. Anal. Chem. **76**(12), 3429–3444 (2004). https://doi.org/10.1021/AC0400818
95. Winey, M., Meehl, J.B., O'Toole, E.T., Giddings, T.H.: Conventional transmission electron microscopy. Mol. Biol. Cell **25**(3), 319–323 (2014). https://doi.org/10.1091/mbc.E12-12-0863
96. Su, D.S., Perathoner, S., Centi, G.: Nanocarbons for the development of advanced catalysts. Chem. Rev. **113**(8), 5782–5816 (2013). https://doi.org/10.1021/CR300367D

Chapter 5
Electrochemical Sensors

Electrochemical sensors are pivotal in modern detection technologies, offering high sensitivity and versatility across diverse applications. This chapter provides a comprehensive overview of electrochemical sensors, detailing their operational principles, configurations, and material choices. It explores various sensor architectures, including traditional three-electrode setups, microelectrode arrays, screen-printed electrodes, and flexible and wearable sensors. Special attention is given to the role of electrolytes in sensor performance, with discussions on aqueous, non-aqueous, and solid-state electrolytes. The chapter also highlights the application of electrochemical sensors in detecting environmental pollutants and corrosion products, showcasing their relevance in environmental monitoring and infrastructure maintenance. Future trends and advancements, such as the integration of nanomaterials and smart technologies, are discussed, emphasizing the ongoing evolution and potential of electrochemical sensors in addressing contemporary challenges.

5.1 Introduction

Electrochemical sensors have emerged as vital tools in modern science and technology, providing a powerful means to detect and quantify a wide range of chemical and biological species [1, 2]. These sensors operate on the principle of converting chemical information into an electrical signal, which can then be measured and analyzed. This ability to directly link chemical phenomena with electronic output has made electrochemical sensors indispensable in various fields, including environmental monitoring [3], healthcare [4], industrial process control [5], and national security [6]. The simplicity, sensitivity, and versatility of these sensors [7], combined with their potential for miniaturization and integration into portable devices, have driven their widespread adoption and ongoing development. Figure 5.1a illustrates

P. Phogat et al., *Electrochemical Devices*, Engineering Materials,
https://doi.org/10.1007/978-981-96-0527-9_5

the ready to use a paper-based electrochemical sensor. The fundamental operation of electrochemical sensors is based on the interaction of an analyte with an electrode surface, leading to a change in an electrical parameter such as current, voltage, or impedance. This change is often proportional to the concentration of the analyte, allowing for quantitative analysis. Different electrochemical techniques, such as potentiometry, amperometry, voltammetry, and electrochemical impedance spectroscopy, are employed depending on the nature of the analyte and the desired sensitivity and selectivity of the sensor. The versatility of these techniques enables the detection of a broad spectrum of substances, ranging from simple ions and molecules to complex biological entities. The choice of electrode material is crucial in determining the performance of electrochemical sensors. Traditionally, noble metals such as gold and platinum have been used due to their excellent conductivity and chemical stability. However, recent advances in materials science have introduced a variety of new materials, including carbon-based nanomaterials [8, 9], conducting polymers [10], and semiconductor materials [11, 12]. These materials offer unique properties that can enhance the sensitivity, selectivity, and response time of sensors. For instance, the high surface area and tunable electronic properties of nanomaterials allow for a greater interaction between the electrode and the analyte, leading to improved sensor performance.

One of the most exciting developments in the field of electrochemical sensing is the use of semiconductor materials as electrodes [11, 13]. Semiconductors such as silicon, metal oxides, and transition metal dichalcogenides have been widely explored

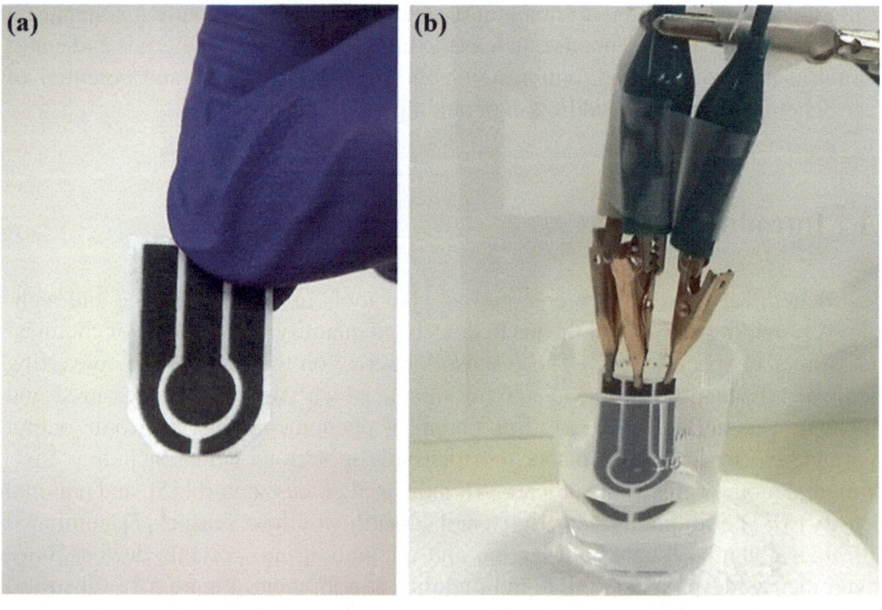

Fig. 5.1 **a** Electrochemical sensor device and **b** electrochemical sensor immersed into iron sulphate solution reproduced from Ref. [27]. Copyright 2024 IOP Publishing

due to their ability to facilitate charge transfer processes at the electrode-electrolyte interface. The electronic properties of these materials can be precisely controlled through doping, surface modification, and nano structuring, enabling the detection of specific analytes with high sensitivity. The integration of semiconductor electrodes with microfabrication technologies has also paved the way for the development of miniaturized and flexible sensors that can be used in wearable and implantable devices, opening new avenues for real-time health monitoring and personalized medicine. The detection mechanism in electrochemical sensors is another critical aspect that influences their performance. The interaction between the analyte and the electrode surface can occur through various mechanisms, including adsorption, redox reactions, or complexation. These interactions lead to measurable changes in the electrochemical properties of the system, such as the potential difference [14], current flow, or impedance, which are then used to quantify the concentration of the analyte. The nature of the electrolyte also plays a significant role in these processes, providing the necessary ionic medium for charge transfer and influencing the stability and selectivity of the sensor [15]. The development of novel electrolytes, such as ionic liquids and solid-state electrolytes, has further expanded the range of conditions under which electrochemical sensors can operate, enabling their use in extreme environments and complex matrices.

The architecture and configuration of electrochemical sensors have also evolved significantly over the years. Traditional electrochemical sensors typically employ a three-electrode system, consisting of a working electrode, a reference electrode, and a counter electrode as shown in Fig. 5.1b [16]. This configuration allows for precise control and measurement of the electrochemical processes occurring at the working electrode. However, advances in microfabrication and nanotechnology have led to the development of more sophisticated sensor architectures, including integrated sensor arrays, lab-on-a-chip devices [17], and flexible and stretchable sensors [18]. These innovations have not only improved the performance of electrochemical sensors but have also expanded their application range to include areas such as point-of-care diagnostics, environmental monitoring, and wearable technology.

Electrochemical sensors have found extensive applications in pH sensing [19], metal ion detection [20], and the monitoring of biological molecules [21], gases, redox-active species [22], environmental pollutants [23], and corrosion products [24]. pH sensors, for instance, are among the most widely used electrochemical sensors, employed in diverse fields ranging from environmental science to biomedicine [25]. The ability to accurately measure the acidity or basicity of a solution is crucial for many chemical processes, and advances in pH sensing technology have led to the development of solid-state pH sensors that are more robust and capable of miniaturization. Metal ion detection is another critical application of electrochemical sensors, particularly in environmental monitoring and public health [26]. The presence of toxic metal ions in water and soil can have severe consequences for both human health and the environment.

Electrochemical sensors offer a sensitive and selective method for detecting trace levels of metal ions, often employing techniques such as anodic stripping voltammetry to achieve low detection limits. The development of sensors capable of detecting multiple metal ions simultaneously has further enhanced their utility in complex environmental and biological samples [27]. In the field of healthcare, the detection of biological molecules using electrochemical sensors has revolutionized medical diagnostics. Biosensors, which combine a biological recognition element with an electrochemical transducer, are capable of detecting specific biomolecules with high sensitivity and specificity. These sensors have been widely used for glucose monitoring in diabetes management [28], detection of biomarkers for various diseases, and even in the development of personalized medicine [29]. The integration of electrochemical sensors with digital technologies has also enabled the real-time monitoring of physiological parameters [30], providing valuable data for the management of chronic diseases and the early detection of health conditions [31]. Gas sensing is another important application of electrochemical sensors [32], particularly in industrial safety, environmental monitoring, and air quality assessment. Electrochemical gas sensors operate by detecting the oxidation or reduction of gas molecules at the electrode surface, resulting in a measurable electrical signal. These sensors are widely used for detecting toxic gases such as carbon monoxide, nitrogen oxides, and hydrogen sulphide [33]. The development of sensors with enhanced sensitivity and selectivity has been driven by the use of advanced materials such as metal oxides and conducting polymers, as well as the optimization of sensor design to control gas diffusion and reaction kinetics. The detection of redox-active species [34], environmental pollutants [35], and corrosion products represents further areas where electrochemical sensors have made significant contributions. Redox-active species, including organic compounds and transition metal complexes, are often detected using voltammetric techniques, which provide detailed information about the redox properties of the analyte. Environmental pollutants, such as pesticides and organic contaminants, are detected using sensors that exploit the specific interactions between the pollutant and the electrode surface [36]. Corrosion monitoring, on the other hand, relies on electrochemical techniques such as impedance spectroscopy to detect the onset of corrosion and monitor the progress of corrosion processes in real-time. Electrochemical sensors represent a dynamic and rapidly evolving field with a wide range of applications across different sectors.

The ongoing advancements in materials science, microfabrication, and electrochemical techniques continue to drive the development of more sensitive, selective, and versatile sensors. As these sensors become increasingly integrated with digital technologies and other analytical tools, their role in addressing global challenges such as environmental protection, healthcare [37], and industrial safety is set to expand even further. This chapter will explore the principles, materials, configurations, and applications of electrochemical sensors in greater detail, providing a comprehensive overview of the state of the art in this exciting and impactful field.

5.2 Fundamentals of Electrochemical Sensing

Electrochemical sensing is a powerful analytical technique that relies on the direct conversion of chemical information into an electrical signal as shown in Fig. 5.2. This conversion is achieved through interactions between the analyte of interest and an electrode surface, where electrochemical reactions take place. The resulting electrical signal, which can be in the form of current, voltage, or impedance, is then measured and correlated with the concentration or presence of the analyte. The versatility, sensitivity, and specificity of electrochemical sensors make them suitable for a wide range of applications, from environmental monitoring to biomedical diagnostics.

5.2.1 Detection of Redox-Active Species

The detection of redox-active species is a pivotal aspect of electrochemical sensing, with applications spanning environmental monitoring, clinical diagnostics, and industrial process control [38]. Redox-active species are substances that undergo reduction and oxidation reactions, which can be detected and quantified using electrochemical sensors. These sensors leverage the principles of electrochemical reactions to provide sensitive and selective measurements of various redox-active analytes. This section explores the fundamental principles, techniques, and applications involved in detecting redox-active species, highlighting the advancements and challenges in this field.

Detection of redox-active species in electrochemical sensors is based on the measurement of current changes resulting from redox reactions at the electrode surface [39, 40]. These reactions involve the transfer of electrons between the analyte and the electrode, which can be quantified to determine the concentration of the species.

Redox-active species participate in electrochemical reactions where they either gain electrons (reduction) or lose electrons (oxidation). The general reaction at the electrode surface can be represented as:

$$\mathrm{Ox} + \mathrm{e}^- \leftrightarrow \mathrm{Red}$$

where "Ox" is the oxidized form and "Red" is the reduced form of the species. The current generated during these reactions is proportional to the concentration of the redox-active species, allowing for its quantification [40]. The electrode potential, controlled by applying a voltage, dictates the redox reaction occurring at the electrode. By scanning the electrode potential across a range of values, the electrochemical behavior of the redox-active species can be studied. Techniques like cyclic voltammetry (CV) and electrochemical impedance spectroscopy (EIS) are commonly used to characterize these reactions and extract quantitative information [41, 42].

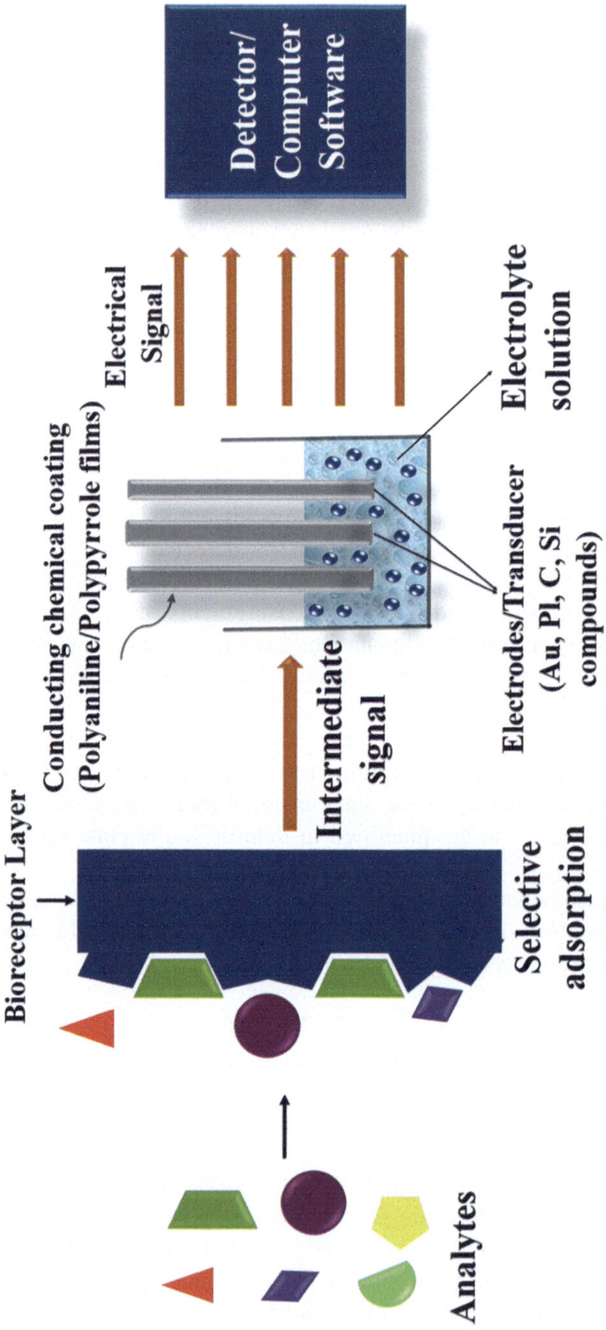

Fig. 5.2 Fundamental principle of electrochemical sensing

5.2.2 Techniques for Detection

Several electrochemical techniques are employed to detect and analyze redox-active species, each offering unique advantages and suited to different applications.

Cyclic Voltammetry (**CV**): CV is a widely used technique where the electrode potential is cycled between two values, and the resulting current is measured. The shape and magnitude of the current–voltage curve as shown in Fig. 5.3a provide information about the redox properties of the species, including its concentration, reaction kinetics, and thermodynamics. CV is useful for identifying redox-active species and studying their electrochemical behavior [43].

Differential Pulse Voltammetry (**DPV**): DPV as shown in Fig. 5.3b involves applying a series of voltage pulses to the electrode and measuring the current response

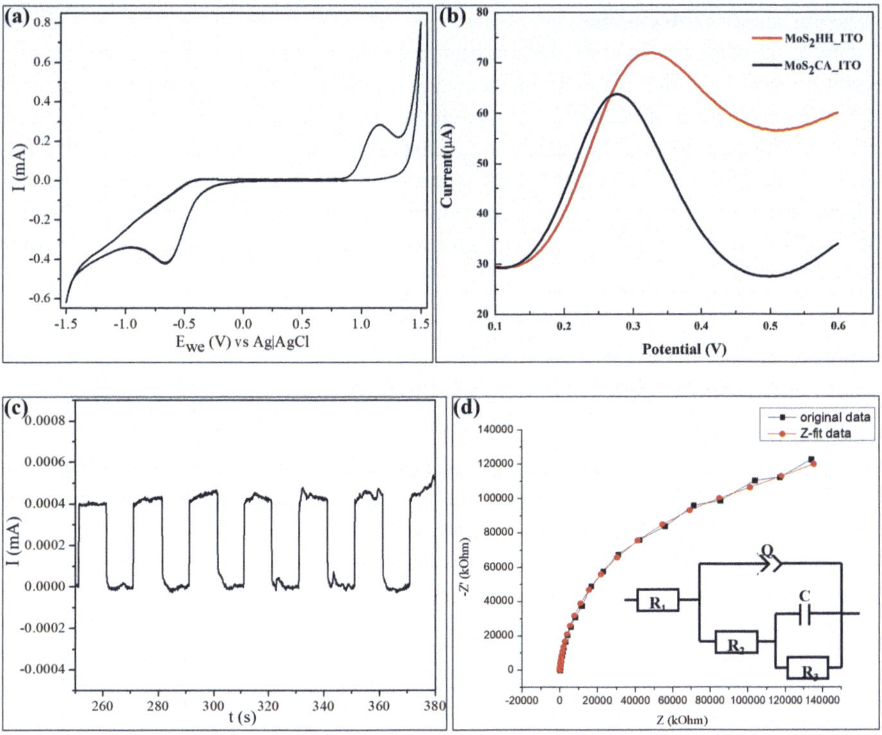

Fig. 5.3 **a** Cyclic voltammogram, **b** differential pulse voltammetry of two different samples; Reproduced with permission from Ref. [47]. Copyright © 2020 Published by Elsevier Ltd., **c** chronoamperometry and **d** Nyquist plot along with Randle circuit as inset; Reproduced with permission from Ref. [48]. Copyright © 2024 Springer Nature

[44]. This technique enhances the sensitivity of detection by minimizing the background current and improving the signal-to-noise ratio. DPV is particularly useful for detecting low concentrations of redox-active species in complex matrices.

Chronoamperometry: In chronoamperometry, a constant potential is applied to the electrode, and the resulting current is measured over time as illustrated in Fig. 5.3c. This technique is used to study the kinetics of redox reactions and monitor the concentration of redox-active species in real time. It is commonly employed in applications where continuous monitoring is required [45].

Impedance Spectroscopy: Electrochemical impedance spectroscopy (EIS) measures the impedance of the electrode as a function of frequency. Changes in impedance can be correlated with the presence and concentration of redox-active species [46]. The obtained data is then quantitatively analysed by Nyquist plot and Randle circuit as obtained by its Z-fitting as shown in Fig. 5.3d. EIS is useful for studying the interactions between redox species and electrode materials, as well as assessing sensor performance.

The basic principle behind electrochemical sensing lies in the redox (reduction-oxidation) reactions that occur at the electrode-electrolyte interface. When an analyte undergoes oxidation or reduction at the electrode surface, electrons are transferred between the analyte and the electrode. This electron transfer generates an electrical signal, typically a current, that is proportional to the concentration of the analyte. The electrode serves as both a conduit for electron transfer and a site for the chemical reaction, making its material and surface properties critical to the sensor's performance.

5.2.3 Electrode Materials and Surface Chemistry

The choice of electrode material is a critical factor in determining the sensitivity, selectivity, and stability of an electrochemical sensor. Traditional electrodes are made from noble metals like gold, platinum, and silver due to their excellent conductivity and chemical stability. However, the rise of nanotechnology and materials science has expanded the range of materials available for electrode fabrication.

(a) **Carbon-Based Materials**: Carbon materials, including graphite, glassy carbon, carbon nanotubes, and graphene, are widely used in electrochemical sensors due to their high surface area, excellent conductivity, and chemical stability. The surface of carbon electrodes can be easily modified with functional groups or nanomaterials to enhance their sensitivity and selectivity. For example, graphene-based electrodes have been employed in a variety of sensors for detecting small molecules, metal ions, and biological species.

(b) **Metal Oxides and Semiconductors**: Metal oxides such as titanium dioxide, zinc oxide, and iron oxide are popular choices for electrochemical sensors, particularly in gas sensing and biosensing applications [49]. These materials exhibit unique electronic properties, such as wide band gaps and high electron mobility, which can be tuned through doping or nanostructuring. Semiconductor materials like silicon and metal dichalcogenides are also gaining attention for their ability to facilitate charge transfer processes at the electrode interface, making them suitable for detecting a wide range of analytes.

(c) **Conducting Polymers**: Conducting polymers, such as polyaniline, polypyrrole, and PEDOT (poly(3,4-ethylenedioxythiophene)), are another class of materials used in electrochemical sensors. These polymers combine the electronic properties of metals with the flexibility and processability of plastics. They can be synthesized with various functional groups that enhance their affinity for specific analytes, making them ideal for use in biosensors and environmental sensors [50].

The surface chemistry of the electrode is equally important in electrochemical sensing. Surface modifications, such as the attachment of functional groups, nanomaterials, or biological molecules, can significantly enhance the performance of a sensor by increasing its sensitivity, selectivity, and response time. For example, the immobilization of enzymes on an electrode surface can create a biosensor capable of specifically detecting a target molecule, such as glucose, through an enzymatic reaction that produces a measurable electrical signal.

5.2.4 Electrolytes and Their Role in Sensing

Electrolytes play a crucial role in electrochemical sensing by providing the ionic medium necessary for charge transfer between the electrode and the analyte. The choice of electrolyte can significantly impact the sensor's performance, including its sensitivity, stability, and selectivity.

(a) **Aqueous Electrolytes**: Aqueous electrolytes, such as phosphate buffer saline (PBS), are commonly used in electrochemical sensors due to their compatibility with biological systems and their ability to support efficient charge transfer [51]. The pH and ionic strength of the electrolyte can be adjusted to optimize sensor performance for specific analytes.

(b) **Non-aqueous Electrolytes**: Non-aqueous electrolytes, such as organic solvents or ionic liquids, are used in situations where water-based electrolytes may not be suitable, such as in the detection of hydrophobic analytes or in high-temperature applications [52]. Ionic liquids, in particular, offer unique advantages, including a wide electrochemical window, high ionic conductivity, and low volatility, making them ideal for specialized sensing applications.

(c) **Solid-State Electrolytes**: Solid-state electrolytes, such as polymer electrolytes or solid ionic conductors, are used in the development of miniaturized and portable sensors. These electrolytes provide the necessary ionic conductivity while eliminating the need for liquid components, making the sensors more robust and suitable for use in harsh environments or wearable devices.

5.2.5 Sensor Design and Configuration

The design and configuration of an electrochemical sensor play a critical role in its functionality and application. Traditional sensors often employ a three-electrode system, consisting of a working electrode, a reference electrode, and a counter electrode as depicted in Fig. 5.4. This configuration allows for precise control and measurement of the electrochemical processes occurring at the working electrode.

(a) **Working Electrode**: The working electrode is the site where the redox reaction of the analyte occurs, and it is typically made from materials such as gold, platinum, carbon, or semiconductor materials [53]. The surface area, shape, and surface chemistry of the working electrode can be tailored to optimize sensitivity and selectivity for a specific analyte.
(b) **Reference Electrode**: The reference electrode provides a stable and known potential against which the potential of the working electrode is measured. Common reference electrodes include the silver/silver chloride (Ag/AgCl) electrode and the saturated calomel electrode (SCE) [54]. The stability and reproducibility of the reference electrode are crucial for accurate and reliable sensor performance.
(c) **Counter Electrode**: The counter electrode, also known as the auxiliary electrode, completes the electrochemical circuit by allowing the flow of current between the working and reference electrodes [55, 56]. It is typically made from a conductive material such as platinum or carbon and is designed to have a large surface area to minimize its impedance.

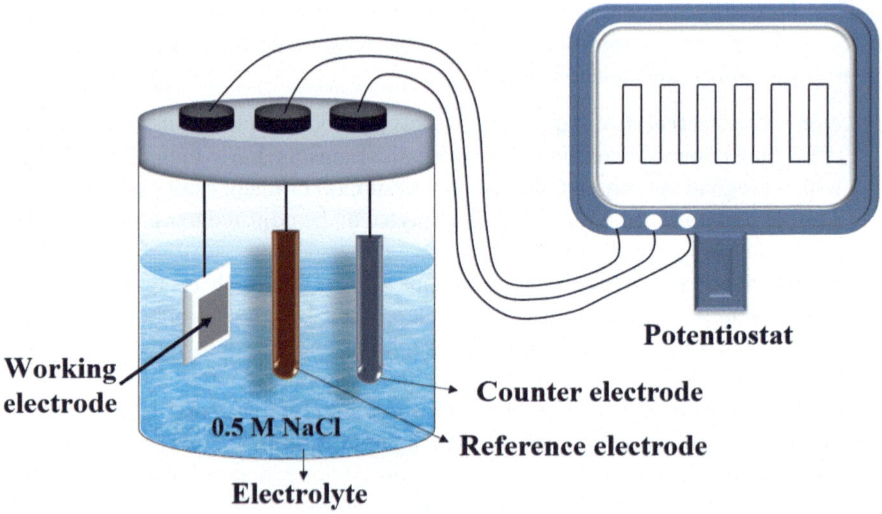

Fig. 5.4 Schematic diagram of three electrode electrochemical workstation setup

In modern electrochemical sensor design, there is a growing trend towards miniaturization and integration of sensors into compact and portable devices. Advances in microfabrication and nanotechnology have enabled the development of microelectrodes, sensor arrays, and lab-on-a-chip devices, which offer the potential for multiplexed detection and high-throughput analysis. Flexible and stretchable sensors, made from materials like conductive polymers and nanomaterials, are also being developed for wearable and implantable applications, providing new opportunities for real-time health monitoring and personalized medicine.

5.3 Semiconductor Electrodes for EC Sensing

Semiconductor electrodes have garnered significant attention in the field of electrochemical sensing due to their unique electronic properties and versatility [57]. Unlike traditional metal electrodes, semiconductors offer tunable electronic characteristics, such as band gap, carrier concentration, and surface states, which can be precisely controlled through doping, nanostructuring, and surface modifications [58]. These properties make semiconductor materials particularly suitable for developing highly sensitive and selective sensors for a wide range of chemical and biological analytes. One of the primary advantages of using semiconductor electrodes in electrochemical sensing is their ability to facilitate efficient charge transfer processes at the electrode-electrolyte interface.

The semiconductor's band structure plays a crucial role in determining the kinetics of these processes. For instance, when a semiconductor electrode comes into contact with an electrolyte, a space-charge region is formed near the surface, where the electric field affects the movement of charge carriers (electrons and holes). This region significantly influences the electron transfer reactions that occur at the interface, making semiconductor electrodes particularly sensitive to changes in the local environment caused by the presence of an analyte [59]. A wide variety of semiconductor materials have been explored for electrochemical sensing applications, including metal oxides, silicon, transition metal dichalcogenides (TMDs), and organic semiconductors. Metal oxides, such as titanium dioxide (TiO_2), zinc oxide (ZnO), and tin oxide (SnO_2), are among the most widely used semiconductor materials due to their stability, wide band gaps, and strong catalytic activity [60]. These materials are particularly effective in gas sensing applications, where they can detect a range of gases like oxygen, nitrogen oxides, and carbon monoxide. The high sensitivity of metal oxide-based sensors is often attributed to the large surface area of nanostructured films, which provides more active sites for gas adsorption and reaction.

Silicon, a cornerstone of the semiconductor industry, has also been extensively studied as an electrode material for electrochemical sensors [61]. Silicon's well-understood electronic properties, along with the mature fabrication techniques developed by the semiconductor industry, make it an attractive material for sensor applications. Silicon-based electrodes can be easily integrated into microelectronic devices, enabling the development of lab-on-a-chip systems for point-of-care diagnostics and environmental monitoring [62]. Additionally, silicon can be functionalized with various chemical groups or nanomaterials to enhance its selectivity for specific analytes, further broadening its application in electrochemical sensing. Transition metal dichalcogenides (TMDs) [63, 64], such as molybdenum disulfide (MoS_2) [64] and tungsten disulfide (WS_2), have recently emerged as promising materials for electrochemical sensors due to their layered structure and tunable electronic properties. TMDs exhibit a range of band gaps that can be adjusted by controlling their thickness, from bulk to monolayer, making them suitable for detecting a variety of analytes [65]. The unique two-dimensional structure of TMDs provides a high surface-to-volume ratio, which enhances their interaction with target molecules and improves sensor performance. Moreover, TMDs can be engineered to exhibit catalytic activity, further enhancing their sensitivity in applications like hydrogen evolution reaction (HER) sensing and biosensing.

Organic semiconductors, such as conjugated polymers, have also been explored for electrochemical sensing due to their flexibility, processability, and ability to undergo reversible redox reactions [66]. Conducting polymers like polyaniline (PANI) and polypyrrole (PPy) are particularly attractive because their electronic properties can be easily tuned by doping or chemical modification. These materials are often used in biosensors, where their biocompatibility and ability to interact with biological molecules, such as enzymes or antibodies, enable the selective detection of biomolecules like glucose, DNA, and proteins [67]. Surface modification and nanostructuring are key strategies for enhancing the performance of semiconductor electrodes in electrochemical sensing. By modifying the surface with functional groups, nanoparticles, or molecular recognition elements, the sensitivity and selectivity of the sensor can be significantly improved [68]. For example, decorating the surface of a ZnO electrode with noble metal nanoparticles like gold or platinum can enhance its catalytic activity and enable the detection of low concentrations of analytes. Similarly, the immobilization of biological molecules on a semiconductor surface can create biosensors capable of highly specific interactions with target analytes, leading to improved detection limits and faster response times.

The integration of semiconductor electrodes into advanced sensor architectures has also opened new possibilities for electrochemical sensing. For instance, the development of semiconductor-based field-effect transistors (FETs) [69] as sensors has enabled the detection of analytes through changes in the electrical conductivity of the semiconductor channel. These FET-based sensors offer the advantage of label-free detection and can be easily miniaturized for incorporation into portable or wearable devices [70]. Additionally, semiconductor electrodes are increasingly being used in combination with other sensing modalities, such as optical or piezoelectric sensors, to create hybrid sensors that offer enhanced performance through multimodal detection. Despite the numerous advantages of semiconductor electrodes, challenges remain in optimizing their performance for specific sensing applications. Issues such as long-term stability, reproducibility, and the influence of environmental factors on sensor response need to be addressed to fully realize the potential of semiconductor-based electrochemical sensors [71]. Nevertheless, the ongoing research and development in this field continue to drive innovations, paving the way for the next generation of high-performance sensors that can meet the growing demands of modern analytical applications.

Semiconductor electrodes represent a versatile and powerful platform for electrochemical sensing, offering unique advantages in terms of sensitivity, selectivity, and integration with electronic devices [72]. As research in materials science and nanotechnology progresses, the development of new semiconductor materials and fabrication techniques will likely lead to further enhancements in sensor performance, expanding their applications across various domains, including environmental monitoring, healthcare, and industrial process control.

5.4 Analyte Detection Mechanisms and Electrolytes

In electrochemical sensing, the accurate detection of analytes relies on the interplay between the electrode materials, the nature of the analyte, and the electrolyte used in the system. The detection mechanisms involve complex electrochemical processes, such as electron transfer, ion exchange, and redox reactions, which occur at the interface between the electrode and the electrolyte. Understanding these mechanisms is crucial for designing sensors with high sensitivity, selectivity, and stability. Additionally, the choice of electrolyte plays a vital role in facilitating these electrochemical processes, influencing the overall performance of the sensor.

5.4.1 Analyte Detection Mechanisms

Direct Electron Transfer is a fundamental mechanism in electrochemical sensing, where the analyte directly participates in a redox reaction at the electrode surface. This process is commonly observed in the detection of small molecules, metal ions, and certain gases. For example, in the detection of hydrogen peroxide (H_2O_2), the analyte undergoes reduction or oxidation at the electrode, generating a current proportional to its concentration [73]. This current response, often measured using amperometric sensors, is directly related to the analyte concentration, making this mechanism highly effective for straightforward and sensitive detection [13]. Direct electron transfer is particularly advantageous for applications where the analyte is electroactive and can readily donate or accept electrons without the need for additional mediators.

Mediated Electron Transfer comes into play when the analyte does not readily undergo direct redox reactions. In such cases, a mediator, which is a redox-active species, facilitates the transfer of electrons between the analyte and the electrode. This mechanism is widely used in biosensors, where biological molecules like enzymes are involved. For instance, in glucose sensing, the enzyme glucose oxidase catalyzes the oxidation of glucose, producing hydrogen peroxide. A mediator like ferrocyanide then shuttles electrons from the hydrogen peroxide to the electrode, generating a measurable current. This approach is crucial in biosensors as it allows the detection of analytes that otherwise would not be electroactive under normal conditions [74].

Adsorptive Stripping is a detection mechanism that involves preconcentrating the analyte on the electrode surface through adsorption, followed by a stripping step where the analyte is electrochemically desorbed, generating a signal. This technique is particularly effective in detecting trace levels of metal ions using voltammetric sensors. For example, in anodic stripping voltammetry (ASV), metal ions are initially reduced and deposited on the electrode as a thin film. The potential is then swept in the positive direction, causing the metal to oxidize and strip off the electrode, producing a peak in the current that corresponds to the metal ion concentration. Adsorptive stripping enhances the sensitivity of sensors, making it possible to detect analytes at very low concentrations.

Capacitive and Impedance-Based Detection mechanisms measure changes in the capacitance or impedance at the electrode-electrolyte interface, which are indicative of analyte binding events [75]. This approach is particularly useful for detecting biomolecules, such as proteins or DNA, where the binding of the analyte to a receptor on the electrode surface alters the charge distribution or impedance [76, 77]. Electrochemical impedance spectroscopy (EIS) is a common technique used in these applications, providing detailed information about interfacial processes, including charge transfer resistance and double-layer capacitance. This method offers high sensitivity and is capable of detecting small changes in analyte concentration, making it suitable for applications requiring precise measurements.

Electrocatalysis enhances the redox reaction of an analyte by utilizing a catalytic material on the electrode surface. This mechanism is often employed in sensors designed for gas detection or the detection of small molecules like hydrogen and oxygen. For instance, platinum or palladium nanoparticles deposited on an electrode can catalyze the reduction of oxygen, leading to an amplified current response [78]. The catalytic activity of the electrode material significantly boosts the sensor's sensitivity, allowing for the detection of analytes at lower concentrations. Electrocatalytic mechanisms are essential in applications where high sensitivity and low detection limits are required.

Field-Effect Sensing is a mechanism employed in sensors based on field-effect transistors (FETs), where the detection of analytes involves changes in the conductivity of a semiconductor channel. This change occurs when an analyte binds to the gate electrode, altering the electric field at the interface and modulating the charge carrier concentration in the channel. This modulation results in a change in current flow, which is measured as the sensor's output [79]. Field-effect sensing is particularly useful for detecting ions, gases, and biomolecules, offering the advantage of label-free detection and high sensitivity. This mechanism is widely used in modern electrochemical sensors, especially in applications that require integration with electronic devices.

5.4.2 Electrolytes in Electrochemical Sensing

Electrolytes are an integral component of electrochemical sensors, providing the ionic medium necessary for charge conduction between the electrode and the analyte. The choice of electrolyte can significantly influence the sensor's performance, affecting its sensitivity, stability, and selectivity. Electrolytes can be broadly categorized into three main types: aqueous, non-aqueous, and solid-state electrolytes. Each category possesses distinct characteristics and is suited to different applications.

5.4.2.1 Aqueous Electrolytes

Aqueous electrolytes, such as phosphate-buffered saline (PBS), sodium chloride (NaCl) solutions, and potassium chloride (KCl) solutions, are among the most commonly used in electrochemical sensors. Their popularity is largely due to their compatibility with biological systems and their high ionic conductivity [80]. In biosensors, where the accurate detection of biological molecules like enzymes and antibodies is critical, maintaining a physiological pH and ionic strength is essential. Aqueous electrolytes are particularly effective in these contexts, as they provide a stable environment that preserves the activity and integrity of biological molecules. The performance of an electrochemical sensor in an aqueous medium is closely tied to the pH of the electrolyte. Many redox reactions are pH-dependent, meaning that changes in pH can directly impact the sensor's response. For instance, in pH sensing, the Nernst equation governs the relationship between the electrode potential and the pH of the solution. A precise understanding of this relationship allows for the accurate detection of pH changes, which is critical in various applications, including environmental monitoring and biomedical diagnostics. Another crucial factor in aqueous electrolytes is their ionic strength. Ionic strength influences the double-layer capacitance and the charge transfer resistance at the electrode surface, both of which are key parameters in determining the sensitivity and selectivity of the sensor. For example, in sensors designed to detect ions or small molecules, the ionic strength of the electrolyte can enhance or inhibit the interaction between the analyte and the electrode surface, thereby affecting the overall sensor performance.

5.4.2.2 Non-aqueous Electrolytes

Non-aqueous electrolytes, which include organic solvents like acetonitrile and dimethyl sulfoxide, as well as ionic liquids, are used in electrochemical sensors when water-based electrolytes are unsuitable. These electrolytes are particularly advantageous for detecting hydrophobic analytes, performing high-temperature sensing, or in situations where a wide electrochemical window is required. Ionic liquids, a subset of non-aqueous electrolytes, are salts that remain in a liquid state at room temperature and offer unique benefits such as low volatility, high thermal stability, and tunable viscosity and conductivity [81]. The use of non-aqueous electrolytes allows electrochemical sensors to detect a broader range of analytes, including those that might be unstable or unreactive in aqueous environments. For instance, in organic molecule detection or gas sensing, non-aqueous electrolytes can prevent side reactions like water electrolysis, which can interfere with the sensor's accuracy and sensitivity. Additionally, the chemical stability of non-aqueous electrolytes makes them ideal for use in harsh environments, where aqueous electrolytes might degrade or evaporate. Ionic liquids, in particular, have garnered interest for their ability to improve sensor stability and performance in extreme conditions. Their tunable properties allow researchers to design electrolytes that are tailored to specific sensing applications, enhancing both the sensitivity and the lifespan of the sensor. This flexibility makes non-aqueous electrolytes a valuable tool in the development of next-generation electrochemical sensors.

5.4.2.3 Solid-State Electrolytes

Solid-state electrolytes represent a significant advancement in the field of electrochemical sensing, particularly in the development of miniaturized, portable, and flexible sensors. These electrolytes include polymer electrolytes and solid ionic conductors, which provide the necessary ionic conductivity while eliminating the need for liquid components. This solid-state configuration makes sensors more robust and durable, particularly for applications in wearable devices, implantable sensors, and remote monitoring systems [82].

One of the primary advantages of solid-state electrolytes is their stability, which allows for long-term, maintenance-free operation. This characteristic is particularly valuable in applications where the sensor needs to function reliably over extended periods, such as in environmental monitoring or medical implants. Solid-state electrolytes also offer ease of integration into microfabricated devices, enabling the creation of compact, high-performance sensors that can be mass-produced for commercial use. In the context of ion-selective electrodes (ISEs), solid-state electrolytes play a crucial role in detecting ions like sodium, potassium, and calcium in biological fluids. These electrolytes provide a stable environment that supports the selective interaction between the ion-selective membrane and the target ion, resulting in accurate and reliable measurements.

However, despite these advantages, challenges remain in achieving high ionic conductivity in solid-state electrolytes, which is essential for ensuring consistent sensor performance. Electrolytes are a fundamental component of electrochemical sensors, with each type offering distinct advantages depending on the application. Aqueous electrolytes are ideal for biological sensing, non-aqueous electrolytes extend the range of detectable analytes and improve stability in harsh environments, and solid-state electrolytes enable the development of robust, portable sensors. As electrochemical sensing technology continues to evolve [83], the careful selection and optimization of electrolytes will remain a key factor in enhancing sensor performance and expanding their applications across various fields.

5.5 Microelectrode Arrays and Miniaturized Sensors

As the demand for portable and miniaturized sensing devices has grown, there has been significant interest in developing microelectrode arrays (MEAs) and other miniaturized sensor architectures [84]. MEAs consist of a dense array of small electrodes, typically with diameters ranging from micrometers to nanometers, integrated on a single substrate. These microelectrodes offer several advantages over traditional larger electrodes, including enhanced mass transport, reduced ohmic drop, and lower capacitance, which contribute to improved sensitivity and faster response times.

Enhanced Sensitivity: The small size of microelectrodes allows for the establishment of a steady-state current more rapidly than in larger electrodes, leading to increased sensitivity [85]. This is particularly beneficial in applications where the analyte concentration is very low, such as in the detection of trace metals or low-abundance biomolecules.

Faster Response Time: The reduced size of microelectrodes also shortens the diffusion path for analytes, resulting in faster sensor response times [86]. This feature is critical in applications requiring real-time monitoring, such as glucose sensing in diabetic patients or continuous environmental monitoring of pollutants.

Array Configurations: MEAs can be configured in various patterns, such as linear arrays, circular arrays, or random distributions, depending on the application. These configurations allow for multiplexed sensing, where multiple analytes can be detected simultaneously using different microelectrodes within the same array [87]. This capability is particularly useful in complex biological samples or environmental matrices where multiple contaminants may be present.

The miniaturization of EC sensors has also led to the development of lab-on-a-chip (LOC) devices, where entire electrochemical systems, including electrodes, microfluidics, and signal processing components, are integrated onto a single chip. These LOC devices are highly portable, require minimal sample volumes, and can perform complex analyses in situ, making them ideal for point-of-care diagnostics and on-site environmental testing.

5.5.1 Screen-Printed Electrodes (SPEs)

Screen-printed electrodes (SPEs) represent a versatile and cost-effective approach to EC sensor fabrication [88]. SPEs are produced by depositing conductive inks, typically containing carbon, silver, or gold, onto a substrate using a screen-printing process. This method allows for the mass production of disposable, low-cost sensors that can be customized for specific applications. Figure 5.5a depicts the screen-printed electrode and Fig. 5.5b shows three different printing layers of screen printed electrode.

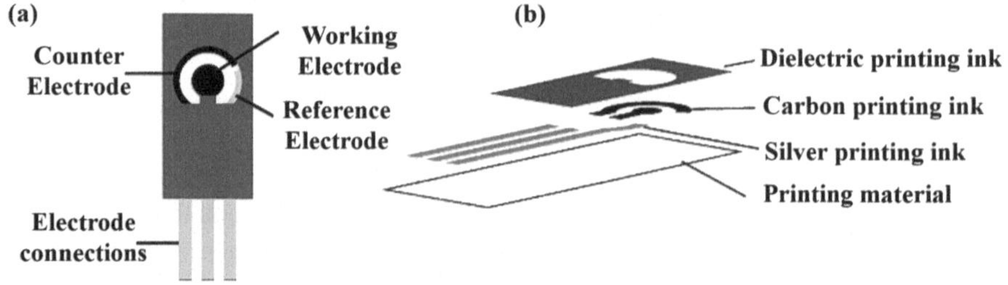

Fig. 5.5 **a** Screen-printed electrode and **b** three different printing layers of screen printed electrode

Customization and Flexibility: SPEs can be easily modified with various recognition elements, such as enzymes, nanoparticles, or molecularly imprinted polymers, to tailor the sensor for specific analytes [89]. The ability to print multiple layers also allows for the incorporation of additional functionalities, such as reference electrodes, protective layers, or electrochemical mediators, directly onto the sensor.

Disposable Sensors: The low cost and ease of production make SPEs ideal for single-use applications, where sensor contamination or fouling is a concern. This is particularly relevant in clinical diagnostics, where SPEs can be used for rapid, on-site testing of blood glucose, cholesterol levels, or infectious diseases [90].

Scalability and Portability: The screen-printing process is highly scalable, enabling the production of large batches of sensors with consistent quality. The compact size and portability of SPEs make them suitable for field testing, environmental monitoring, and wearable sensor applications.

5.5.2 Flexible and Wearable Sensors

The advent of flexible electronics has paved the way for the development of flexible and wearable EC sensors. These sensors are designed to conform to the shape of the human body or other curved surfaces [91], enabling continuous monitoring of physiological parameters, environmental conditions, or industrial processes. Figure 5.6 shows a wearable and flexible sensor.

Flexible EC sensors are typically fabricated using materials such as flexible polymers (e.g., polyimide, polyethylene terephthalate), paper, or textiles. These substrates provide mechanical flexibility while maintaining the necessary electrical properties for reliable sensing [92]. Conductive materials like graphene, carbon nanotubes, or metallic nanowires are often used to create the electrode structures on these flexible substrates.

Wearable EC sensors can be integrated into clothing, accessories, or directly onto the skin to monitor a range of physiological parameters, including sweat composition, pH levels, electrolyte balance, or glucose levels. These sensors offer the potential

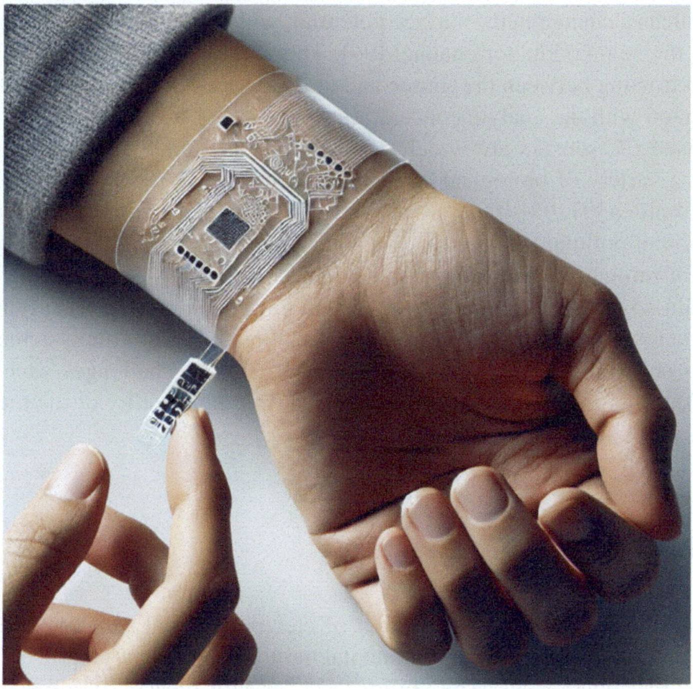

Fig. 5.6 Flexible and wearable sensor created through lexica art (AI)

for continuous, non-invasive health monitoring, providing real-time data that can be used to manage chronic conditions or detect early signs of illness [93].

While flexible and wearable EC sensors hold great promise, they also present unique challenges, such as ensuring long-term stability, biocompatibility, and durability under repeated mechanical stress [94]. Advances in materials science, such as the development of stretchable conductive inks or self-healing materials, are expected to address these challenges and expand the range of applications for flexible and wearable sensors.

5.5.3 Field-Effect Transistor (FET)-Based Sensors

Field-effect transistor (FET)-based EC sensors represent a unique class of devices where the sensing mechanism is based on the modulation of the electrical current in a semiconductor channel by the target analyte [95]. These sensors offer high sensitivity, rapid response times, and the ability to operate at low power, making them attractive for a wide range of applications.

In FET-based sensors, the gate electrode is typically functionalized with a recognition element that interacts with the target analyte. When the analyte binds to the gate, it induces a change in the surface potential, which modulates the charge carrier density in the semiconductor channel [96]. This modulation results in a change in the current flowing between the source and drain electrodes, which can be measured and correlated with the analyte concentration.

FET-based EC sensors are widely used in biosensing applications, where they can detect a variety of biological molecules, including proteins, nucleic acids, and small metabolites [97, 98]. They are also employed in environmental monitoring for the detection of pollutants, gases, and heavy metals, offering high sensitivity and the potential for miniaturization into portable devices.

The main advantages of FET-based sensors include their high sensitivity, ability to operate at low power, and potential for integration with electronic systems for real-time data processing [99]. However, these sensors also face challenges, such as maintaining selectivity in complex samples and ensuring the stability of the recognition elements over time.

5.6 pH Sensing

pH sensing is a fundamental application of electrochemical sensors, crucial for a wide range of fields including environmental monitoring, industrial processes [100], and biomedical diagnostics. The ability to accurately measure the pH of a solution—essentially the concentration of hydrogen ions (H^+) in a liquid—provides critical information about the acidity or alkalinity of the sample. This section delves into the principles of pH sensing [101], the technologies employed, and the various applications and challenges associated with pH measurement.

5.6.1 Principles of pH Sensing

The basic principle behind pH sensing involves measuring the electrochemical potential difference that arises between two electrodes when they are immersed in a solution. This potential difference correlates with the pH of the solution, which is a logarithmic measure of the concentration of hydrogen ions [102]. The most common method for pH sensing utilizes a glass electrode paired with a reference electrode, forming a two-electrode system that measures the potential difference.

Glass Electrode: The glass electrode, often referred to as the pH electrode, is constructed from a special type of glass that is sensitive to hydrogen ions. The glass membrane is typically a thin, porous layer of specialized glass material that allows hydrogen ions to diffuse through it. When the electrode is placed in a solution, a potential difference is generated across the glass membrane due to the interaction of hydrogen ions with the glass surface. This potential difference is then measured and converted into a pH value [103].

Reference Electrode: The reference electrode provides a stable reference potential against which the potential of the glass electrode is measured. Common reference electrodes include the silver/silver chloride (Ag/AgCl) electrode and the saturated calomel electrode (SCE). The stability and consistency of the reference electrode are crucial for accurate pH measurements [104], as any fluctuation in its potential can lead to erroneous readings.

Nernst Equation: The relationship between the pH of the solution and the potential difference measured by the electrodes is described by the Nernst equation. According to this equation, the potential difference (E) is linearly related to the pH of the solution, with a sensitivity of approximately 59.16 mV per pH unit at 25 °C. This relationship allows for precise conversion of the measured potential into a pH value.

5.6.2 Technologies and Methods in pH Sensing

Several technologies and methods are employed in pH sensing, each with its own advantages and limitations [105, 106]. The choice of technology depends on the specific requirements of the application, such as the range of pH values, the sample environment, and the desired sensitivity and accuracy.

Glass Membrane Electrodes: Glass membrane electrodes are the most commonly used technology for pH sensing due to their wide pH range, high accuracy, and robustness. These electrodes are suitable for measuring pH in aqueous solutions, including biological fluids, industrial processes, and environmental samples. However, glass electrodes can be susceptible to fouling and require regular calibration to maintain accuracy.

Ion-Selective Field-Effect Transistors (ISFETs): ISFET-based pH sensors utilize a semiconductor field-effect transistor with a gate electrode that is sensitive to hydrogen ions. The gate is typically coated with a proton-sensitive material, such as silicon nitride or polyvinyl chloride (PVC), which interacts with the hydrogen ions in the solution [107]. Changes in the gate potential alter the current flowing through the transistor, providing a measure of pH. ISFETs offer advantages such as miniaturization, rapid response times, and the ability to integrate with electronic systems, making them suitable for portable and wearable pH sensors. However, they may have limited pH range and require careful calibration.

Optical pH Sensors: Optical pH sensors use pH-sensitive dyes or indicators that change color in response to changes in pH. These sensors typically involve a light source, a detector, and a pH-sensitive coating or membrane. The color change of the dye is detected by the optical system and correlated with the pH of the solution. Optical pH sensors offer advantages such as non-contact measurement, suitability for opaque or turbid samples, and resistance to interference from electromagnetic fields [108]. However, they may have lower sensitivity compared to electrochemical sensors and can be affected by the stability of the dye.

Solid-State pH Sensors: Solid-state pH sensors utilize solid-state materials, such as conductive polymers or metal oxides, to detect changes in pH. These sensors often operate based on changes in electrical properties, such as conductivity or impedance, in response to pH variations. Solid-state pH sensors offer advantages such as robustness, miniaturization, and the potential for integration into flexible or wearable devices [109]. However, they may require careful material selection and calibration to achieve high accuracy.

5.6.3 Applications of pH Sensors

pH sensors find applications across a wide range of industries and scientific fields as depicted in Fig. 5.7, reflecting their versatility and importance in monitoring and controlling chemical processes.

Environmental Monitoring: In environmental monitoring, pH sensors are used to measure the acidity or alkalinity of water bodies, soils, and atmospheric samples. Accurate pH measurement is essential for assessing water quality, detecting pollution, and studying ecological impacts [110]. For example, monitoring the pH of rivers and lakes helps in tracking the effects of acid rain, while soil pH measurement is crucial for agriculture and land management.

Industrial Processes: In industrial settings, pH sensors play a critical role in controlling and optimizing chemical processes, such as wastewater treatment, food and beverage production, and pharmaceuticals. pH control is vital for maintaining product quality, ensuring safety, and maximizing efficiency. For instance, in the food industry,

Fig. 5.7 Various applications of pH sensors

pH sensors are used to monitor and adjust the acidity of products like dairy, beverages, and processed foods [111].

Biomedical Diagnostics: In biomedical applications, pH sensors are used to monitor physiological parameters, such as blood pH, urine pH, and gastric pH [112]. These measurements provide valuable information about metabolic status, renal function, and gastrointestinal health [113]. pH sensors are also employed in the development of diagnostic tests and monitoring devices for conditions like acid-base imbalances, respiratory disorders, and diabetes.

Chemical Research: pH sensors are essential tools in chemical research and analysis, where they provide precise measurements of pH during experiments and reactions. Accurate pH control is crucial for studying chemical kinetics, equilibrium, and reaction mechanisms. pH sensors are also used in laboratory settings for titrations, quality control, and method development [114].

5.7 Detection of Metal Ions

The detection of metal ions is a critical aspect of electrochemical sensing due to the wide range of applications spanning environmental monitoring, industrial process control, and health diagnostics. Metal ions, such as lead (Pb^{2+}), mercury (Hg^{2+}), cadmium (Cd^{2+}), and others, are often present in trace amounts and can have significant health and environmental impacts [115]. Electrochemical sensors offer an effective and sensitive method for detecting these ions, leveraging various electrochemical techniques and sensor configurations.

5.7.1 Electrochemical Techniques for Metal Ion Detection

Anodic Stripping Voltammetry (ASV): One of the most widely used techniques for metal ion detection is anodic stripping voltammetry. This method involves preconcentrating the metal ions onto a working electrode surface by applying a deposition potential. Following the accumulation period, the potential is swept in the positive direction, leading to the oxidation of the metal ions [116]. The resulting stripping current, which is proportional to the concentration of the metal ions, is then measured. ASV is highly sensitive and capable of detecting metal ions at very low concentrations, making it ideal for environmental monitoring and water quality assessment.

Cyclic Voltammetry (CV): Cyclic voltammetry is another technique used for metal ion detection. It involves sweeping the electrode potential back and forth and recording the resulting current. The peak current and potential observed in the cyclic voltammogram can provide information about the metal ion concentration and the electrochemical behavior of the metal [117]. CV is often used in conjunction with other techniques to gain insights into the redox properties of metal ions and to optimize sensor performance.

Differential Pulse Voltammetry (DPV): Differential pulse voltammetry is a sensitive technique that applies a series of potential pulses to the electrode and measures the current response. This method enhances sensitivity and resolution by reducing background noise and improving the detection limit [118]. DPV is particularly useful for detecting metal ions in complex matrices, such as biological samples or industrial effluents.

5.7.2 Electrode Materials and Modifications

Carbon-Based Electrodes: Carbon-based electrodes, including glassy carbon and carbon nanotubes, are commonly used in metal ion detection due to their wide electrochemical window, high conductivity, and resistance to fouling. These electrodes can be modified with various substances, such as nanoparticles or organic ligands, to enhance selectivity and sensitivity towards specific metal ions.

Metal Electrodes: Metal electrodes, such as gold and platinum, are also used for detecting metal ions. These electrodes can form stable complexes with metal ions or facilitate redox reactions, enhancing the sensitivity of the sensor. For example, gold electrodes are often used in the detection of heavy metals due to their ability to form stable complexes with thiol groups, which can improve the detection limits.

Composite Electrodes: Composite electrodes, which combine different materials to create a synergistic effect, are becoming increasingly popular. These electrodes can incorporate metal nanoparticles, conducting polymers, or other functional materials to achieve enhanced sensitivity, selectivity, and stability [119]. For example, electrodes modified with silver nanoparticles can provide improved sensitivity for detecting lead ions.

Advancements in sensor materials and technologies are expected to further enhance the performance of metal ion detectors. Emerging technologies, such as nanomaterials, molecularly imprinted polymers, and hybrid sensor platforms, hold the potential to improve sensitivity, selectivity, and portability. Additionally, the integration of electrochemical sensors with digital technologies, such as smartphones and wearable devices, could enable real-time monitoring and data analysis, expanding the applications of metal ion detection in both research and practical settings.

5.8 Detection of Biological Molecules

The detection of biological molecules using electrochemical sensors is a rapidly evolving field with significant implications for medical diagnostics, environmental monitoring, and food safety. Biological molecules, such as glucose, cholesterol, DNA, and proteins, play critical roles in various biological processes and can serve as biomarkers for disease, exposure, or contamination. Electrochemical sensors offer a powerful and versatile platform for the selective and sensitive detection of these molecules. Figure 5.8 shows a sensor to detect a highly selective peptide which is derived from HarmOBP7 aldehyde binding site.

5.8.1 Electrochemical Techniques for Biological Molecule Detection

(a) **Enzyme-Linked Electrochemical Detection**: Enzyme-based electrochemical sensors utilize specific enzymes that catalyze reactions involving target biological molecules. For example, glucose sensors often use glucose oxidase to catalyze the oxidation of glucose, producing hydrogen peroxide as a byproduct. The resulting electrochemical reaction is monitored to determine glucose concentration [121]. This technique is widely used in glucose meters for diabetes management and has applications in clinical diagnostics and point-of-care testing.

(b) **Immunoassays**: Electrochemical immunoassays employ antibodies or antigen-binding molecules to selectively bind to target biological molecules. The binding

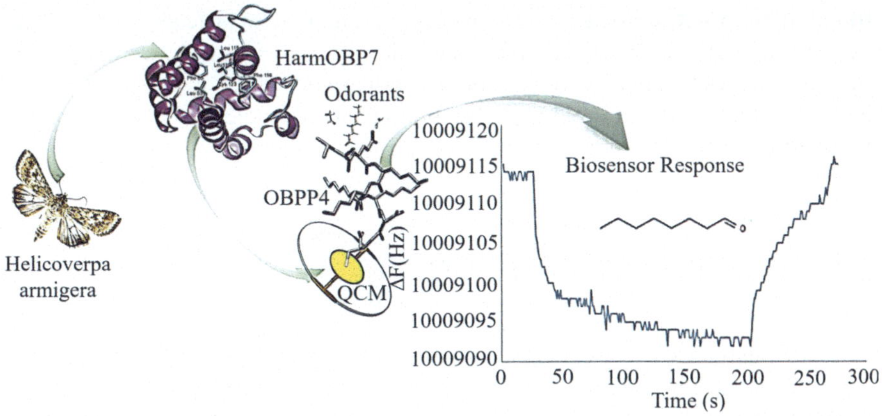

Fig. 5.8 Sensor to detect a highly selective peptide which is derived from HarmOBP7 aldehyde binding site reproduced from Ref. [120]. Copyright © 1996–2024 MDPI (Basel, Switzerland)

event is detected electrochemically, providing information about the concentration of the target molecule. For example, electrochemical immunosensors can detect biomarkers for cancer or infectious diseases by measuring the current response resulting from antibody-antigen interactions [122]. This approach offers high specificity and sensitivity for detecting low-abundance biomolecules.

(c) **DNA Sensors**: Electrochemical DNA sensors are designed to detect specific DNA sequences through hybridization with complementary DNA probes. The binding of DNA molecules induces changes in the electrochemical properties of the sensor, such as changes in current or impedance [123, 124]. These sensors are used in genetic testing, pathogen detection, and environmental monitoring. The ability to detect specific genetic sequences enables early diagnosis of genetic disorders and identification of pathogens.

5.8.2 Electrode Materials and Modifications

Carbon Nanomaterials: Carbon-based materials, including graphene and carbon nanotubes, are frequently used in electrochemical sensors for their high surface area, electrical conductivity, and biocompatibility. These materials enhance the sensitivity and performance of sensors by providing a large active surface for biomolecule interactions and facilitating efficient charge transfer.

Conductive Polymers: Conductive polymers, such as polyaniline and polypyrrole, are used to modify electrodes and improve sensor performance. These polymers can provide additional functionality, such as enzyme immobilization or redox activity, which enhances the sensitivity and selectivity of the sensor for specific biological molecules.

Nanoparticle Functionalization: The incorporation of nanoparticles, such as gold, silver, or magnetic nanoparticles, into electrode materials can enhance the performance of electrochemical sensors. These nanoparticles offer high surface areas and can facilitate the immobilization of biological molecules or improve signal amplification, leading to more sensitive and accurate detection.

5.8.3 Applications and Challenges

Medical Diagnostics: Electrochemical sensors are extensively used in medical diagnostics for monitoring glucose levels in diabetic patients, detecting cholesterol levels, and diagnosing various diseases [125]. These sensors offer rapid, on-site analysis with minimal sample preparation, making them suitable for point-of-care testing and home diagnostics.

Environmental Monitoring: In environmental monitoring, electrochemical sensors can detect biological contaminants, such as pathogens or toxins, in water and air samples. The ability to perform real-time, on-site analysis allows for timely intervention and management of environmental risks [126].

Food Safety: Electrochemical sensors are also employed in food safety testing to detect contaminants, allergens, or spoilage indicators [127]. The ability to provide quick and accurate results ensures the safety and quality of food products and helps prevent foodborne illnesses [128].

The development of advanced materials, improved sensor designs, and integrated technologies are expected to drive the future of biological molecule detection. Innovations in nanotechnology, bioengineering, and data analytics are likely to enhance the sensitivity, selectivity, and versatility of electrochemical sensors. Additionally, the integration of electrochemical sensors with wearable devices and smartphone applications could enable continuous, real-time monitoring of biological molecules, providing valuable insights into health and environmental conditions.

5.9 Gas Sensing

Gas sensing is a critical application of electrochemical sensors, with widespread use in environmental monitoring, industrial safety, and indoor air quality control [129]. The ability to detect and quantify gases with high sensitivity and selectivity is essential for addressing challenges related to pollution, safety, and health. Gas sensors are designed to respond to specific gases by leveraging various electrochemical principles and technologies [130]. This section explores the principles, technologies, and applications of gas sensing, highlighting the advances and challenges associated with this field. Figure 5.9a depicts the mechanism of resistive sensing and Fig. 5.9b shows metal-oxide-semiconductor gas sensor.

5.9.1 Principles of Gas Sensing

Gas sensing in electrochemical sensors typically relies on the interaction between the target gas and the electrode materials, resulting in measurable changes in electrical signals. The most common principles used in gas sensing include potentiometric, amperometric, and conductometric methods.

Potentiometric Gas Sensors: In potentiometric gas sensors, the potential difference between the working electrode and a reference electrode is measured. The potential change is related to the concentration of the gas, which interacts with the electrode surface and alters its electrochemical properties [131]. For instance, in the case of carbon dioxide (CO_2) sensing, a solid-state electrolyte reacts with CO_2 to form a change in potential, which is then measured.

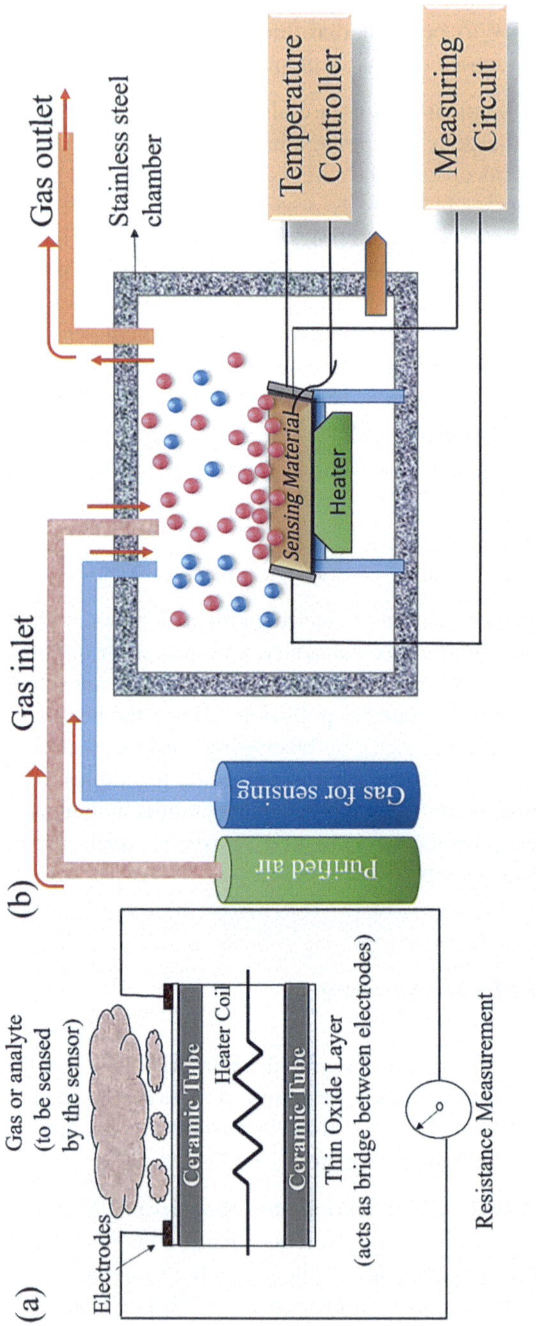

Fig. 5.9 **a** Mechanism of resistive sensing and **b** metal-oxide-semiconductor gas sensor

Amperometric Gas Sensors: Amperometric gas sensors measure the current generated by the electrochemical reactions occurring at the working electrode in response to the target gas. The current is proportional to the concentration of the gas, allowing for quantitative measurement. For example, in nitrogen dioxide (NO_2) sensing, NO_2 reacts with the electrode surface to produce a current signal, which correlates with the gas concentration [132].

Conductometric Gas Sensors: Conductometric gas sensors detect changes in the electrical conductivity of a sensing material caused by the adsorption of the target gas. The change in conductivity is used to infer the concentration of the gas [133]. Metal-oxide semiconductors (MOS) are commonly used in conductometric gas sensors due to their high sensitivity and rapid response times.

5.9.2 Technologies and Materials for Gas Sensing

The choice of materials and technologies is crucial for the performance of gas sensors. Various materials and sensor designs are employed to enhance sensitivity, selectivity, and stability.

Metal-Oxide Semiconductors (MOS): MOS sensors are widely used for gas detection due to their high sensitivity and relatively low cost [134]. Materials such as tin dioxide (SnO_2), zinc oxide (ZnO), and titanium dioxide (TiO_2) are commonly used in MOS sensors. These materials change their electrical resistance in the presence of target gases, such as carbon monoxide (CO) or hydrogen (H_2), which allows for detection. MOS sensors are particularly useful in detecting combustible gases and pollutants [135].

Electrochemical Cells: Electrochemical gas sensors often use a combination of electrodes and electrolytes to detect gases. Common configurations include electrochemical cells with a working electrode, a reference electrode, and a counter electrode [136]. The working electrode is typically coated with a catalyst or sensing material that interacts with the gas. For example, palladium (Pd) or platinum (Pt) can be used as catalysts for detecting hydrogen gas.

Conducting Polymers: Conducting polymers, such as polyaniline (PANI) and polypyrrole (PPy), are used in gas sensors due to their tunable electrical properties and high surface area. These polymers can be functionalized to enhance selectivity for specific gases. When a gas interacts with the polymer, it changes the polymer's conductivity, which is measured to determine the gas concentration.

5.9.3 Applications of Gas Sensors

Gas sensors have a wide range of applications across various industries, including environmental monitoring, industrial safety, and healthcare.

Environmental Monitoring: Gas sensors play a vital role in monitoring air quality and detecting pollutants in the environment. They are used to measure concentrations of gases such as carbon monoxide (CO), nitrogen dioxide (NO_2), and ozone (O_3) in urban areas. Continuous monitoring of these gases helps in assessing the impact of pollution on public health and formulating strategies to improve air quality.

Industrial Safety: In industrial settings, gas sensors are essential for detecting hazardous gases and ensuring workplace safety. Sensors for gases like methane (CH_4) and hydrogen sulfide (H_2S) are used to monitor potential leaks and prevent explosions or toxic exposures [137]. Gas sensors are integrated into safety systems to provide early warnings and trigger safety measures when gas concentrations exceed permissible limits.

Healthcare: Gas sensors are increasingly used in healthcare applications, including breath analysis and monitoring of respiratory gases [138]. Sensors can detect biomarkers in exhaled breath, providing insights into metabolic processes and disease states. For example, sensors can be used to monitor blood alcohol levels, detect volatile organic compounds (VOCs) related to diseases, and assess lung function.

5.10 Detection of Environmental Pollutants

Environmental pollutants are substances that contaminate the natural environment and pose risks to human health, wildlife, and ecosystems [139]. Effective detection and monitoring of these pollutants are crucial for environmental protection and public health. Electrochemical sensors have emerged as powerful tools for detecting a wide range of environmental pollutants due to their high sensitivity, selectivity, and the ability to operate in various environmental conditions.

Electrochemical sensors can detect various environmental pollutants, including heavy metals, pesticides, volatile organic compounds (VOCs), and gases like nitrogen dioxide (NO_2), sulfur dioxide (SO_2), and carbon monoxide (CO). Each type of pollutant presents unique challenges for detection [140]. For example, heavy metals like lead and mercury are toxic even at low concentrations and require sensors with high sensitivity and selectivity. Pesticides, which often exist in complex mixtures, demand sensors capable of distinguishing between different chemical species. VOCs and gases, which can be present in trace amounts, necessitate sensors with fast response times and high sensitivity.

Different electrochemical techniques are employed for the detection of environmental pollutants, including amperometry, voltammetry, and potentiometry. In amperometric sensors, the current produced by the oxidation or reduction of the pollutant at the electrode surface is measured. This technique is particularly effective for detecting gases and dissolved species in aqueous environments [141]. Voltammetric sensors, which involve sweeping the potential of the working electrode and measuring the resulting current, are useful for analyzing the electrochemical behavior of pollutants and their concentrations. Potentiometric sensors measure the potential difference between the working and reference electrodes, which can be correlated to the concentration of certain pollutants.

Recent advancements in electrochemical sensors have improved their performance in environmental monitoring. Nanomaterials, such as graphene, carbon nanotubes, and metal nanoparticles, have been incorporated into sensor designs to enhance their sensitivity and selectivity. These materials offer high surface areas, excellent electrical conductivity, and catalytic properties that improve the detection of low-concentration pollutants. Portable and wearable sensors have also been developed for real-time monitoring of environmental pollutants [142]. These devices are used in field applications, such as monitoring air quality in urban areas, assessing water contamination in rivers and lakes, and tracking industrial emissions. They provide valuable data that can inform regulatory decisions, guide pollution mitigation efforts, and protect public health [143].

5.11 Detection of Corrosion Products

Corrosion is the deterioration of materials, usually metals, due to chemical reactions with their environment [144]. This process can lead to significant economic losses, structural failures, and safety hazards. Detecting corrosion products is essential for monitoring the condition of infrastructure, machinery, and other metal components, and for implementing effective corrosion control measures. Electrochemical sensors play a crucial role in the detection and analysis of corrosion products, providing real-time information about the extent of corrosion and the effectiveness of preventive measures. Corrosion can result in various products, including metal oxides, hydroxides, and salts [145, 146]. Common corrosion products include iron oxides (rust), copper oxides, and lead compounds. These products can accumulate on metal surfaces and influence the material's integrity and performance. Monitoring these corrosion products helps assess the severity of corrosion, evaluate the effectiveness of protective coatings, and predict the remaining service life of metal structures.

Electrochemical sensors for detecting corrosion products typically employ techniques such as electrochemical impedance spectroscopy (EIS), potentiodynamic polarization, and electrochemical noise analysis. EIS measures the impedance of the metal surface as a function of frequency, providing information about the corrosion rate and the condition of protective coatings [147]. Potentiodynamic polarization involves sweeping the electrode potential and measuring the resulting current to evaluate the corrosion resistance and identify the corrosion mechanisms. Electrochemical noise analysis monitors fluctuations in the current or potential of the electrode, which can be correlated to the rate of corrosion and the presence of corrosion products.

Advancements in electrochemical sensor technology have significantly improved the detection of corrosion products [146, 148]. The development of advanced electrode materials, such as composites and nanomaterials, has enhanced the sensitivity and selectivity of these sensors [149]. For example, the use of nanostructured electrodes can increase the surface area and improve the interaction with corrosion products, leading to more accurate measurements. In practical applications, electrochemical sensors are used in various industries, including construction, transportation, and energy. They are employed to monitor the condition of bridges, pipelines, storage tanks, and other critical infrastructure. By providing real-time data on corrosion processes, these sensors enable timely maintenance and repair, reducing the risk of catastrophic failures and extending the lifespan of metal components.

5.12 Conclusion

In this chapter, we have explored the multifaceted world of electrochemical sensors, delving into their principles, configurations, and applications across various domains. Electrochemical sensors stand at the forefront of detection technology, offering unparalleled sensitivity, selectivity, and versatility. Their ability to transform complex chemical information into measurable electrical signals makes them indispensable tools in fields ranging from environmental monitoring and biomedical diagnostics to industrial process control.

The principles of electrochemical sensing, involving interactions between electrodes and analytes, provide a foundation for understanding how these sensors operate. The choice of electrode materials and the configuration of the sensing setup, including the use of microelectrode arrays and screen-printed electrodes, play crucial roles in enhancing performance and meeting specific application needs. Innovations such as flexible and wearable sensors and FET-based devices highlight the ongoing advancements in the field, pushing the boundaries of what electrochemical sensors can achieve. Electrolytes, as a vital component in electrochemical sensing, significantly impact sensor performance. Aqueous, non-aqueous, and solid-state electrolytes each offer unique benefits and challenges, influencing the sensitivity, stability, and applicability of sensors in different environments. The selection of an appropriate electrolyte is crucial for optimizing sensor performance and ensuring accurate and reliable measurements. The chapter also covered specialized applications of electrochemical sensors in detecting environmental pollutants and corrosion products. Environmental sensors have become essential for monitoring pollutants and ensuring environmental protection. Advances in sensor technology, such as the use of nanomaterials and portable devices, have enhanced their capabilities and expanded their applications. Similarly, electrochemical sensors for corrosion detection provide valuable insights into material degradation, enabling timely maintenance and improving the longevity of critical infrastructure.

As we look to the future, several trends and developments are poised to shape the evolution of electrochemical sensors. Continued advancements in materials science, including the development of novel electrode materials and advanced electrolytes, will further enhance sensor performance. Innovations in sensor design, such as the integration of wireless communication and data analytics, will enable more sophisticated and accessible monitoring solutions. The integration of electrochemical sensors with emerging technologies, such as artificial intelligence and machine learning, holds the potential to revolutionize data analysis and interpretation. These technologies can enhance the accuracy of sensor measurements, facilitate real-time decision-making, and uncover new insights into complex systems. Moreover, the growing emphasis on sustainability and environmental responsibility will drive the development of sensors that are not only more effective but also more environmentally friendly. This includes designing sensors with recyclable materials and minimizing their environmental impact throughout their lifecycle.

Electrochemical sensors represent a dynamic and evolving field with broad applications and significant potential for future advancements. By harnessing the principles of electrochemical reactions and leveraging innovative materials and designs, these sensors continue to provide critical insights across various domains. As technology progresses, electrochemical sensors will play an increasingly vital role in addressing global challenges, from environmental protection to healthcare and beyond. The ongoing research and development in this field promise to unlock new possibilities, making electrochemical sensors a cornerstone of modern detection technology.

References

1. Maduraiveeran, G., Sasidharan, M., Ganesan, V.: Electrochemical sensor and biosensor platforms based on advanced nanomaterials for biological and biomedical applications. Biosens. Bioelectron. **103**, 113–129 (2018). https://doi.org/10.1016/j.bios.2017.12.031
2. Baranwal, J., Barse, B., Gatto, G., Broncova, G., Kumar, A.: Electrochemical sensors and their applications: a review. Chemosensors **10**(9) (2022). https://doi.org/10.3390/chemosensors10090363
3. Khanmohammadi, A., Jalili Ghazizadeh, A., Hashemi, P., Afkhami, A., Arduini, F., Bagheri, H.: An overview to electrochemical biosensors and sensors for the detection of environmental contaminants. J. Iran. Chem. Soc. **17**(10), 2429–2447 (2020). https://doi.org/10.1007/s13738-020-01940-z
4. Hernández-Rodríguez, J.F., Rojas, D., Escarpa, A.: Electrochemical sensing directions for next-generation healthcare: trends, challenges, and frontiers. Anal. Chem. **93**(1), 167–183 (2021). https://doi.org/10.1021/acs.analchem.0c04378
5. Manikandan, V.S., Adhikari, B., Chen, A.: Nanomaterial based electrochemical sensors for the safety and quality control of food and beverages. Analyst **143**(19), 4537–4554 (2018). https://doi.org/10.1039/C8AN00497H
6. Apak, R., Üzer, A., Sağlam, Ş., Arman, A.: Selective electrochemical detection of explosives with nanomaterial based electrodes. Electroanalysis **35**(1), e202200175 (2023). https://doi.org/10.1002/elan.202200175
7. Kajal, N., Singh, V., Gupta, R., Gautam, S.: Metal organic frameworks for electrochemical sensor applications: a review. Environ. Res. **204**, 112320 (2022). https://doi.org/10.1016/j.envres.2021.112320

8. Kothandam, G., et al.: Recent advances in carbon-based electrodes for energy storage and conversion. Adv. Sci. **10**(18), 2301045 (2023). https://doi.org/10.1002/advs.202301045
9. Pathak, R., Punetha, V.D., Bhatt, S., Punetha, M.: Multifunctional role of carbon dot-based polymer nanocomposites in biomedical applications: a review. J Mater Sci **58**(15), 6419–6443 (2023). https://doi.org/10.1007/s10853-023-08408-4
10. Jadoun, S., Fuentes, J.P., Urbano, B.F., Yáñez, J.: A review on adsorption of heavy metals from wastewater using conducting polymer-based materials. J. Environ. Chem. Eng. **11**(1), 109226 (2023). https://doi.org/10.1016/j.jece.2022.109226
11. Leau, S.-A., Lete, C., Lupu, S.: Nanocomposite materials based on metal nanoparticles for the electrochemical sensing of neurotransmitters. Chemosensors **11**(3) (2023). https://doi.org/10.3390/chemosensors11030179
12. Ahmed, S., Sinha, S.K.: Studies on nanomaterial-based p-type semiconductor gas sensors. Environ. Sci. Pollut. Res. **30**(10), 24975–24986 (2023). https://doi.org/10.1007/s11356-022-21218-6
13. Saputra, H.A.: Electrochemical sensors: basic principles, engineering, and state of the art. Monatsh. Chem. Chem. Mon. **154**(10), 1083–1100 (2023). https://doi.org/10.1007/s00706-023-03113-z
14. Meskher, H., et al.: A review on CNTs-based electrochemical sensors and biosensors: unique properties and potential applications. Crit. Rev. Anal. Chem. 1–24. https://doi.org/10.1080/10408347.2023.2171277
15. Qian, L., Durairaj, S., Prins, S., Chen, A.: Nanomaterial-based electrochemical sensors and biosensors for the detection of pharmaceutical compounds. Biosens. Bioelectron. **175**, 112836 (2021). https://doi.org/10.1016/j.bios.2020.112836
16. Cardoso, R.M., et al.: 3D printing for electroanalysis: from multiuse electrochemical cells to sensors. Anal. Chim. Acta. **1033**, 49–57 (2018). https://doi.org/10.1016/j.aca.2018.06.021
17. Dkhar, D.S., Kumari, R., Malode, S.J., Shetti, N.P., Chandra, P.: Integrated lab-on-a-chip devices: fabrication methodologies, transduction system for sensing purposes. J. Pharm. Biomed. Anal. **223**, 115120 (2023). https://doi.org/10.1016/j.jpba.2022.115120
18. Jeon, S., Lim, S.-C., Trung, T.Q., Jung, M., Lee, N.-E.: Flexible multimodal sensors for electronic skin: principle, materials, device, array architecture, and data acquisition method. Proc. IEEE **107**(10), 2065–2083 (2019). https://doi.org/10.1109/JPROC.2019.2930808
19. Manjakkal, L., Szwagierczak, D., Dahiya, R.: Metal oxides based electrochemical pH sensors: current progress and future perspectives. Prog. Mater. Sci. **109**, 100635 (2020). https://doi.org/10.1016/j.pmatsci.2019.100635
20. Liu, X., Yao, Y., Ying, Y., Ping, J.: Recent advances in nanomaterial-enabled screen-printed electrochemical sensors for heavy metal detection. TrAC Trends Anal. Chem. **115**, 187–202 (2019). https://doi.org/10.1016/j.trac.2019.03.021
21. Terán-Alcocer, Á., Bravo-Plascencia, F., Cevallos-Morillo, C., Palma-Cando, A.: Electrochemical sensors based on conducting polymers for the aqueous detection of biologically relevant molecules. Nanomaterials **11**(1) (2021). https://doi.org/10.3390/nano11010252
22. Duanghathaipornsuk, S., Farrell, E.J., Alba-Rubio, A.C., Zelenay, P., Kim, D.-S.: Detection technologies for reactive oxygen species: fluorescence and electrochemical methods and their applications. Biosensors (Basel) **11**(2) (2021). https://doi.org/10.3390/bios11020030
23. Tajik, S., et al.: Recent developments in polymer nanocomposite-based electrochemical sensors for detecting environmental pollutants. Ind. Eng. Chem. Res. **60**(3), 1112–1136 (2021). https://doi.org/10.1021/acs.iecr.0c04952
24. Xia, D.-H., et al.: Electrochemical measurements used for assessment of corrosion and protection of metallic materials in the field: a critical review. J. Mater. Sci. Technol. **112**, 151–183 (2022). https://doi.org/10.1016/j.jmst.2021.11.004
25. Ghoneim, M.T., et al.: Recent progress in electrochemical pH-sensing materials and configurations for biomedical applications. Chem. Rev. **119**(8), 5248–5297 (2019). https://doi.org/10.1021/acs.chemrev.8b00655
26. Boselli, E., Wu, Z., Friedman, A., Claus Henn, B., Papautsky, I.: Validation of electrochemical sensor for determination of manganese in drinking water. Environ. Sci. Technol. **55**(11), 7501–7509 (2021). https://doi.org/10.1021/acs.est.0c05929

27. da Costa, T.H., Song, E., Tortorich, R.P., Choi, J.-W.: A paper-based electrochemical sensor using inkjet-printed carbon nanotube electrodes. ECS J. Solid State Sci. Technol. **4**(10), S3044–S3047 (2015). https://doi.org/10.1149/2.0121510JSS/XML
28. Teymourian, H., Barfidokht, A., Wang, J.: Electrochemical glucose sensors in diabetes management: an updated review (2010–2020). Chem. Soc. Rev. **49**(21), 7671–7709 (2020). https://doi.org/10.1039/D0CS00304B
29. Dong, Q., Ryu, H., Lei, Y.: Metal oxide based non-enzymatic electrochemical sensors for glucose detection. Electrochim. Acta **370**, 137744 (2021). https://doi.org/10.1016/j.electacta.2021.137744
30. Pang, X., Shaw, M.D., Gillot, S., Lewis, A.C.: The impacts of water vapour and co-pollutants on the performance of electrochemical gas sensors used for air quality monitoring. Sens. Actuators B Chem. **266**, 674–684 (2018). https://doi.org/10.1016/j.snb.2018.03.144
31. Nsabimana, A., Lan, Y., Du, F., Wang, C., Zhang, W., Xu, G.: Alkaline phosphatase-based electrochemical sensors for health applications. Anal. Methods **11**(15), 1996–2006 (2019). https://doi.org/10.1039/C8AY02793E
32. Nasri, A., Pétrissans, M., Fierro, V., Celzard, A.: Gas sensing based on organic composite materials: review of sensor types, progresses and challenges. Mater. Sci. Semicond. Process. **128**, 105744 (2021). https://doi.org/10.1016/j.mssp.2021.105744
33. Khan, M.A.H., Rao, M.V., Li, Q.: Recent advances in electrochemical sensors for detecting toxic gases: NO_2, SO_2 and H_2S. Sensors **19**(4) (2019). https://doi.org/10.3390/s19040905
34. Monisha, S., Kumar, A.S.: A sustainable redox-active electrocatalyst utilizing natural quercetin-infused MWCNT for ultra-sensitive detection of environmental Cr(VI) contaminants. Chem. Eng. J. **487**, 150381 (2024). https://doi.org/10.1016/j.cej.2024.150381
35. Debsharma, K., Dutta, B., Dey, S., Sinha, C.: Metal–organic coordination polymers: a review of electrochemical sensing of environmental pollutants. Cryst. Growth Des. **24**(15), 6503–6530 (2024). https://doi.org/10.1021/acs.cgd.4c00421
36. Zare-Shehneh, N., Mollarasouli, F., Ghaedi, M.: Recent advances in carbon nanostructure-based electrochemical biosensors for environmental monitoring. Crit. Rev. Anal. Chem. **53**(3), 520–536 (2023). https://doi.org/10.1080/10408347.2021.1967719
37. Yadav, N., Mishra, A., Narang, J.: 31—Electrochemical sensor method for food quality evaluation. In: Zhong, J., Wang, X. (eds.) Evaluation Technologies for Food Quality, pp. 793–815. Woodhead Publishing (2019). https://doi.org/10.1016/B978-0-12-814217-2.00031-7
38. Zhang, Y., Dai, M., Yuan, Z.: Methods for the detection of reactive oxygen species. Anal. Methods **10**(38), 4625–4638 (2018). https://doi.org/10.1039/C8AY01339J
39. Kalyanaraman, B., Cheng, G., Hardy, M., Ouari, O., Bennett, B., Zielonka, J.: Teaching the basics of reactive oxygen species and their relevance to cancer biology: mitochondrial reactive oxygen species detection, redox signaling, and targeted therapies. Redox. Biol. **15**, 347–362 (2018). https://doi.org/10.1016/j.redox.2017.12.012
40. Wang, H., et al.: Redox flow batteries: how to determine electrochemical kinetic parameters. ACS Nano **14**(3), 2575–2584 (2020). https://doi.org/10.1021/acsnano.0c01281
41. Ganiyu, S.O., Martínez-Huitle, C.A., Oturan, M.A.: Electrochemical advanced oxidation processes for wastewater treatment: advances in formation and detection of reactive species and mechanisms. Curr. Opin. Electrochem. **27**, 100678 (2021). https://doi.org/10.1016/j.coelec.2020.100678
42. Nemčeková, K., Labuda, J.: Advanced materials-integrated electrochemical sensors as promising medical diagnostics tools: a review. Mater. Sci. Eng. C **120**, 111751 (2021). https://doi.org/10.1016/j.msec.2020.111751
43. Rafiee, M., Abrams, D.J., Cardinale, L., Goss, Z., Romero-Arenas, A., Stahl, S.S.: Cyclic voltammetry and chronoamperometry: mechanistic tools for organic electrosynthesis. Chem. Soc. Rev. **53**(2), 566–585 (2024). https://doi.org/10.1039/D2CS00706A
44. Cruz-Navarro, J.A., Hernandez-Garcia, F., Alvarez Romero, G.A.: Novel applications of metal-organic frameworks (MOFs) as redox-active materials for elaboration of carbon-based electrodes with electroanalytical uses. Coord. Chem. Rev. **412**, 213263 (2020). https://doi.org/10.1016/j.ccr.2020.213263

45. Kinyua Muthuri, L., Nagy, L., Nagy, G.: Chronopotentiometric method for assessing antioxidant activity: a reagentless measuring technique. Electrochem. Commun. **122**,106907 (2021). https://doi.org/10.1016/j.elecom.2020.106907
46. Donadt, T.B., Lilio, A.M., Stinson, T.A., Lama, B., Luca, O.R.: DOSY NMR and normal pulse voltammetry for the expeditious determination of number of electrons exchanged in redox events. ChemistrySelect **3**(25), 7410–7415 (2018). https://doi.org/10.1002/slct.201801231
47. Khatri, R., Puri, N.K.: Electrochemical study of hydrothermally synthesised reduced MoS_2 layered nanosheets. Vacuum **175** (2020). https://doi.org/10.1016/J.VACUUM.2020.109250
48. Sharma, S., Phogat, P., Jha, R., Singh, S.: Microwave-synthesized γ-WO_3 nanorods exhibiting high current density and diffusion characteristics. Transit. Metal Chem. **1**, 1–17 (2023). https://doi.org/10.1007/S11243-023-00533-Y/FIGURES/16
49. Fazio, E., et al.: Metal-oxide based nanomaterials: synthesis, characterization and their applications in electrical and electrochemical sensors. Sensors **21**(7) (2021). https://doi.org/10.3390/s21072494
50. Wang, G., Morrin, A., Li, M., Liu, N., Luo, X.: Nanomaterial-doped conducting polymers for electrochemical sensors and biosensors. J. Mater. Chem. B **6**(25), 4173–4190 (2018). https://doi.org/10.1039/C8TB00817E
51. Fu, C.-C., et al.: Electrochemical sensing of mercury ions in electrolyte solutions by nitrogen-doped graphene quantum dot electrodes at ultralow concentrations. J. Mol. Liq. **302**, 112593 (2020). https://doi.org/10.1016/j.molliq.2020.112593
52. Benoudjit, A., Bader, M.M., Wan Salim, W.W.A.: Study of electropolymerized PEDOT: PSS transducers for application as electrochemical sensors in aqueous media. Sens. Biosensing. Res. **17**, 18–24 (2018). https://doi.org/10.1016/j.sbsr.2018.01.001
53. Sharma, S., Phogat, P., Jha, R., Singh, S.: Electrochemical study of cerium and iron doped MoO_3 nanoparticles showing potential for supercapacitor application. Next Mater. **5**, 100260 (2024). https://doi.org/10.1016/j.nxmate.2024.100260
54. Ning, Z.H., Huang, J.Q., Guo, S.X., Wang, L.H.: A portable potentiostat for three-electrode electrochemical sensor. J. Phys. Conf. Ser. **1550**(4), 42049 (2020). https://doi.org/10.1088/1742-6596/1550/4/042049
55. Wang, Y., Liu, Y., Wang, X., Cao, X., Xia, J., Wang, Z.: A flexible and wearable three-electrode electrochemical sensing system consisting of a two-in-one enzyme-mimic working electrode. Anal. Chim. Acta. **1278**, 341688 (2023). https://doi.org/10.1016/j.aca.2023.341688
56. Sharma, S., Phogat, P., Jha, R., Singh, S.: Emerging advances and future prospects of two dimensional nanomaterials based solar cells. J. Alloys Compd. **1001**, 175063 (2024). https://doi.org/10.1016/j.jallcom.2024.175063
57. Vogel, Y.B., Gooding, J.J., Ciampi, S.: Light-addressable electrochemistry at semiconductor electrodes: redox imaging, mask-free lithography and spatially resolved chemical and biological sensing. Chem. Soc. Rev. **48**(14), 3723–3739 (2019). https://doi.org/10.1039/C8CS00762D
58. Sassa, F., Biswas, G.C., Suzuki, H.: Microfabricated electrochemical sensing devices. Lab Chip **20**(8), 1358–1389 (2020). https://doi.org/10.1039/C9LC01112A
59. Raju, P., Li, Q.: Review—semiconductor materials and devices for gas sensors. J. Electrochem. Soc. **169**(5), 57518 (2022). https://doi.org/10.1149/1945-7111/ac6e0a
60. Shu, J., Tang, D.: Recent advances in photoelectrochemical sensing: from engineered photoactive materials to sensing devices and detection modes. Anal. Chem. **92**(1), 363–377 (2020). https://doi.org/10.1021/acs.analchem.9b04199
61. Guo, K., et al.: Porous silicon nanostructures as effective faradaic electrochemical sensing platforms. Adv. Funct. Mater. **29**(24), 1809206 (2019). https://doi.org/10.1002/adfm.201809206
62. Xu, Y., et al.: Silicon-based sensors for biomedical applications: a review. Sensors **19**(13) (2019). https://doi.org/10.3390/s19132908
63. Tajik, S., Dourandish, S., Garkani Nejad, F., Beitollahi, H., Jahani, P.M., Di Bartolomeo, A.: Transition metal dichalcogenides: synthesis and use in the development of electrochemical sensors and biosensors. Biosens. Bioelectron. **216**, 114674 (2022). https://doi.org/10.1016/j.bios.2022.114674

64. Mphuthi, N., Sikhwivhilu, L., Ray, S.S.: Functionalization of 2D MoS_2 nanosheets with various metal and metal oxide nanostructures: their properties and application in electrochemical sensors. Biosensors (Basel) **12**(6) (2022). https://doi.org/10.3390/bios12060386
65. Ahmed, J., Faisal, M., Harraz, F.A., Jalalah, M., Alsareii, S.A.: Porous silicon-mesoporous carbon nanocomposite based electrochemical sensor for sensitive and selective detection of ascorbic acid in real samples. J. Taiwan Inst. Chem. Eng. **125**, 360–371 (2021). https://doi.org/10.1016/j.jtice.2021.06.018
66. Yao, Y., et al.: Flexible and stretchable organic electrochemical transistors for physiological sensing devices. Adv. Mater. **35**(35), 2209906 (2023). https://doi.org/10.1002/adma.202209906
67. Borges-González, J., Kousseff, C.J., Nielsen, C.B.: Organic semiconductors for biological sensing. J. Mater. Chem. C **7**(5), 1111–1130 (2019). https://doi.org/10.1039/C8TC05900D
68. Kumar, P., Kim, K.-H., Mehta, P.K., Ge, L., Lisak, G.: Progress and challenges in electrochemical sensing of volatile organic compounds using metal-organic frameworks. Crit. Rev. Environ. Sci. Technol. **49**(21), 2016–2048 (2019). https://doi.org/10.1080/10643389.2019.1601489
69. Sinha, S., Pal, T.: A comprehensive review of FET-based pH sensors: materials, fabrication technologies, and modeling. Electrochem. Sci. Adv. **2**(5), e2100147 (2022). https://doi.org/10.1002/elsa.202100147
70. Rollo, S., Rani, D., Olthuis, W., Pascual García, C.: High performance Fin-FET electrochemical sensor with high-k dielectric materials. Sens. Actuators B Chem. **303**, 127215 (2020). https://doi.org/10.1016/j.snb.2019.127215
71. Syedmoradi, L., Ahmadi, A., Norton, M.L., Omidfar, K.: A review on nanomaterial-based field effect transistor technology for biomarker detection. Microchim. Acta **186**(11), 739 (2019). https://doi.org/10.1007/s00604-019-3850-6
72. Sadighbayan, D., Hasanzadeh, M., Ghafar-Zadeh, E.: Biosensing based on field-effect transistors (FET): recent progress and challenges. TrAC Trends Anal. Chem. **133**, 116067 (2020). https://doi.org/10.1016/j.trac.2020.116067
73. Bollella, P., Katz, E.: Enzyme-based biosensors: tackling electron transfer issues Sensors **20**(12) (2020). https://doi.org/10.3390/s20123517
74. Martins, M.V.A.: Direct and mediated electron transfer in enzyme electrodes. In: Crespilho, F.N. (ed.) Advances in Bioelectrochemistry. Volume 1: Surface, Electron Transfer and Techniques, pp. 25–34. Springer International Publishing, Cham (2022). https://doi.org/10.1007/978-3-030-94988-4_2
75. Strong, M.E., Richards, J.R., Torres, M., Beck, C.M., La Belle, J.T.: Faradaic electrochemical impedance spectroscopy for enhanced analyte detection in diagnostics. Biosens. Bioelectron. **177**, 112949 (2021). https://doi.org/10.1016/j.bios.2020.112949
76. Tasić, Ž., Petrović Mihajlović, M.B., Simonović, A.T., Radovanović, M.B., Antonijević, M.M.: Recent advances in electrochemical sensors for caffeine determination. Sensors **22**(23) (2022). https://doi.org/10.3390/s22239185
77. Rashed, M.Z., et al.: Rapid detection of SARS-CoV-2 antibodies using electrochemical impedance-based detector. Biosens. Bioelectron. **171**, 112709 (2021). https://doi.org/10.1016/j.bios.2020.112709
78. Lu, L.: Nanoporous noble metal-based alloys: a review on synthesis and applications to electrocatalysis and electrochemical sensing. Microchim. Acta **186**(9), 664 (2019). https://doi.org/10.1007/s00604-019-3772-3
79. Gibi, C., Liu, C.-H., Barton, S.C., Anandan, S., Wu, J.J.: Carbon materials for electrochemical sensing application—a mini review. J. Taiwan Inst. Chem. Eng. **154**, 105071 (2024). https://doi.org/10.1016/j.jtice.2023.105071
80. Zhang, W., Wang, R., Luo, F., Wang, P., Lin, Z.: Miniaturized electrochemical sensors and their point-of-care applications. Chin. Chem. Lett. **31**(3), 589–600 (2020). https://doi.org/10.1016/j.cclet.2019.09.022
81. Murtada, K., Moreno, V.: Nanomaterials-based electrochemical sensors for the detection of aroma compounds—towards analytical approach. J. Electroanal. Chem. **861**, 113988 (2020). https://doi.org/10.1016/j.jelechem.2020.113988

82. Gorbova, E., Tzorbatzoglou, F., Molochas, C., Chloros, D., Demin, A., Tsiakaras, P.: fundamentals and principles of solid-state electrochemical sensors for high temperature gas detection. Catalysts **12**(1) (2022). https://doi.org/10.3390/catal12010001
83. Patella, B., et al.: Electrochemical detection of chloride ions using Ag-based electrodes obtained from compact disc. Anal. Chim. Acta. **1190**, 339215 (2022). https://doi.org/10.1016/j.aca.2021.339215
84. Steins, H., et al.: A flexible protruding microelectrode array for neural interfacing in bioelectronic medicine. Microsyst. Nanoeng. **8**(1), 131 (2022). https://doi.org/10.1038/s41378-022-00466-z
85. Hannah, S., Blair, E., Corrigan, D.K.: Developments in microscale and nanoscale sensors for biomedical sensing. Curr. Opin. Electrochem. **23**, 7–15 (2020). https://doi.org/10.1016/j.coelec.2020.02.012
86. Obien, M.E.J., Frey, U.: Large-scale, high-resolution microelectrode arrays for interrogation of neurons and networks. In: Chiappalone, M., Pasquale, V., Frega, M. (eds.) Vitro Neuronal Networks: From Culturing Methods to Neuro-technological Applications, pp. 83–123. Springer International Publishing, Cham (2019). https://doi.org/10.1007/978-3-030-11135-9_4
87. Kosri, E., Ibrahim, F., Thiha, A., Madou, M.: Micro and nano interdigitated electrode array (IDEA)-based MEMS/NEMS as electrochemical transducers: a review. Nanomaterials**12**(23) (2022). https://doi.org/10.3390/nano12234171
88. Costa-Rama, E., Fernández-Abedul, M.T.: Paper-based screen-printed electrodes: a new generation of low-cost electroanalytical platforms. Biosensors (Basel) **11**(2) (2021). https://doi.org/10.3390/bios11020051
89. Mincu, N.-B., Lazar, V., Stan, D., Mihailescu, C.M., Iosub, R., Mateescu, A.L.: Screen-printed electrodes (SPE) for in vitro diagnostic purpose. Diagnostics **10**(8) (2020). https://doi.org/10.3390/diagnostics10080517
90. Suresh, R.R., et al.: Fabrication of screen-printed electrodes: opportunities and challenges. J. Mater. Sci. **56**(15), 8951–9006 (2021). https://doi.org/10.1007/s10853-020-05499-1
91. Khan, S., Ali, S., Bermak, A.: Recent developments in printing flexible and wearable sensing electronics for healthcare applications. Sensors **19**(5) (2019). https://doi.org/10.3390/s19051230
92. Gao, W., Ota, H., Kiriya, D., Takei, K., Javey, A.: Flexible electronics toward wearable sensing. Acc. Chem. Res. **52**(3), 523–533 (2019). https://doi.org/10.1021/acs.accounts.8b00500
93. Wang, C., Xia, K., Wang, H., Liang, X., Yin, Z., Zhang, Y.: Advanced carbon for flexible and wearable electronics. Adv. Mater. **31**(9), 1801072 (2019). https://doi.org/10.1002/adma.201801072
94. Yoon, Y., Truong, P.L., Lee, D., Ko, S.H.: Metal-oxide nanomaterials synthesis and applications in flexible and wearable sensors. ACS Nanosci. Au **2**(2), 64–92 (2022). https://doi.org/10.1021/acsnanoscienceau.1c00029
95. Hong, S., et al.: FET-type gas sensors: a review. Sens. Actuators B Chem. **330**, 129240 (2021). https://doi.org/10.1016/j.snb.2020.129240
96. Kwon, J., et al.: Nanoscale FET-based transduction toward sensitive extended-gate biosensors. ACS Sens. **4**(6), 1724–1729 (2019). https://doi.org/10.1021/acssensors.9b00731
97. Wang, J., et al.: Field effect transistor-based tactile sensors: from sensor configurations to advanced applications. InfoMat **5**(1), e12376 (2023). https://doi.org/10.1002/inf2.12376
98. Dai, C., Liu, Y., Wei, D.: Two-dimensional field-effect transistor sensors: the road toward commercialization. Chem. Rev. **122**(11), 10319–10392 (2022). https://doi.org/10.1021/acs.chemrev.1c00924
99. Sharma, B., Sharma, A., Kim, J.-S.: Recent advances on H_2 sensor technologies based on MOX and FET devices: a review. Sens. Actuators B Chem. **262**, 758–770 (2018). https://doi.org/10.1016/j.snb.2018.01.212
100. Moradi, M., Tajik, H., Almasi, H., Forough, M., Ezati, P.: A novel pH-sensing indicator based on bacterial cellulose nanofibers and black carrot anthocyanins for monitoring fish freshness. Carbohydr. Polym. **222**, 115030 (2019). https://doi.org/10.1016/j.carbpol.2019.115030

101. Wang, S., et al.: Anti-quenching NIR-II molecular fluorophores for in vivo high-contrast imaging and pH sensing. Nat. Commun. **10**(1), 1058 (2019). https://doi.org/10.1038/s41467-019-09043-x
102. Bai, J., et al.: Solvent-controlled and solvent-dependent strategies for the synthesis of multi-color carbon dots for pH sensing and cell imaging. J. Mater. Chem. C **7**(31), 9709–9718 (2019). https://doi.org/10.1039/C9TC02422K
103. Nakata, S., Shiomi, M., Fujita, Y., Arie, T., Akita, S., Takei, K.: A wearable pH sensor with high sensitivity based on a flexible charge-coupled device. Nat. Electron. **1**(11), 596–603 (2018). https://doi.org/10.1038/s41928-018-0162-5
104. Steinegger, A., Wolfbeis, O.S., Borisov, S.M.: Optical sensing and imaging of pH values: spectroscopies, materials, and applications. Chem. Rev. **120**(22), 12357–12489 (2020). https://doi.org/10.1021/acs.chemrev.0c00451
105. Hassan Akhtar, M., et al.: Advances in pH sensing: from traditional approaches to next-generation sensors in biological contexts. Chem. Rec. **24**(7), e202300369 (2024). https://doi.org/10.1002/tcr.202300369
106. Salvo, P., et al.: Graphene-based devices for measuring pH. Sens. Actuators B Chem. **256**, 976–991 (2018). https://doi.org/10.1016/j.snb.2017.10.037
107. Vivaldi, F., et al.: Recent advances in optical, electrochemical and field effect pH sensors. Chemosensors **9**(2) (2021). https://doi.org/10.3390/chemosensors9020033
108. Zhao, Y., Lei, M., Liu, S.-X., Zhao, Q.: Smart hydrogel-based optical fiber SPR sensor for pH measurements. Sens. Actuators B Chem. **261**, 226–232 (2018). https://doi.org/10.1016/j.snb.2018.01.120
109. Lonsdale, W., Wajrak, M., Alameh, K.: Manufacture and application of RuO_2 solid-state metal-oxide pH sensor to common beverages. Talanta **180**, 277–281 (2018). https://doi.org/10.1016/j.talanta.2017.12.070
110. Dincer, C., et al.: Disposable sensors in diagnostics, food, and environmental monitoring. Adv. Mater. **31**(30), 1806739 (2019). https://doi.org/10.1002/adma.201806739
111. Jiang, Y., Yin, S., Dong, J., Kaynak, O.: A review on soft sensors for monitoring, control, and optimization of industrial processes. IEEE Sens. J. **21**(11), 12868–12881 (2021). https://doi.org/10.1109/JSEN.2020.3033153
112. Correia, R., James, S., Lee, S.-W., Morgan, S.P., Korposh, S.: Biomedical application of optical fibre sensors. J. Opt. **20**(7), 73003 (2018). https://doi.org/10.1088/2040-8986/aac68d
113. Alam, A.U., et al.: Polymers and organic materials-based pH sensors for healthcare applications. Prog. Mater. Sci. **96**, 174–216 (2018). https://doi.org/10.1016/j.pmatsci.2018.03.008
114. Di Costanzo, L., Panunzi, B.: Visual pH sensors: from a chemical perspective to new bioengineered materials. Molecules **26**(10) (2021). https://doi.org/10.3390/molecules26102952
115. Shellaiah, M., Sun, K.W.: Diamond-based electrodes for detection of metal ions and anions. Nanomaterials **12**(1) (2022). https://doi.org/10.3390/nano12010064
116. Durai, L., Badhulika, S.: Stripping voltammetry and chemometrics assisted ultra-selective, simultaneous detection of trace amounts of heavy metal ions in aqua and blood serum samples. Sens. Actuators Rep. **4**, 100097 (2022). https://doi.org/10.1016/j.snr.2022.100097
117. Lu, Y., Liang, X., Niyungeko, C., Zhou, J., Xu, J., Tian, G.: A review of the identification and detection of heavy metal ions in the environment by voltammetry. Talanta **178**, 324–338 (2018). https://doi.org/10.1016/j.talanta.2017.08.033
118. Saylan, Y., Akgönüllü, S., Yavuz, H., Ünal, S., Denizli, A.: Molecularly imprinted polymer based sensors for medical applications. Sensors **19**(6) (2019). https://doi.org/10.3390/s19061279
119. Zheng, J., Rahim, Md.A., Tang, J., Allioux, F.-M., Kalantar-Zadeh, K.: Post-transition metal electrodes for sensing heavy metal ions by stripping voltammetry. Adv. Mater. Technol. **7**(1), 2100760 (2022). https://doi.org/10.1002/admt.202100760
120. Wasilewski, T., Szulczyński, B., Wojciechowski, M., Kamysz, W., Gębicki, J.: A highly selective biosensor based on peptide directly derived from the HarmOBP7 Aldehyde binding site. Sensors **19**(19), 4284 (2019). https://doi.org/10.3390/S19194284

121. Mahmudunnabi, R.G., Farhana, F.Z., Kashaninejad, N., Firoz, S.H., Shim, Y.-B., Shiddiky, M.J.A.: Nanozyme-based electrochemical biosensors for disease biomarker detection. Analyst **145**(13), 4398–4420 (2020). https://doi.org/10.1039/D0AN00558D
122. Oberhaus, F.V., Frense, D., Beckmann, D.: Immobilization techniques for aptamers on gold electrodes for the electrochemical detection of proteins: a review. Biosensors (Basel) **10**(5) (2020). https://doi.org/10.3390/bios10050045
123. Manzano, M., Viezzi, S., Mazerat, S., Marks, R.S., Vidic, J.: Rapid and label-free electrochemical DNA biosensor for detecting hepatitis A virus. Biosens. Bioelectron. **100**, 89–95 (2018). https://doi.org/10.1016/j.bios.2017.08.043
124. Rasouli, E., et al.: Advancements in electrochemical DNA sensor for detection of human papilloma virus—a review. Anal. Biochem. **556**, 136–144 (2018). https://doi.org/10.1016/j.ab.2018.07.002
125. Simoska, O., Stevenson, K.J.: Electrochemical sensors for rapid diagnosis of pathogens in real time. Analyst **144**(22), 6461–6478 (2019). https://doi.org/10.1039/C9AN01747J
126. He, Q., et al.: Research on the construction of portable electrochemical sensors for environmental compounds quality monitoring. Mater. Today Adv. **17**, 100340 (2023). https://doi.org/10.1016/j.mtadv.2022.100340
127. Wang, K., Lin, X., Zhang, M., Li, Y., Luo, C., Wu, J.: Review of electrochemical biosensors for food safety detection. Biosensors (Basel) **12**(11) (2022). https://doi.org/10.3390/bios12110959
128. Curulli, A.: Electrochemical biosensors in food safety: challenges and perspectives. Molecules **26**(10) (2021). https://doi.org/10.3390/molecules26102940
129. Dhall, S., Mehta, B.R., Tyagi, A.K., Sood, K.: A review on environmental gas sensors: materials and technologies. Sen. Inter. **2**, 100116 (2021). https://doi.org/10.1016/j.sintl.2021.100116
130. Majhi, S.M., Mirzaei, A., Kim, H.W., Kim, S.S., Kim, T.W.: Recent advances in energy-saving chemiresistive gas sensors: a review. Nano Energy **79**, 105369 (2021). https://doi.org/10.1016/j.nanoen.2020.105369
131. Zhang, H., Zhang, Z., Li, Z., Han, H., Song, W., Yi, J.: A chemiresistive-potentiometric multivariate sensor for discriminative gas detection. Nat. Commun. **14**(1), 3495 (2023). https://doi.org/10.1038/s41467-023-39213-x
132. Sedlák, P., Kuberský, P., Mívalt, F.: Effect of various flow rate on current fluctuations of amperometric gas sensors. Sens. Actuators B Chem. **283**, 321–328 (2019). https://doi.org/10.1016/j.snb.2018.12.006
133. Vázquez-López, A., Bartolomé, J., Cremades, A., Maestre, D.: High-performance room-temperature conductometric gas sensors: materials and strategies. Chemosensors **10**(6) (2022). https://doi.org/10.3390/chemosensors10060227
134. Zappa, D., Galstyan, V., Kaur, N., Munasinghe Arachchige, H.M.M., Sisman, O., Comini, E.: 'Metal oxide -based heterostructures for gas sensors'—a review. Anal. Chim. Acta **1039**, 1–23 (2018). https://doi.org/10.1016/j.aca.2018.09.020
135. Dey, A.: Semiconductor metal oxide gas sensors: a review. Mater. Sci. Eng. B **229**, 206–217 (2018). https://doi.org/10.1016/j.mseb.2017.12.036
136. Alrammouz, R., Podlecki, J., Abboud, P., Sorli, B., Habchi, R.: A review on flexible gas sensors: from materials to devices. Sens. Actuators A Phys. **284**, 209–231 (2018). https://doi.org/10.1016/j.sna.2018.10.036
137. Zhou, T., Zhang, T.: Recent progress of nanostructured sensing materials from 0D to 3D: overview of structure–property-application relationship for gas sensors. Small Methods **5**(9), 2100515 (2021). https://doi.org/10.1002/smtd.202100515
138. Das, S., Mojumder, S., Saha, D., Pal, M.: Influence of major parameters on the sensing mechanism of semiconductor metal oxide based chemiresistive gas sensors: a review focused on personalized healthcare. Sens. Actuators B Chem. **352**, 131066 (2022). https://doi.org/10.1016/j.snb.2021.131066
139. Shi, L., Yin, Y., Zhang, L.-C., Wang, S., Sillanpää, M., Sun, H.: Design and engineering heterojunctions for the photoelectrochemical monitoring of environmental pollutants: a review. Appl. Catal. B **248**, 405–422 (2019). https://doi.org/10.1016/j.apcatb.2019.02.044

140. Sousa, J.C.G., Ribeiro, A.R., Barbosa, M.O., Pereira, M.F.R., Silva, A.M.T.: A review on environmental monitoring of water organic pollutants identified by EU guidelines. J. Hazard Mater. **344**, 146–162 (2018). https://doi.org/10.1016/j.jhazmat.2017.09.058
141. Hernandez-Vargas, G., Sosa-Hernández, J.E., Saldarriaga-Hernandez, S., Villalba-Rodríguez, A.M., Parra-Saldivar, R., Iqbal, H.M.N.: Electrochemical biosensors: a solution to pollution detection with reference to environmental contaminants. Biosensors (Basel) **8**(2) (2018). https://doi.org/10.3390/bios8020029
142. Sohrabi, H., et al.: Recent advances on portable sensing and biosensing assays applied for detection of main chemical and biological pollutant agents in water samples: a critical review. TrAC Trends Anal. Chem. **143**, 116344 (2021). https://doi.org/10.1016/j.trac.2021.116344
143. Shetty, S.S., et al.: Environmental pollutants and their effects on human health. Heliyon **9**(9) (2023). https://doi.org/10.1016/j.heliyon.2023.e19496
144. Hu, J.Y., Zhang, S.S., Chen, E., Li, W.G.: A review on corrosion detection and protection of existing reinforced concrete (RC) structures. Constr. Build Mater. **325**, 126718 (2022). https://doi.org/10.1016/j.conbuildmat.2022.126718
145. Reddy, M.S.B., et al.: Sensors in advancing the capabilities of corrosion detection: a review. Sens. Actuators A Phys. **332**, 113086 (2021). https://doi.org/10.1016/j.sna.2021.113086
146. Refait, P., Grolleau, A.-M., Jeannin, M., Rémazeilles, C., Sabot, R.: Corrosion of carbon steel in marine environments: role of the corrosion product layer. Corros. Mater. Degrad. **1**(1), 198–218 (2020). https://doi.org/10.3390/cmd1010010
147. Liu, L., Xu, Y., Xu, C., Wang, X., Huang, Y.: Detecting and monitoring erosion–corrosion using ring pair electrical resistance sensor in conjunction with electrochemical measurements. Wear **428–429**, 328–339 (2019). https://doi.org/10.1016/j.wear.2019.03.025
148. Tan, X., Fan, L., Huang, Y., Bao, Y.: Detection, visualization, quantification, and warning of pipe corrosion using distributed fiber optic sensors. Autom. Constr. **132**, 103953 (2021). https://doi.org/10.1016/j.autcon.2021.103953
149. Shi, J., Ming, J., Wu, M.: Electrochemical behavior and corrosion products of Cr-modified reinforcing steels in saturated $Ca(OH)_2$ solution with chlorides. Cem. Concr. Compos. **110**, 103587 (2020). https://doi.org/10.1016/j.cemconcomp.2020.103587

Chapter 6
Electrochemical Capacitors: EDLCs and Pseudocapacitors

Electrochemical capacitors, comprising Electric Double-Layer Capacitors (EDLCs) and pseudocapacitors, are crucial components in advanced energy storage systems due to their high power density, rapid charge–discharge capabilities, and long cycle life. This chapter provides a comprehensive overview of EDLCs and pseudocapacitors, including their underlying principles, materials, and design considerations. It delves into the specifics of electrode materials such as activated carbon, carbon nanotubes, and graphene, as well as the various types of electrolytes used to optimize performance. The chapter further explores pseudocapacitive materials, including metal oxides and conducting polymers, and their mechanisms of energy storage. Additionally, it examines hybrid capacitor systems and composite electrodes that combine the benefits of EDLCs and pseudocapacitors to enhance performance. Characterization techniques, including cyclic voltammetry, electrochemical impedance spectroscopy, galvanostatic charge–discharge, and electron microscopy, are discussed for evaluating these devices' performance. Applications across different sectors, from transportation and consumer electronics to renewable energy systems and grid storage, highlight the versatility and importance of these capacitors. The chapter concludes with an overview of current challenges and future directions, emphasizing the need for material innovation, advanced design, and practical integration to meet the evolving demands of energy storage.

6.1 Introduction to Electrochemical Capacitors

Electrochemical capacitors, commonly known as supercapacitors [1], have emerged as a vital class of energy storage devices, occupying a unique niche between conventional capacitors and batteries [2]. As the demand for efficient, reliable, and sustainable energy storage systems escalates across various sectors, the role of electrochemical capacitors [3] in addressing this demand has become increasingly significant.

P. Phogat et al., *Electrochemical Devices*, Engineering Materials,
https://doi.org/10.1007/978-981-96-0527-9_6

These devices are celebrated for their exceptional power density [4], rapid charge and discharge capabilities [5], and remarkable cycling stability [6], which make them indispensable in applications requiring high power output and longevity. The advent of electrochemical capacitors represents a paradigm shift in energy storage technology, offering solutions to the limitations posed by traditional batteries and capacitors [7]. The origins of electrochemical capacitors can be traced back to the mid-twentieth century, but it was only in the past few decades that they have gained widespread attention, driven by advancements in materials science and nanotechnology. The core principle behind electrochemical capacitors is their ability to store energy through the electrostatic and electrochemical processes occurring at the interface between the electrode and the electrolyte. This is in contrast to batteries, where energy is stored and released through bulk electrochemical reactions. This distinction allows electrochemical capacitors to deliver much higher power densities, albeit with lower energy densities compared to batteries [8].

Electrochemical capacitors are broadly categorized into two main types: Electric Double-Layer Capacitors (EDLCs) [9] and Pseudocapacitors [10]. EDLCs store energy through the formation of an electric double layer at the electrode–electrolyte interface, a purely electrostatic process. The amount of energy stored in EDLCs is directly related to the surface area of the electrode material and the dielectric properties of the electrolyte. Consequently, materials with high surface area, such as activated carbon [11], carbon nanotubes, and graphene [12], are extensively used in EDLCs. The simplicity of the charge storage mechanism in EDLCs imparts them with excellent cycling stability, often exceeding one million charge–discharge cycles. Pseudocapacitors, on the other hand, involve faradaic processes [13] where charge transfer occurs through redox reactions, intercalation, or electrosorption at the electrode surface. This mechanism allows pseudocapacitors to achieve higher capacitance [14] and energy density [14] compared to EDLCs. Pseudocapacitive materials include transition metal oxides [15], such as ruthenium oxide (RuO_2) [16] and manganese oxide (MnO_2) and conducting polymers like polyaniline (PANI) [17] and polypyrrole (PPy). The faradaic processes in pseudocapacitors, while providing higher energy storage, are often accompanied by challenges such as lower cycling stability and potential degradation of the electrode materials. The distinction between EDLCs and pseudocapacitors is not always clear-cut, as many materials exhibit characteristics of both types, leading to the development of hybrid capacitors [18]. These hybrid systems combine the advantages of both EDLCs and pseudocapacitors, aiming to enhance both energy and power density while maintaining good cycling stability [19]. Hybrid capacitors typically employ different materials for the positive and negative electrodes, creating an asymmetric configuration that extends the operating voltage and improves the overall performance of the device. Materials science plays a critical role in the advancement of electrochemical capacitors. The performance of these devices is heavily dependent on the properties of the electrode materials, which must exhibit high surface area [20], good electrical conductivity, chemical stability [21], and appropriate porosity to facilitate ion transport. Carbon-based materials, such as activated carbon, carbon nanotubes, and graphene, dominate the landscape of EDLCs due to their favorable properties. In pseudocapacitors, transition metal oxides and conducting polymers are the materials of choice, each offering unique advantages and challenges. The development of composite materials [22], where different materials are combined to leverage their individual strengths, has also emerged as a promising approach to enhance the performance of electrochemical capacitors. Beyond the electrodes, the electrolyte is another crucial component that influences the performance of electrochemical capacitors. The electrolyte must provide high ionic conductivity, a wide electrochemical stability window, and compatibility with the electrode materials. Aqueous electrolytes [23], organic electrolytes [24], and ionic liquids [25] are the primary types of electrolytes used in these devices, each offering distinct advantages and limitations. Aqueous electrolytes are known for their high ionic conductivity and low cost but are limited by a narrow electrochemical stability window. Organic electrolytes offer a wider voltage range, enabling higher energy densities, but they are more expensive and have lower ionic conductivity. Ionic liquids, with their non-volatile [26] and non-flammable nature, offer the potential for very high energy densities, but their high viscosity and cost pose challenges [27].

The applications of electrochemical capacitors are as diverse as the challenges they address. In the transportation sector, supercapacitors are employed in electric vehicles (EVs) [28] and hybrid electric vehicles (HEVs) [29] for energy recovery and power assistance during acceleration. They are also used in public transportation systems, such as buses and trains, to provide power stabilization and energy recovery during braking. In consumer electronics, supercapacitors are finding increasing use in portable devices where they can provide quick bursts of power, extend battery life, and improve energy efficiency. Renewable energy systems benefit from supercapacitors [30] by using them as energy buffers to smooth out fluctuations in power output from intermittent sources like solar panels [31] and wind turbines [32]. In industrial applications, supercapacitors are used in uninterruptible power supplies (UPS), cranes [33], and drilling rigs [33], where they provide reliable and instantaneous power. The military sector [34] also relies on supercapacitors for various applications, including power systems for armored vehicles, aircraft, and missiles, where their reliability and rapid response time are critical. Despite their numerous advantages, electrochemical capacitors face several challenges that must be addressed to fully realize their potential. One of the most significant challenges is the relatively low energy density compared to batteries, which limits their use in applications where long-term energy storage is required. Researchers are actively exploring new materials and hybrid systems to overcome this limitation. The cost of electrochemical capacitors, driven by the expensive materials and complex manufacturing processes, is another barrier to widespread adoption. Developing cost-effective materials and scalable production methods is crucial for making these devices more accessible. Additionally, the cycling stability of pseudocapacitive materials remains a concern, as repeated faradaic reactions can lead to the degradation of the electrode materials [35]. Addressing these stability issues is essential for the long-term reliability of pseudocapacitors. Looking to the future, the field of electrochemical capacitors is poised for significant advancements. The ongoing research into novel materials, such as two-dimensional (2D) materials [36] and metal–organic frameworks (MOFs)

[37], holds promise for improving the performance of both EDLCs and pseudocapacitors. These materials offer unique properties that could lead to higher energy and power densities, better cycling stability, and enhanced safety. Additionally, the integration of supercapacitors with other energy storage systems, such as batteries and fuel cells, could create hybrid energy storage [9, 38] solutions that capitalize on the strengths of each technology. As the demand for efficient and sustainable energy storage continues to grow, electrochemical capacitors are expected to play an increasingly important role in the energy landscape, driving innovation and enabling new applications across a wide range of industries.

6.2 Principles of EDLCs

Electric Double-Layer Capacitors (EDLCs) [9] are a type of electrochemical capacitor that store energy through the electrostatic accumulation of charges at the interface between an electrode and an electrolyte. Unlike batteries, which store energy via chemical reactions within the bulk of the material, EDLCs operate based on purely physical processes, allowing them to deliver rapid charge and discharge [39] cycles with minimal degradation over time. This fundamental principle underlies the high power density and long cycle life that characterize [39] EDLCs, making them particularly suitable for applications requiring quick bursts of energy. The core of an EDLC is its two electrodes, usually made from highly porous carbon-based materials such as activated carbon, carbon nanotubes (CNTs), or graphene. These materials are chosen for their high surface area, which directly influences the capacitance of the device [40]. When a voltage is applied across the electrodes, positive and negative charges accumulate at the surfaces of the respective electrodes as depicted in Fig. 6.1. These charges attract oppositely charged ions from the electrolyte, which adhere to the electrode surfaces, forming what is known as the electric double layer. The electric double layer comprises two layers of charges: one on the surface of the electrode and the other in the electrolyte, separated by a nanometer-scale distance. This separation of charge across a small distance creates a strong electric field, enabling the storage of energy. The amount of energy stored in an EDLC is proportional to the surface area of the electrodes and inversely proportional to the thickness of the double layer. Therefore, the performance of EDLCs is highly dependent on the properties of the electrode material and the electrolyte used. The capacitance C of an EDLC can be described by the equation:

$$C = \frac{\epsilon_r \epsilon_0 A}{d}$$

where ϵ_r is the relative permittivity of the electrolyte, ϵ_0 is the permittivity of free space, A is the surface area of the electrode, d is the effective thickness of the double layer. This equation highlights the importance of maximizing the surface area A and minimizing the separation distance d to achieve high capacitance.

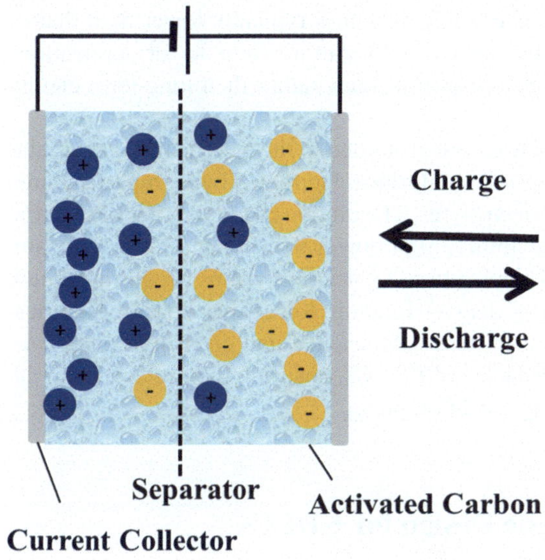

Fig. 6.1 Schematic diagram shows the working principle of EDLC reconstructure from Ref. [43]

The choice of electrolyte plays a crucial role in determining the performance of EDLCs [41]. The electrolyte must be compatible with the electrode material, provide high ionic conductivity, and possess a wide electrochemical stability window. Aqueous electrolytes, such as sulfuric acid and potassium hydroxide, are commonly used due to their high ionic conductivity. However, their electrochemical stability window is limited to approximately 1.23 V, restricting the energy density of the EDLC. Organic electrolytes, on the other hand, offer a broader stability window (up to 3 V) but have lower ionic conductivity and are more expensive. Ionic liquids, with their wide stability window and non-volatility, offer the potential for higher energy densities, though their higher viscosity can hinder ion mobility. The energy E stored in an EDLC is given by the equation [42]:

$$E = \frac{1}{2}CV^2$$

where V is the voltage across the capacitor. This equation indicates that increasing the operating voltage V significantly enhances the energy storage capability, which is why optimizing the electrolyte is critical for maximizing the energy density of EDLCs. One of the most remarkable features of EDLCs is their ability to undergo millions of charge–discharge cycles with minimal degradation. This longevity is attributed to the absence of chemical reactions in the energy storage process. The electrostatic nature of charge storage ensures that the electrode materials do not undergo significant structural changes during operation, unlike in batteries, where repetitive redox reactions can lead to material degradation and capacity fading. However, the reliance on surface area and the physical charge separation mechanism also imposes limitations on the energy density of EDLCs, which is typically lower than that of batteries. This lower energy density makes EDLCs more suitable for applications requiring high power density and fast response times rather than long-term energy storage.

In summary, the principles of EDLCs are grounded in the electrostatic accumulation of charges at the electrode–electrolyte interface, facilitated by high-surface-area electrode materials and optimized electrolytes. The unique characteristics of EDLCs, such as rapid charge–discharge capability, high power density, and long cycle life, make them ideal for a variety of applications, particularly those requiring quick energy delivery. While their energy density remains lower than that of batteries, ongoing research and development in materials science and electrolyte optimization continue to enhance the performance of EDLCs, ensuring their relevance in the evolving landscape of energy storage technologies.

6.3 Materials and Electrode Design for EDLCs

The performance of EDLCs is heavily influenced by the choice of electrode materials and their design. Ideal electrode materials for EDLCs should possess a high surface area, good electrical conductivity, chemical stability, and porosity to facilitate ion transport.

6.3.1 Activated Carbon

Activated carbon [44] is one of the most widely used materials in Electric Double-Layer Capacitors (EDLCs) due to its high surface area, tunable porosity, and relatively low cost. Derived from various carbonaceous materials such as coconut shells, wood, or synthetic precursors, activated carbon undergoes processes like pyrolysis and activation to develop a porous structure [45] ideal for electrochemical applications. The large surface area, often exceeding 2000 m^2/g, provides abundant sites for ion adsorption, directly contributing to the capacitance of the EDLC. The performance of activated carbon as an electrode material is influenced by its pore size distribution. Micropores (pores < 2 nm) are critical for maximizing capacitance, as they offer the highest surface area for double-layer formation. However, the accessibility of these pores to electrolyte ions can be limited by pore size, especially in the case of larger solvated ions. Mesopores (2–50 nm) and macropores (> 50 nm), though providing lower surface area, play an essential role in facilitating ion transport within the electrode structure, enhancing the rate capability of the EDLC. Therefore, a hierarchical pore structure, combining micropores for high capacitance and meso/macropores [46] for efficient ion transport, is often targeted in the design of activated carbon for EDLCs. Additionally, the surface chemistry of activated carbon influences its electrochemical performance. The presence of oxygen-containing functional groups, such as hydroxyl, carbonyl, and carboxyl groups, can enhance wettability and ion transport in aqueous electrolytes. However, these functional groups may also introduce faradaic reactions, potentially affecting the cycling stability of the device. Balancing surface functionalization to optimize electrochemical performance without compromising stability is a key challenge in the use of activated carbon in EDLCs. Despite its advantages, activated carbon has limitations, particularly in terms of energy density. The relatively low conductivity and the limited ability to control pore size distribution can restrict performance. Nevertheless, ongoing research into the optimization of precursor materials, activation processes, and post-treatment methods continues to advance the capabilities of activated carbon, ensuring its continued prominence in the field of EDLCs [47].

6.3.2 Carbon Nanotubes (CNTs)

Carbon Nanotubes (CNTs) [48] have garnered significant attention as electrode materials for EDLCs due to their unique structural, electrical, and mechanical properties. CNTs are cylindrical structures composed of one or more layers of graphene, rolled into a tube with diameters typically in the nanometer range. They can be classified into single-walled (SWCNTs) and multi-walled (MWCNTs) nanotubes, with each type offering distinct advantages for energy storage applications. One of the most compelling features of CNTs is their high electrical conductivity [49], which facilitates efficient electron transport within the electrode, reducing internal resistance and enhancing the power density of EDLCs. This property, combined with their high surface area (up to 1000 m^2/g), allows CNTs to serve as effective electrode materials capable of storing significant amounts of charge. The tubular morphology of CNTs provides open, accessible pathways for electrolyte ions, enabling rapid ion transport and contributing to the fast charge–discharge cycles characteristic of EDLCs [50]. In addition to their conductivity and surface area, CNTs possess exceptional mechanical strength and flexibility, which contribute to the structural stability of the electrode during cycling. This is particularly beneficial in flexible and wearable energy storage devices, where mechanical durability is crucial. CNTs can also be functionalized with various chemical groups (depicted in Fig. 6.2) or combined with other materials, such as metal oxides or conducting polymers, to enhance their electrochemical performance. However, the practical application of CNTs in EDLCs faces challenges [51]. The aggregation of CNTs due to van der Waals forces can lead to a reduction in accessible surface area and pore availability, limiting the overall capacitance. Additionally, the cost and scalability of CNT production remain barriers to their widespread adoption in commercial supercapacitors. Efforts to overcome these challenges include the development of dispersion techniques, the synthesis of CNT composites, and the exploration of cost-effective production methods. Despite these hurdles, CNTs continue to represent a promising avenue for the development of high-performance EDLCs.

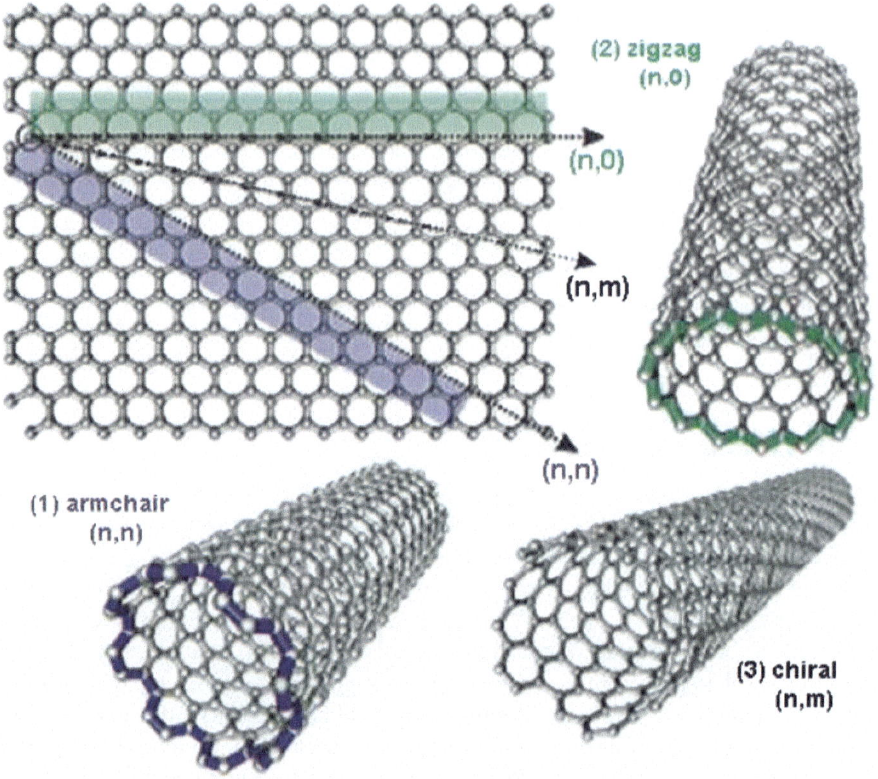

Fig. 6.2 Chirality of single wall and multiwall carbon nanotube reproduced from Ref. [52] copyright

6.3.3 Graphene

Graphene [53], a single layer of carbon atoms arranged in a two-dimensional honeycomb lattice, has revolutionized the field of materials science and holds immense potential as an electrode material for EDLCs. Its exceptional properties [54], including high surface area (theoretically up to 2630 m^2/g), excellent electrical conductivity, mechanical strength, and chemical stability, make graphene an ideal candidate for high-performance energy storage devices. The high surface area of graphene allows for a significant amount of charge to be stored within a compact structure, contributing to the high capacitance of graphene-based EDLCs [55]. The two-dimensional structure of graphene ensures that nearly every atom is exposed to the electrolyte, maximizing the electrostatic interaction between the electrode and the ions. Additionally, the conductivity of graphene facilitates rapid electron transport across the electrode, which is crucial for the fast charge–discharge cycles and high-power density [56] of EDLCs. Graphene's mechanical flexibility and strength further enhance its appeal for EDLC applications, particularly in flexible and wearable devices. Its ability to maintain structural integrity under mechanical deformation ensures reliable performance even under repeated cycling. Moreover, the chemical stability of graphene in various electrolytes, including aqueous, organic, and ionic liquids, broadens the range of operating conditions for graphene-based EDLCs. Despite these advantages, challenges remain in harnessing the full potential of graphene for EDLCs [57]. One significant challenge is the restacking of graphene layers during electrode fabrication, which reduces the accessible surface area and limits ion transport. To address this, strategies such as creating graphene oxide (GO) or reduced graphene oxide (rGO) [58], introducing spacers between graphene layers, or developing three-dimensional (3D) graphene structures have been explored. These approaches aim to preserve the high surface area and improve ion accessibility, thereby enhancing the performance of graphene-based EDLCs. Another challenge is the scalable production of high-quality graphene at a low cost. While methods such as chemical vapor deposition (CVD) [59] and chemical exfoliation have been developed, they often involve trade-offs between quality, yield, and cost. Ongoing research is focused on optimizing these methods to make graphene-based EDLCs commercially viable. Despite these challenges, the unique properties of graphene make it one of the most promising materials for next-generation EDLCs, with the potential to significantly improve energy storage performance [60].

6.3.4 Electrode Design

The design of the electrode is critical in determining the performance of EDLCs [61], as it directly influences the capacitance, energy density, power density, and cycling stability of the device. An ideal electrode design must optimize the surface area for charge storage, minimize internal resistance, and ensure efficient ion transport within the electrode structure. One of the key considerations in electrode design is the material's porosity. A hierarchical pore structure [62], incorporating micropores, mesopores, and macropores, is often desirable. Micropores provide the surface area needed for double-layer formation, while mesopores and macropores facilitate ion transport and electrolyte diffusion throughout the electrode. This multi-scale porosity ensures that the electrode can store a large amount of charge while maintaining high rate capability, which is essential for the rapid charge–discharge cycles characteristic of EDLCs. The thickness of the electrode also plays a significant role in performance. While thicker electrodes may store more charge due to increased surface area, they can also suffer from increased internal resistance and slower ion diffusion, leading to lower power density and longer charge times [63]. Therefore, optimizing electrode thickness is crucial for balancing energy density and power density. Electrode composition is another critical factor. Composite electrodes, combining carbon-based materials (such as activated carbon, CNTs, or graphene) with other materials like metal oxides or conducting polymers, can leverage the strengths of each component. For example, adding a small amount of metal oxide can enhance pseudocapacitance, increasing the overall capacitance of the electrode. However, care must be taken to ensure that the addition of such materials does not compromise the stability or conductivity of the electrode. Additionally, the binder used in the electrode fabrication can impact performance. Binders are necessary to hold the active materials together and ensure good contact with the current collector. However, they should be electrically conductive and chemically inert to avoid adding unnecessary resistance or degrading over time. Research into binder-free electrode designs [64], where the active material adheres directly to the current collector, is ongoing and holds promise for further improving EDLC performance.

6.4 Electrolytes for EDLCs

The choice of electrolyte is crucial in determining the performance of EDLCs. Electrolytes should possess high ionic conductivity, wide electrochemical stability, and compatibility with the electrode material. There are three main types of electrolytes used in EDLCs:

6.4.1 Aqueous Electrolytes

Aqueous electrolytes are among the most commonly used electrolytes in Electric Double-Layer Capacitors (EDLCs) due to their high ionic conductivity, low cost, and environmental friendliness. These electrolytes [65] are typically composed of a salt, such as potassium hydroxide (KOH), sulfuric acid (H_2SO_4), or sodium sulfate (Na_2SO_4), dissolved in water. The high ionic conductivity of aqueous electrolytes, often in the range of 1–2 S/cm, facilitates rapid ion transport within the electrode, enabling fast charge–discharge cycles and high power density in EDLCs. One of the primary advantages of aqueous electrolytes is their ability to operate at room temperature without significant loss of performance. The simple preparation and handling of aqueous electrolytes also make them attractive for large-scale applications, where ease of manufacturing and cost-effectiveness are critical. Additionally, aqueous electrolytes are non-flammable, reducing the risk of fire hazards in energy storage systems, which is a significant safety consideration compared to organic or ionic liquid electrolytes. However, the electrochemical stability window of aqueous electrolytes is limited, typically to about 1.23 V [66], due to the decomposition of water via electrolysis. This relatively narrow voltage window constrains the energy density of EDLCs using aqueous electrolytes, as the energy stored is proportional to the square of the operating voltage. Efforts to expand the voltage window include using neutral electrolytes, such as Na_2SO_4, which can reduce water electrolysis by operating at lower voltages, though this often comes at the cost of reduced ionic conductivity compared to strong acids or bases. The interaction between aqueous electrolytes and electrode materials is another critical factor affecting the performance and stability of EDLCs. For instance, carbon-based electrodes, such as activated carbon or graphene, exhibit excellent wettability and compatibility with aqueous electrolytes, leading to efficient ion adsorption and desorption as illustrated in Fig. 6.3. However, the presence of oxygen-containing functional groups on the carbon surface can induce faradaic reactions, particularly in the presence of acidic or basic electrolytes, which can lead to self-discharge and reduced cycling stability. This issue is often addressed through surface modification techniques to minimize unwanted reactions while maintaining high capacitance. Despite these challenges, aqueous electrolytes remain a popular choice for EDLCs in applications where high power density, low cost, and safety are prioritized over energy density. They are particularly well-suited for applications such as regenerative braking systems in vehicles, backup power supplies, and grid stabilization, where rapid energy delivery and robustness are crucial. Ongoing research is focused on optimizing the composition and concentration of aqueous electrolytes to improve their electrochemical stability and compatibility with advanced electrode materials, aiming to push the performance limits of EDLCs.

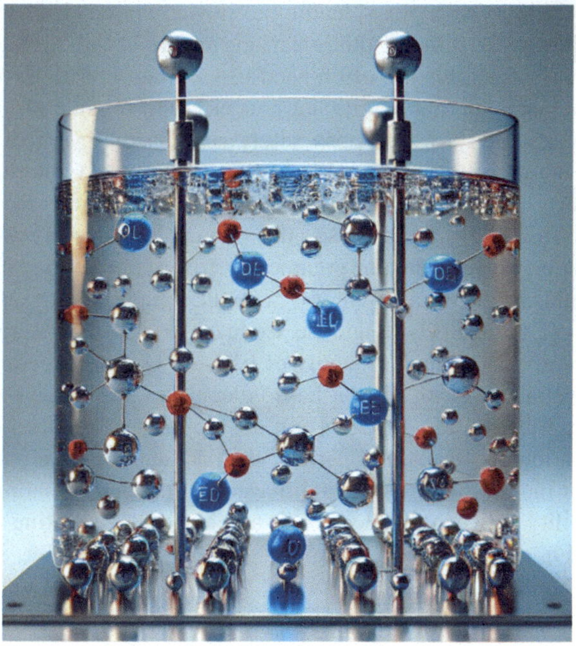

Fig. 6.3 This image is generated by AI lexica. Art shows the aqueous electrolyte

6.4.2 Organic Electrolytes

Organic electrolytes [67] are widely used in EDLCs due to their broad electrochemical stability window [68], which can range from 2.5 to 3.5 V, significantly higher than that of aqueous electrolytes. This wider voltage window allows for greater energy storage, as the energy density of an EDLC is proportional to the square of the operating voltage. Organic electrolytes typically consist of a salt, such as tetraethylammonium tetrafluoroborate ($TEABF_4$), dissolved in an organic solvent like acetonitrile (ACN) or propylene carbonate (PC). The higher voltage window offered by organic electrolytes makes them ideal for applications where energy density is a critical factor, such as in electric vehicles, portable electronics, and renewable energy storage systems. The use of organic electrolytes can increase the energy density of EDLCs by a factor of 2–3 compared to those using aqueous electrolytes, providing a competitive edge in markets where both energy and power density are important. However, organic electrolytes have several drawbacks [69] that must be addressed to fully capitalize on their advantages. One of the main challenges is their relatively low ionic conductivity, which is typically in the range of 10^{-3}–10^{-2} S/cm, much lower than that of aqueous electrolytes. This lower conductivity can limit the rate capability of EDLCs, leading to slower charge–discharge cycles and reduced power density. To mitigate this, electrode materials with highly conductive and well-aligned pore structures are often used to facilitate ion transport and maintain high performance. Another significant concern with organic electrolytes is their flammability and toxicity, which pose safety risks in the event of a leak or short circuit. The volatility of organic solvents like ACN and PC increases the risk of fire, particularly under high-temperature operating conditions or in large-scale energy storage systems. To address these safety concerns, research efforts are focused on developing safer, non-flammable organic electrolytes or solid-state alternatives that can maintain a high voltage window without compromising safety. The interaction between organic electrolytes and electrode materials is also a crucial factor in determining the performance and longevity of EDLCs. Organic electrolytes generally exhibit lower wettability with carbon-based electrodes compared to aqueous electrolytes, which can lead to less efficient ion adsorption and reduced capacitance. Surface modification of electrode materials, such as functionalizing with specific chemical groups or incorporating polar additives, can enhance wettability and improve the compatibility between the electrolyte and electrode, resulting in better overall performance (Fig. 6.4).

6.4.3 Ionic Liquids

Ionic liquids [70] are a class of electrolytes that have gained considerable interest for use in EDLCs due to their unique combination of properties, including a wide electrochemical stability window, non-volatility, and thermal stability. Ionic liquids

Fig. 6.4 This image is generated by lexica. Art shows organic electrolyte

are salts that are liquid at or near room temperature, typically composed of bulky, asymmetric organic cations and inorganic or organic anions. The most commonly used ionic liquids in EDLCs include imidazolium-based, pyrrolidinium-based, and ammonium-based ionic liquids, each offering distinct advantages depending on the application. One of the primary advantages of ionic liquids is their wide electrochemical stability window [71], which can exceed 4 V, far surpassing that of both aqueous and organic electrolytes. This high voltage window enables the storage of significantly more energy, making ionic liquids ideal for applications requiring both high energy density and long cycle life. The non-volatile and non-flammable nature of ionic liquids also addresses the safety concerns associated with organic electrolytes, reducing the risk of fire and making them suitable for use in high-temperature environments. The ionic conductivity of ionic liquids, while generally lower than that of aqueous electrolytes, is often comparable to or slightly lower than that of organic electrolytes, typically in the range of 10^{-3}–10^{-2} S/cm. This makes ionic liquids suitable for EDLCs where the balance between energy density and power density is essential. The high viscosity [72] of many ionic liquids can, however, limit ion mobility, particularly at lower temperatures, which can negatively impact the rate capability and power performance of the EDLC. To address this, researchers are exploring the development of ionic liquid mixtures or the addition of small amounts of co-solvents to reduce viscosity and enhance ion transport. The interaction between ionic liquids and electrode materials is another critical factor in determining the performance of EDLCs. Ionic liquids tend to form stable electric double layers with carbon-based electrodes, such as activated carbon, CNTs, and graphene, leading to high capacitance and efficient energy storage. However, the large size and asymmetry of the ions in ionic liquids can result in slower ion diffusion and limited access to micropores within the electrode, reducing the overall capacitance as shown in Fig. 6.5. Designing electrode materials with a hierarchical pore structure that includes a combination of micropores, mesopores, and macropores can help mitigate this issue by providing pathways for efficient ion transport and maximizing the accessible surface area. Ionic liquids also offer the potential for tenability [73], as their physical and chemical properties can be tailored by modifying the cation and anion components. This allows for the design of ionic liquids with specific characteristics suited to particular applications, such as low viscosity for high power density or enhanced thermal stability for use in extreme environments. However, the high cost of ionic liquids compared to traditional aqueous or organic electrolytes remains a barrier to their widespread adoption, particularly in commercial EDLCs. In conclusion, ionic liquids present a promising electrolyte option for EDLCs, offering a unique combination of a wide electrochemical stability window, non-volatility, and thermal stability. While challenges related to ionic conductivity, viscosity, and cost persist, ongoing research and development continue to improve the performance and economic viability of ionic liquids, positioning them as a key component in the future of high-performance EDLC.

6.5 Pseudocapacitive Materials and Mechanisms

Pseudocapacitors [75] store energy through faradaic reactions involving charge transfer between the electrode and electrolyte, rather than just electrostatic accumulation. This mechanism allows pseudocapacitors to achieve higher energy densities compared to EDLCs.

6.5.1 Metal Oxides

Metal oxides are among the most prominent pseudocapacitive materials used in electrochemical capacitors due to their ability to store charge through fast and reversible faradaic reactions [76]. Unlike Electric Double-Layer Capacitors (EDLCs), which store energy purely through ion adsorption at the electrode–electrolyte interface,

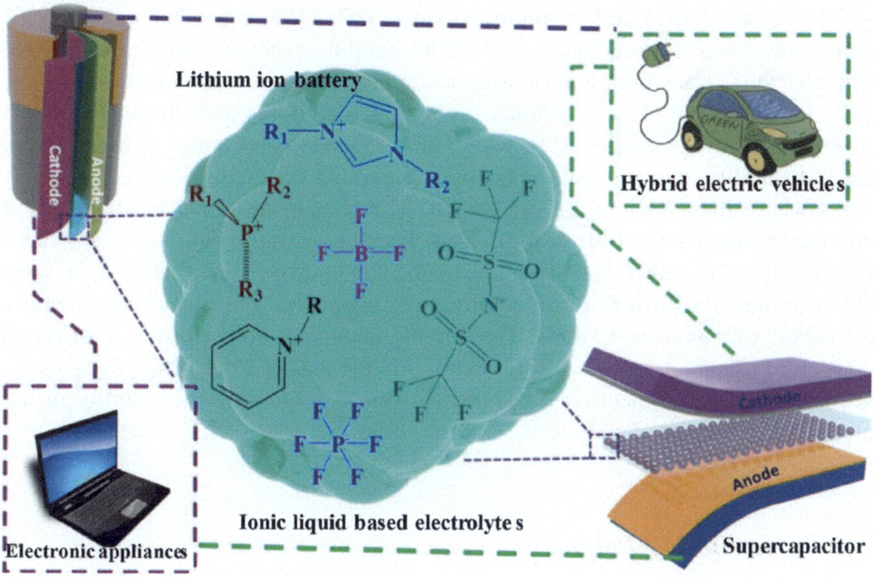

Fig. 6.5 Shows the ionic liquid used for various application reproduced from Ref. [74] copyright © 1996–2024 MDPI (Basel, Switzerland)

pseudocapacitors utilize redox reactions, intercalation, or surface adsorption within the electrode material to achieve higher capacitance and energy density. Commonly used metal oxides [77] include ruthenium oxide (RuO_2), manganese oxide (MnO_2), nickel oxide (NiO), and titanium dioxide (TiO_2), each offering distinct electrochemical properties. Ruthenium oxide (RuO_2) is one of the most studied metal oxides for pseudocapacitors due to its high electrical conductivity, excellent reversibility, and large specific capacitance. RuO_2 exhibits multiple oxidation states, allowing for efficient charge transfer during redox reactions. Its high capacitance, often exceeding 1000 F/g, stems from the rapid faradaic processes occurring at the surface and within the bulk of the material. However, the high cost and toxicity of ruthenium limit its widespread commercial application, prompting the search for more affordable alternatives. Manganese oxide (MnO_2) [78] is another widely explored metal oxide, valued for its abundance, low cost, and environmental friendliness. MnO_2 exhibits a variety of crystallographic forms, such as α-MnO_2, β-MnO_2, and γ-MnO_2, each with distinct electrochemical properties. The pseudocapacitive behavior of MnO_2 is primarily attributed to the redox reactions involving the Mn^{4+}/Mn^{3+} couple, which occur at the surface and in the bulk material. While MnO_2 offers lower capacitance compared to RuO_2, its cost-effectiveness and non-toxic nature make it an attractive candidate for large-scale applications. However, MnO_2 often suffers from poor electrical conductivity and limited cycle life, challenges that are being addressed through material engineering, such as doping with other elements or combining with conductive substrates.

Nickel oxide (NiO) and titanium dioxide (TiO_2) [79] are other examples of metal oxides used in pseudocapacitors. NiO exhibits excellent redox activity due to the Ni^{2+}/Ni^{3+} transition, contributing to its high specific capacitance. However, similar to MnO_2, NiO faces challenges related to electrical conductivity and structural stability during cycling. TiO_2, while offering lower capacitance, is noted for its chemical stability, low cost, and non-toxicity, making it suitable for specific applications where these factors are prioritized. The pseudocapacitive mechanisms in metal oxides involve the diffusion of electrolyte ions into the bulk material, where they participate in redox reactions. This process, known as intercalation, is typically faster than the diffusion-limited processes in batteries but slower than the pure electrostatic processes in EDLCs. The key to enhancing the performance of metal oxide-based pseudocapacitors lies in optimizing the material's structure to facilitate fast ion diffusion and electron transport, as well as improving the stability of the material during repeated cycling.

6.5.2 *Conducting Polymers*

Conducting polymers [80] are another class of pseudocapacitive materials that have gained significant attention for their ability to store charge through faradaic processes. Unlike traditional polymers, which are generally insulating, conducting polymers possess conjugated backbones that enable electron delocalization, making them electrically conductive. This unique property allows conducting polymers to participate in redox reactions, leading to pseudocapacitive behavior. Common conducting polymers used in pseudocapacitors include polyaniline (PANI), polypyrrole (PPy), and polythiophene (PTh). Polyaniline (PANI) is one of the most extensively studied conducting polymers for pseudocapacitive applications due to its high specific capacitance, ease of synthesis, and environmental stability. PANI [81] can exist in various oxidation states, such as leucoemeraldine, emeraldine, and pernigraniline, each offering different electrochemical properties. The pseudocapacitive behavior of PANI arises from the reversible redox transitions between these states, which involve the doping and de-doping of the polymer with electrolyte ions. PANI's high capacitance, often ranging from 400 to 1000 F/g, is attributed to the large number of active sites available for redox reactions. However, the cycling stability of PANI can be a concern, as repeated doping and de-doping can lead to structural degradation, affecting long-term performance depicted in 3D structure as Fig. 6.6. Polypyrrole [17] (PPy) is another widely used conducting polymer known for its high conductivity, ease of processing, and good environmental stability. The pseudocapacitive behavior of PPy is similar to that of PANI, involving redox reactions associated with the polymer backbone. PPy offers moderate capacitance, typically in the range of 200–500 F/g, and is often used in combination with other materials to enhance its performance. One of the challenges associated with PPy is its relatively poor mechanical stability during cycling, as the polymer tends to undergo swelling and shrinking during ion exchange, leading to reduced capacitance over time. Polythiophene (PTh) and its derivatives represent another important class of conducting polymers for pseudocapacitors. PTh offers good environmental stability, high conductivity, and moderate capacitance, typically ranging from 100 to 300 F/g. The pseudocapacitive behavior of PTh is based on the reversible redox reactions of the thiophene rings, which involve the movement of counterions into and out of the polymer during charging and discharging. PTh is often modified or copolymerized with other monomers to improve its electrochemical properties and stability. One of the key advantages of conducting polymers is their flexibility, which makes them suitable for use in flexible and wearable energy storage devices. However, the long-term stability of conducting polymers remains a challenge due to the mechanical stresses associated with repeated redox cycling. Strategies to improve the performance of conducting polymers in pseudocapacitors include the development of composite materials, incorporating carbon-based nanostructures, or doping with metal oxides to enhance conductivity and mechanical stability.

6.5.3 Composite Materials

Composite materials [83] combine different types of pseudocapacitive materials, such as metal oxides, conducting polymers, and carbon-based nanostructures, to leverage the strengths of each component and overcome their individual limitations. The use of composite materials in pseudocapacitors aims to achieve higher capacitance, better cycling stability, and improved conductivity by synergistically integrating the desirable properties of each constituent material. One common approach to composite materials is the combination of metal oxides with carbon-based materials, such as activated carbon, carbon nanotubes (CNTs), or graphene. In these composites, the carbon material provides a high surface area and excellent conductivity, facilitating fast electron transport and ion diffusion. The metal oxide component, such as MnO_2, RuO_2, or NiO, contributes to the overall capacitance through faradaic redox reactions. The integration of these materials can result in a composite with enhanced electrochemical performance compared to the individual components. For instance, a MnO_2/graphene composite can offer both the high capacitance of MnO_2 and the excellent conductivity and mechanical strength of graphene, leading to improved energy density and cycling stability. Conducting polymer-based composites [84] are another important class of materials used in pseudocapacitors. These composites often involve the incorporation of conducting polymers like PANI, PPy, or PTh with carbon nanostructures or metal oxides. The conducting polymer provides a large number of active sites for redox reactions, while the carbon or metal oxide component enhances conductivity and mechanical stability. For example, a PANI/CNT composite can offer high capacitance due to the redox activity of PANI, combined with the high conductivity and structural support provided by the CNTs. This synergistic effect can lead to improved capacitance, rate capability, and long-term stability given in Fig. 6.7. Another approach to composite materials involves the use of hybrid structures, where different pseudocapacitive materials are combined at

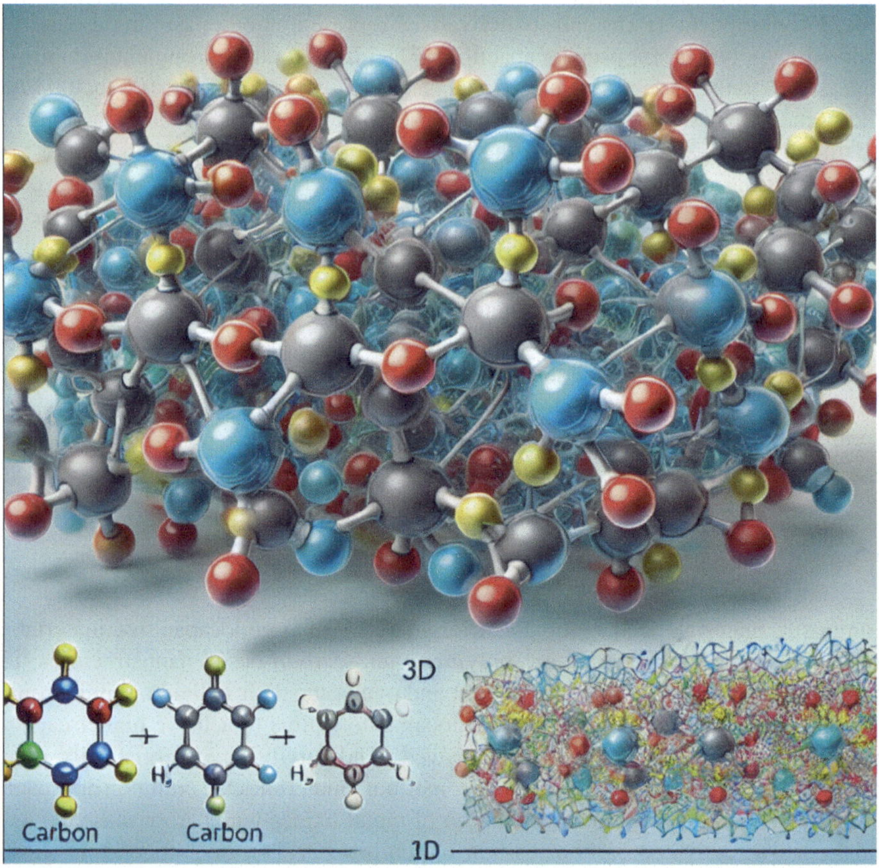

Fig. 6.6 **a** 3D structure of polyaniline. The carbon atoms (cyan balls), nitrogen atoms (red balls), hydrogen atoms (small yellow balls), and clouds (molecular orbitals). **b** Polyaniline 2D structure replicated from Ref. [82]

the nanoscale to create a more uniform and interconnected network. For example, a hybrid composite of TiO_2 nanowires and PPy can provide a three-dimensional (3D) structure with a high surface area, facilitating nanowires offer chemical stability and mechanical support, while the PPy provides pseudocapacitive behavior, resulting in a composite with enhanced electrochemical performance. The development of composite materials for pseudocapacitors is a rapidly evolving field, with ongoing research focused on optimizing the composition, structure, and fabrication methods to achieve the best possible performance. Challenges remain [85], such as ensuring uniform dispersion of the components, achieving strong interfacial bonding, and scaling up the production of these composites for commercial applications. However, the potential of composite materials to combine high capacitance, good conductivity, and excellent stability makes them a promising avenue for the next generation of high-performance pseudocapacitors.

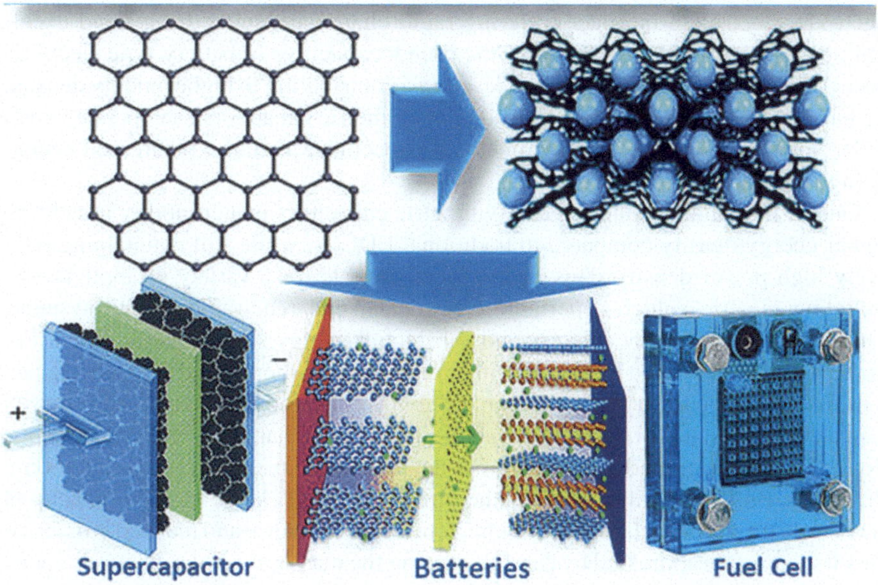

Fig. 6.7 The image shows the graphene nanocomposite enhancing the capacitive performance of supercapacitor reproduced from Ref. [86] copyright © 1996–2024 MDPI (Basel, Switzerland)

6.6 Hybrid Capacitors and Composite Electrodes

Characterization of EDLCs and pseudocapacitors is essential for understanding their performance and guiding the development of new materials and devices. Key techniques include:

6.6.1 Asymmetric Capacitors

Asymmetric capacitors, also known as hybrid capacitors, represent a significant advancement in the field of electrochemical capacitors by combining the advantages of both Electric Double-Layer Capacitors (EDLCs) and pseudocapacitors [87]. These devices typically consist of two different types of electrodes: one with high surface area for double-layer capacitance (such as activated carbon) and the other with pseudocapacitive or battery-like behavior (such as metal oxides or conducting polymers). The combination of these materials allows asymmetric capacitors to achieve a balance between high energy density and high power density [88], addressing some of the limitations of traditional capacitors. The fundamental principle behind asymmetric capacitors lies in the distinct operating mechanisms of the two electrodes. The EDLC electrode stores energy through electrostatic ion adsorption at the electrode–electrolyte interface, offering rapid charge–discharge cycles and excellent power performance. In contrast, the pseudocapacitive or battery-type electrode stores energy through faradaic redox reactions, contributing to higher energy density. By pairing these two different electrodes, asymmetric capacitors can operate over a wider voltage range [89] than symmetrical capacitors, leading to increased energy storage capacity.

One of the main advantages of asymmetric capacitors is their ability to achieve higher energy density compared to traditional EDLCs, while still maintaining relatively high power density. This makes them suitable for a variety of applications, including electric vehicles, portable electronics, and renewable energy systems, where both quick energy delivery and substantial energy storage are required. For example, an asymmetric capacitor with an activated carbon negative electrode and a nickel oxide positive electrode can exhibit significantly higher energy density than a conventional EDLC, without sacrificing the fast charge–discharge capability. However, the design and optimization of asymmetric capacitors come with several challenges [90]. One critical issue is the mismatch in the charge-storage capacity of the two electrodes, which can lead to imbalanced performance and reduced efficiency. This issue is often addressed by carefully tuning the mass ratio of the electrodes or by using electrode materials with complementary properties. Additionally, the cycling stability of the pseudocapacitive electrode is crucial, as repeated faradaic reactions can lead to degradation over time. Researchers are actively exploring advanced materials and electrode architectures to enhance the durability and performance of asymmetric capacitors.

6.6.2 *Composite Electrodes*

Composite electrodes [91] are a key innovation in the development of advanced electrochemical capacitors, combining multiple materials to enhance the overall performance of the device. By integrating different types of materials—such as carbon-based materials, metal oxides [92], and conducting polymers—composite electrodes can achieve superior capacitance, conductivity, and mechanical stability compared to single-material electrodes. The synergy between the components in a composite electrode allows for the optimization of various properties, leading to improved energy and power density, as well as enhanced cycling stability. One of the most common strategies in composite electrode design is the combination of carbon-based materials, like activated carbon, carbon nanotubes (CNTs), or graphene, with pseudocapacitive materials such as metal oxides or conducting polymers [92]. Carbon materials provide a high surface area, excellent electrical conductivity, and structural support, while the pseudocapacitive materials contribute to the overall capacitance through faradaic reactions. For example, a composite electrode made of graphene and manganese oxide (MnO_2) can leverage the high conductivity [93] and surface area of graphene with the redox activity of MnO_2, resulting in a material that offers both high capacitance and good rate capability. Another approach to composite electrodes involves the use of hybrid structures, where different materials are combined at the nanoscale to create a more interconnected and efficient network. For instance, incorporating metal oxides like RuO_2 or NiO into a carbon nanotube matrix can create a composite with enhanced electron and ion transport pathways, leading to faster charge–discharge cycles and higher power density. These hybrid composites often exhibit superior performance due to the uniform distribution of active materials and the reduced resistance at the electrode interfaces.

Composite electrodes also benefit from the mechanical flexibility and stability provided by the combination of materials. Conducting polymers, for instance, can be combined with carbon materials to create flexible and durable electrodes suitable for use in wearable or portable energy storage devices. The polymer component adds mechanical flexibility [94] and a large number of active sites for charge storage, while the carbon material ensures good electrical conductivity and structural integrity. Despite their advantages, the fabrication of composite electrodes poses challenges, particularly in achieving uniform dispersion of the components and ensuring strong interfacial bonding. The manufacturing process must be carefully controlled to prevent agglomeration of the active materials and to maintain the integrity of the composite structure. Additionally, the scalability and cost-effectiveness of producing composite electrodes for commercial applications remain areas of active research. Composite electrodes represent a versatile and powerful approach to enhancing the performance of electrochemical capacitors. By combining the strengths of different materials [95], these electrodes can achieve higher capacitance, better conductivity, and improved stability, making them a critical component in the development of high-performance energy storage systems. As research progresses, the design and optimization of composite electrodes will continue to play a pivotal role in the advancement of next-generation electrochemical capacitors.

6.7 Characterization Techniques for EDLCs and Pseudocapacitors

This Table 6.1 summarizes the essential characterization techniques used to evaluate the performance and properties of EDLCs and pseudocapacitors. Each technique provides unique insights into different aspects of these devices, from electrochemical behaviour to structural features, enabling the optimization of materials and designs for improved energy storage.

Table 6.1 Detailed table outlining the characterization techniques for EDLCs and pseudocapacitors

Characterization technique	Description	Purpose	Key parameters	Typical applications
Cyclic voltammetry (CV)	CV is an electrochemical technique where the potential of a working electrode is cycled between two set values while measuring the resulting current	To analyze the electrochemical properties and reversibility of electrode materials, and to identify the presence of redox reactions in pseudocapacitors	• Current versus potential curves (voltammograms) • Peak current and peak separation • Capacitance calculation from the CV curve	• Identifying pseudocapacitive behavior in materials • Determining the stability and reversibility of redox processes • Estimating the capacitance and energy density
Electrochemical impedance spectroscopy (EIS)	EIS measures the impedance of an electrochemical system over a range of frequencies by applying a small AC voltage and analyzing the resulting current response	To study the charge transfer processes, ion diffusion, and resistance within the electrode material and at the electrode–electrolyte interface	• Nyquist plots (real vs. imaginary impedance) • Equivalent circuit modelling • Resistance (Rs) and charge transfer resistance (Rct) • Diffusion characteristics (Warburg impedance)	• Characterizing the internal resistance of EDLCs • Investigating the kinetics of charge storage • Modeling and simulating electrode behavior
Galvanostatic charge–discharge (GCD)	GCD involves applying a constant current to charge and discharge the capacitor, while monitoring the potential change over time	To measure the capacitance, energy density, power density, and cycling stability of electrochemical capacitors	• Charge–discharge curves (potential vs. time) • Capacitance calculation from discharge time • Coulombic efficiency • Energy and power density	• Determining the practical capacitance of EDLCs and pseudocapacitors • Assessing the efficiency and durability of energy storage • Evaluating the performance at different current densities
Scanning electron microscopy (SEM) and transmission electron microscopy (TEM)	SEM and TEM are imaging techniques that use electron beams to provide detailed images of the electrode material's surface morphology and internal structure, respectively	To visualize the morphology, particle size, porosity, and microstructure of electrode materials, and to identify any changes after cycling	• SEM images: surface topography and particle size • TEM images: internal structure and crystallinity • Elemental analysis (with EDS, if available)	• Examining the structural integrity of electrode materials • Analyzing the porosity and surface area of carbon-based materials • Investigating the dispersion and interaction of components in composite electrodes

6.8 Applications of EDLCs and Pseudocapacitors

This Table 6.2 highlights the diverse applications of EDLCs and pseudocapacitors across different sectors, illustrating their versatility and the specific benefits and challenges associated with each application area.

6.9 Conclusion and Future Directions

The field of electrochemical capacitors, encompassing both Electric Double-Layer Capacitors (EDLCs) and pseudocapacitors, has made significant strides in recent years. These advanced energy storage devices are distinguished by their ability to deliver rapid charge and discharge cycles, high power density, and enhanced cycle life compared to traditional batteries. EDLCs, with their high surface area carbon-based electrodes, excel in applications requiring high power and rapid energy delivery, while pseudocapacitors, with their redox-active materials, offer superior energy density through faradaic charge storage mechanisms. Together, they address a wide range of applications, from consumer electronics and transportation to renewable energy systems and grid storage. As the demand for high-performance, reliable energy storage solutions continues to grow, both EDLCs and pseudocapacitors are poised to play a critical role in shaping the future of energy storage technologies. The key to advancing these devices lies in overcoming current limitations and exploring new material and design innovations. For EDLCs, enhancing the specific capacitance and energy density while maintaining high power density and cycling stability remains a primary challenge. Research is focused on developing advanced carbon-based materials [50], optimizing electrode design, and improving electrolyte formulations to address these issues. In the realm of pseudocapacitors, the exploration of new pseudocapacitive materials and composite electrode systems is crucial. Metal oxides, conducting polymers, and hybrid composites have demonstrated promising results, yet challenges such as material cost, scalability, and long-term stability need to be addressed. The integration of pseudocapacitors with advanced carbon materials and the development of novel hybrid structures could lead to breakthroughs in performance and practical applications. The future directions in this field are likely to focus on several key areas:

(a) **Material Innovation**: Continued research into novel materials [96], including advanced carbon nanostructures, new metal oxides, and innovative conducting polymers, will drive improvements in capacitance, energy density, and stability. The development of scalable, cost-effective synthesis methods for these materials will be crucial for widespread adoption.

Table 6.2 Detailed table outlining the applications of EDLCs and pseudocapacitors in various fields

Application area	Description	Benefits of EDLCs and pseudocapacitors	Examples
Transportation	EDLCs and pseudocapacitors are used in electric and hybrid vehicles, including cars, buses, and trains, to provide quick bursts of power and improve overall energy efficiency	• High power density allows for rapid acceleration and braking energy recovery • Improved fuel efficiency and reduced emissions when integrated with conventional power sources	• Regenerative braking systems in electric vehicles • Power assist in hybrid buses • Electric train energy storage systems
Consumer electronics	In consumer electronics, EDLCs and pseudocapacitors are used to power devices requiring short bursts of energy or to extend battery life	• Fast charge–discharge cycles enhance device performance • Compact size and light weight suitable for portable devices • Long cycle life compared to traditional batteries	• Backup power for digital cameras and smartphones • Power supply for wearable electronics • Energy storage in portable electronics like tablets and smartwatches
Renewable energy systems	These capacitors are employed in renewable energy systems such as wind and solar power to smooth out power fluctuations and store transient energy	• Ability to quickly absorb and release energy helps to stabilize intermittent energy sources • Enhances the reliability and efficiency of renewable energy systems	• Energy buffering in solar panel systems • Smoothing power output in wind turbine systems • Energy storage for off-grid renewable setups
Grid energy storage	EDLCs and pseudocapacitors are used in grid energy storage systems to provide rapid response for frequency regulation, voltage support, and load leveling	• Quick response time and high power density help stabilize the grid • Ability to improve grid reliability and reduce the need for peaking power plants	• Frequency regulation in power grids • Voltage support for grid stabilization • Load leveling and peak shaving in urban power grids

(continued)

Table 6.2 (continued)

Application area	Description	Benefits of EDLCs and pseudocapacitors	Examples
Industrial and military applications	In industrial and military applications, these capacitors are used for backup power, energy storage in rugged environments, and high-power pulse applications	• High reliability and performance in extreme conditions • Fast charge–discharge capability for high-power pulses • Long cycle life and durability in harsh environments	• Backup power for critical industrial systems • Energy storage for military equipment and vehicles • Power conditioning and stabilization in industrial processes

(b) **Electrode Design and Engineering**: Advances in electrode design [97], such as the development of 3D architectures, nanostructured materials, and composites, will enhance the performance of both EDLCs and pseudocapacitors. Innovations in electrode fabrication techniques that enable better control over material properties and interfaces will be essential for achieving optimal device performance.

(c) **Hybrid and Asymmetric Systems**: The integration of EDLCs and pseudocapacitors into hybrid or asymmetric capacitor [87] systems offers a pathway to balance energy and power densities. Research into optimizing the combination of different electrode materials and configurations will continue to be a key focus area.

(d) **Advanced Characterization Techniques**: The development of new and improved characterization methods will facilitate a deeper understanding of the electrochemical processes occurring in these devices [98]. Techniques that provide insights into material properties, reaction mechanisms, and performance under real-world conditions will be instrumental in guiding material and design innovations.

(e) **Applications and Integration**: Expanding the application of EDLCs and pseudocapacitors to new and emerging fields, such as flexible electronics, wearable technology, and large-scale grid storage, will drive further research and development [99]. Addressing the integration challenges and optimizing performance for specific applications will be critical for commercialization and practical use.

In conclusion, while significant progress has been made in the development of EDLCs and pseudocapacitors, ongoing research and innovation are essential for advancing these technologies. By addressing current challenges and exploring new materials, designs, and applications, the field is poised to make substantial contributions to the future of energy storage and management. The continued evolution of these devices will play a pivotal role in meeting the growing demands for efficient, reliable, and sustainable energy solutions.

References

1. Raza, W., et al.: Recent advancements in supercapacitor technology. Elsevier. Accessed: 29 Aug 2024. [Online]. Available: https://www.sciencedirect.com/science/article/pii/S2211285518305755
2. Scherson, D.A., Palencsár, A.: Batteries and electrochemical capacitors. Electrochem. Soc. Interface **15**(1), 17–22 (2006). https://doi.org/10.1149/2.F05061IF/META
3. Zhang, Y., et al.: Progress of electrochemical capacitor electrode materials: a review. Elsevier. Accessed: 29 Aug 2024. [Online]. Available: https://www.sciencedirect.com/science/article/pii/S0360319909004984
4. Zhai, T., et al.: An electrochemical capacitor with applicable energy density of 7.4 Wh/kg at average power density of 3000 W/kg. Nano Lett. **15**(5), 3189–3194 (2015). https://doi.org/10.1021/ACS.NANOLETT.5B00321
5. Leng, C., et al.: 3D carbon frameworks for ultrafast charge/discharge rate supercapacitors with high energy-power density. Nano Micro Lett. **13**(1), 3 (2021). https://doi.org/10.1007/s40820-020-00535-w
6. Li, L., Hu, Z., An, N., Yang, Y., Li, Z.M., Wu, H.Y.: Facile synthesis of MnO_2/CNTs composite for supercapacitor electrodes with long cycle stability. J. Phys. Chem. C **118**(40), 22865–22872 (2014). https://doi.org/10.1021/jp505744p
7. Blooming, T., Carnovale, D.J.: Capacitor application issues. ieeexplore.ieee.org. Accessed: 29 Aug 2024. [Online]. Available: https://ieeexplore.ieee.org/abstract/document/4286298/
8. Aifantis, K.E., Hackney, S.A., Vasant Kumar, R.: High energy density lithium batteries (2010). Accessed: 29 Aug 2024. [Online]. Available: https://doi.org/10.1002/9783527630011
9. Fang, B., Binder, L.: A novel carbon electrode material for highly improved EDLC performance. J. Phys. Chem. B **110**(15), 7877–7882 (2006). https://doi.org/10.1021/JP060110D
10. Bhojane, P.: Recent advances and fundamentals of pseudocapacitors: materials, mechanism, and its understanding. Elsevier. Accessed: 29 Aug 2024. [Online]. Available: https://www.sciencedirect.com/science/article/pii/S2352152X21013293
11. Phogat, P., Sharma, S., Jha, R., Singh, S.: Supercapacitive studies of hybrid materials based on cadmium deuteroxide chloride (CdDOCl) with activated carbon. J. Mater. Sci. **59**(26), 11757–11780 (2024). https://doi.org/10.1007/S10853-024-09874-0
12. Yang, W., et al.: Graphene in supercapacitor applications. Elsevier. Accessed: 22 Aug 2024. [Online]. Available: https://www.sciencedirect.com/science/article/pii/S1359029415000734
13. Chen, Y., et al.: The ions storage mechanism of capacitive-faradic coupling effect for pseudo-intercalation electrode MnO_2. Elsevier. Accessed: 22 Aug 2024. [Online]. Available: https://www.sciencedirect.com/science/article/pii/S1383586623024371
14. Li, H.B., et al.: Amorphous nickel hydroxide nanospheres with ultrahigh capacitance and energy density as electrochemical pseudocapacitor materials. Nature. Commun. **4**(1), 1894 (2013). https://doi.org/10.1038/ncomms2932
15. Yi, C.-Q., Zou, J.-P., Yang, H.-Z., Xian, L.: Recent advances in pseudocapacitor electrode materials: transition metal oxides and nitrides. Elsevier. Accessed: 22 Aug 2024. [Online]. Available: https://www.sciencedirect.com/science/article/pii/S1003632618648435
16. Jadon, A., et al.: Rethinking pseudocapacitance: a way to harness charge storage of crystalline RuO_2. ACS Appl. Energy Mater. **3**(5), 4144–4148 (2020). https://doi.org/10.1021/ACSAEM.0C00476
17. Liu, T., et al.: Polyaniline and polypyrrole pseudocapacitor electrodes with excellent cycling stability. Nano Lett. **14**(5), 2522–2527 (2014). https://doi.org/10.1021/NL500255V
18. Dong, L., Yang, W., Yang, W., Li, Y., Wu, W., Wang, G.: Multivalent metal ion hybrid capacitors: a review with a focus on zinc-ion hybrid capacitors. Pubs.rsc.org. Accessed: 29 Aug 2024. [Online]. Available: https://pubs.rsc.org/en/content/articlehtml/2018/k3/c9ta02678a
19. Wu, Y., et al.: Recent advances in potassium-ion hybrid capacitors: electrode materials, storage mechanisms and performance evaluation. Elsevier. Accessed: 29 Aug 2024. [Online]. Available: https://www.sciencedirect.com/science/article/pii/S2405829721002579

20. Connor, P., Schuch, J., Kaiser, B., Jaegermann, W.: The determination of electrochemical active surface area and specific capacity revisited for the system MnOx as an oxygen evolution catalyst. Z. Phys. Chem. **234**(5), 979–994 (2020). https://doi.org/10.1515/ZPCH-2019-1514/HTML
21. Pognon, G., Brousse, T., Demarconnay, L., Bélanger, D.: Performance and stability of electrochemical capacitor based on anthraquinone modified activated carbon. Elsevier. Accessed: 29 Aug 2024. [Online]. Available: https://www.sciencedirect.com/science/article/pii/S0378775310017039
22. A. Jena, Nookala, M., Shivashankar, S.: Carbonaceous nickel oxide nano-composites: as electrode materials in electrochemical capacitor applications. Elsevier. Accessed: 29 Aug 2024. [Online]. Available: https://www.sciencedirect.com/science/article/pii/S0378775313004291
23. Hsieh, C.-T., Hsu, S., Lin, J.-Y.: Electrochemical capacitors based on graphene oxide sheets using different aqueous electrolytes.. J. Phys. Chem. C **115**(25), 12367–12374 (2011). https://doi.org/10.1021/jp2032687
24. Jiang, D.E., Jin, Z., Henderson, D., Wu, J.: Solvent effect on the pore-size dependence of an organic electrolyte supercapacitor. J. Phys. Chem. Lett. **3**(13), 1727–1731 (2012). https://doi.org/10.1021/JZ3004624
25. Salanne, M.: Ionic liquids for supercapacitor applications. Ionic Liq. II 29–53 (2017). https://doi.org/10.1007/978-3-319-89794-3_2
26. Salar-García, M.J., Ortiz-Martínez, V.M., Hernández-Fernández, F.J., De los Ríos, A.P.: Ionic liquid technology to recover volatile organic compounds (VOCs). Elsevier. Accessed: 29 Aug 2024. [Online]. Available: https://www.sciencedirect.com/science/article/pii/S0304389416308627
27. Shamsuri, A., Abdullah, D.K.: Ionic liquids: preparations and limitations. MAKARA Sci. Ser. **14**(2), 101–106. scholarhub.ui.ac.id. Accessed: 29 Aug 2024. [Online]. Available: https://scholarhub.ui.ac.id/science/vol14/iss2/19/
28. Jinrui, N., Zhifu, W., Qinglian, R.: Simulation and analysis of performance of a pure electric vehicle with a super-capacitor. ieeexplore.ieee.org. Accessed: 29 Aug 2024. [Online]. Available: https://ieeexplore.ieee.org/abstract/document/4211259/
29. Paladini, V., Donateo, T., De Risi, A., Laforgia, D.: Super-capacitors fuel-cell hybrid electric vehicle optimization and control strategy development. Elsevier. Accessed: 29 Aug 2024. [Online]. Available: https://www.sciencedirect.com/science/article/pii/S0196890407002178
30. Hu, X., Tseng, K., Srinivasan, M.: Optimization of battery energy storage system with supercapacitor for renewable energy applications. ieeexplore.ieee.org. Accessed: 29 Aug 2024. [Online]. Available: https://ieeexplore.ieee.org/abstract/document/5944515/
31. Nordin, N.A., et al.: Integrating photovoltaic (PV) solar cells and supercapacitors for sustainable energy devices: a review (2021). Mdpi.com. https://doi.org/10.3390/en14217211
32. Pan, T.-L., Wan, H.-S., Ji, Z.-C.: Stand-alone wind power system with battery/supercapacitor hybrid energy storage. Int. J. Sustain. Eng. **7**(2), 103–110 (2013). https://doi.org/10.1080/19397038.2013.779327
33. Kim, S.M., Sul, S.-K.: Control of rubber tyred gantry crane with energy storage based on supercapacitor bank. ieeexplore.ieee.org. Accessed: 29 Aug 2024. [Online]. Available: https://ieeexplore.ieee.org/abstract/document/1687992/
34. Végvári, Z.: Supercapacitors and their military applicability. real.mtak.hu. https://doi.org/10.35926/HDR.2019.1-2.3
35. Mahala, S., Khosravinia, K., Kiani, A.: Unwanted degradation in pseudocapacitors: challenges and opportunities. Elsevier. Accessed: 22 Aug 2024. [Online]. Available: https://www.sciencedirect.com/science/article/pii/S2352152X23009556
36. Tomy, M., Athira, A.R., Vimuna, V.M., Xavier, T.S.: Emergence of novel 2D materials for high-performance supercapacitor electrode applications: a brief review. Energy Fuels **35**(24), 19881–19900 (2021). https://doi.org/10.1021/ACS.ENERGYFUELS.1C02743
37. Salunkhe, R.R., Kaneti, Y.V., Yamauchi, Y.: Metal-organic framework-derived nanoporous metal oxides toward supercapacitor applications: progress and prospects. ACS Nano **11**(6), 5293–5308 (2017). https://doi.org/10.1021/ACSNANO.7B02796

38. Bocklisch, T.: Hybrid energy storage approach for renewable energy applications. Elsevier. Accessed: 29 Aug 2024. [Online]. Available: https://www.sciencedirect.com/science/article/pii/S2352152X16300044
39. Parida, K., Bhavanasi, V., Kumar, V., Wang, J.: Fast charging self-powered electric double layer capacitor. Elsevier. Accessed: 29 Aug 2024. [Online]. Available: https://www.sciencedirect.com/science/article/pii/S0378775316316342
40. Du, X., Guo, P., Song, H.-H., Chen, X.: Graphene nanosheets as electrode material for electric double-layer capacitors. Elsevier. Accessed: Aug. 29, 2024. [Online]. Available: https://www.sciencedirect.com/science/article/pii/S0013468610004494
41. Cheng, Q., Chen, W.: Analysis of the electrolyte characteristics on the performance of electric double layer capacitor. Elsevier. Accessed: 29 Aug 2024. [Online]. Available: https://www.sciencedirect.com/science/article/pii/S0378775320307084
42. Funaki, T., Hikihara, T.: Characterization and modeling of the voltage dependency of capacitance and impedance frequency characteristics of packed EDLCs. ieeexplore.ieee.org. Accessed: 29 Aug 2024. [Online]. Available: https://ieeexplore.ieee.org/abstract/document/4484947/?casa_token=pjCAZWutPUEAAAAA:28Y-RH6TV1KnviT9XaWeSzHkpTf60AFID_BWJENgpzf6udOWknvQfsh_D9qpaBmQmbgRgeQrhH8a
43. Kado, Y., Soneda, Y., Hatori, H., Kodama, M.: Advanced carbon electrode for electrochemical capacitors. J. Solid State Electrochem. **23**(4), 1061–1081 (2019). https://doi.org/10.1007/S10008-019-04211-X
44. Gamby, J., Taberna, P.L., Simon, P., Fauvarque J.F.: Studies and characterisations of various activated carbons used for carbon/carbon supercapacitors. Elsevier. Accessed: 29 Aug 2024. [Online]. Available: https://www.sciencedirect.com/science/article/pii/S0378775301007078
45. Ismadji, S., Sudaryanto, Y., Deitiana, T., Setiawan, T.: Activated carbon from char obtained from vacuum pyrolysis of teak sawdust: pore structure development and characterization. Elsevier. Accessed: 29 Aug 2024. [Online]. Available: https://www.sciencedirect.com/science/article/pii/S0960852404004055
46. Yeh, C.-L., Hsi, H.-C., Li, K.-C., Hou, C.-H.: Improved performance in capacitive deionization of activated carbon electrodes with a tunable mesopore and micropore ratio. Elsevier. Accessed: 29 Aug 2024. [Online]. Available: https://www.sciencedirect.com/science/article/pii/S0011916415002027
47. (1) (PDF) A review on activated carbon: process, application and prospects. Accessed: 29 Aug 2024. [Online]. Available: https://www.researchgate.net/publication/318827940_A_review_on_activated_carbon_process_application_and_prospects/figures?lo=1
48. De Volder, M., Tawfick, S., Baughman, R.H., Hart, A.J.: Carbon nanotubes: present and future commercial applications. Science **339**(6119), 535–539 (2013). https://doi.org/10.1126/SCIENCE.1222453
49. Ebbesen, T.W., Lezec, H., Hiura, H., Bennett, J.W.: Electrical conductivity of individual carbon nanotubes. nature.com. Accessed: 29 Aug 2024. [Online]. Available: https://www.nature.com/articles/382054a0
50. Obreja, V.: On the performance of supercapacitors with electrodes based on carbon nanotubes and carbon activated material—a review. Elsevier. Accessed: 29 Aug 2024. [Online]. Available: https://www.sciencedirect.com/science/article/pii/S138694770700447X
51. Laszczyk, K.U., Futaba, D.N., Kobashi, K., Hata, K.: The limitation of electrode shape on the operational speed of a carbon nanotube based micro-supercapacitor. pubs.rsc.org. Accessed: 29 Aug 2024. [Online]. Available: https://pubs.rsc.org/en/content/articlehtml/2011/se/c7se00101k
52. Monea, B.F., Ionete, E.I., Spiridon, S.I., Ion-Ebrasu, D., Petre, E.: Carbon nanotubes and carbon nanotube structures used for temperature measurement. Sensors **19**, 2464 (2019). https://doi.org/10.3390/s19112464
53. Ullah, K., Kim, I.J., Yang, S.H., Oh, W.C.: Preparation of highly expanded graphene with large surface area and its additional conductive effect for EDLC performance. J. Mater. Sci. Mater. Electron. **26**(9), 6945–6953 (2015). https://doi.org/10.1007/S10854-015-3313-8

54. Geim, A.K.: Graphene: status and prospects. Science **324**, 1530–1534 (2009) https://doi.org/10.1126/SCIENCE.1158877
55. Chen, J., Li, C., Shi, G.: Graphene materials for electrochemical capacitors. ACS Publication. Accessed: 29 Aug 2024. [Online]. Available: https://pubs.acs.org/doi/abs/https://doi.org/10.1021/jz400160k
56. Miller, J.R., Outlaw, R.A., Holloway, B.C.: Graphene electric double layer capacitor with ultra-high-power performance. Elsevier. Accessed: 29 Aug 2024. [Online]. Available: https://www.sciencedirect.com/science/article/pii/S0013468611008668
57. Horn, M., Gupta, B., MacLeod, J.M., Liu, J., Motta, N.: Graphene-based supercapacitor electrodes: addressing challenges in mechanisms and materials. Elsevier. Accessed: 29 Aug 2024. [Online]. Available: https://www.sciencedirect.com/science/article/pii/S2452223618301275
58. Fathy, M., Gomaa, A., Taher, F.A., El-Fass, M.M., Kashyout, A.E.H.B.: Optimizing the preparation parameters of GO and rGO for large-scale production. J. Mater. Sci. **51**(12), 5664–5675 (2016). https://doi.org/10.1007/S10853-016-9869-8
59. Zhang, Y., Zhang, L., Zhou, C.: Review of chemical vapor deposition of graphene and related applications. Acc. Chem. Res. **46**(10), 2329–2339 (2012). https://doi.org/10.1021/ar300203n
60. (1) (PDF) An properties of graphene. Accessed: 29 Aug 2024. [Online]. Available: https://www.researchgate.net/publication/352357442_An_Properties_of_Graphene/figures?lo=1
61. Wang, F., et al.: Latest advances in supercapacitors: from new electrode materials to novel device designs. pubs.rsc.org. Accessed: 29 Aug 2024. [Online]. Available: https://pubs.rsc.org/en/content/articlehtml/2010/wo/c7cs00205j
62. Dai, S., et al.: A high-performance supercapacitor electrode based on N-doped porous graphene. Elsevier. Accessed: 29 Aug 2024. [Online]. Available: https://www.sciencedirect.com/science/article/pii/S0378775318302970
63. Sim, Y.B., Park, B.K., Kim, K.J.: Reasonable design of thick electrodes in lithium-ion batteries. Front. Batteries Electrochem. **2** (2023). https://doi.org/10.3389/FBAEL.2023.1272439/FULL
64. Stoller, M., Ruoff, R.S.: Best practice methods for determining an electrode material's performance for ultracapacitors. pubs.rsc.org. Accessed: 31 Aug 2024. [Online]. Available: https://pubs.rsc.org/en/content/articlehtml/2010/ee/c0ee00074d
65. Panagiotopoulos, A.Z., Yue, S.: Dynamics of aqueous electrolyte solutions: challenges for simulations. J. Phys. Chem. B **127**(2), 430–437 (2023). https://doi.org/10.1021/ACS.JPCB.2C07477
66. Jarvis, N.L., Scheiman, M.A.: Surface potentials of aqueous electrolyte solutions. J. Phys. Chem. **72**(1), 74–78 (1968). https://doi.org/10.1021/J100847A014
67. Mao, J., Wang, C., Lyu, Y., Zhang, R., Wang, Y., Liu, S., Wang, Z., Zhang, S., Guo, Z.: Organic electrolyte design for practical potassium-ion batteries. pubs.rsc.org. Accessed: 31 Aug 2024. [Online]. Available: https://pubs.rsc.org/en/content/articlehtml/2022/ta/d2ta02223k
68. Qiu, X., et al.: A high-voltage Zn–organic battery using a nonflammable organic electrolyte. Angew. Chem. **133**(38), 21193–21200 (2021). https://doi.org/10.1002/ANGE.202108624
69. Poizot, P., Gaubicher, J., Renault, S., Dubois, L., Liang, Y., Yao, Y.: Opportunities and challenges for organic electrodes in electrochemical energy storage. Chem. Rev. **120**(14), 6490–6557 (2020). https://doi.org/10.1021/ACS.CHEMREV.9B00482
70. Holbrey, J.D., Seddon, K.R.: Ionic liquids. Clean Technol. Environ. Policy **1**(4), 223–236 (1999). https://doi.org/10.1007/S100980050036
71. Hayyan, M., Mjalli, F.S., Hashim, M.A., AlNashef, I.M., Mei, T.X.: Investigating the electrochemical windows of ionic liquids. J. Ind. Eng. Chem. **19**(1), 106–112 (2013). https://doi.org/10.1016/J.JIEC.2012.07.011
72. Diaw, M., Chagnes, A., Carre, B., Willmann, P., Lemordant, D.: Mixed ionic liquid as electrolyte for lithium batteries. Elsevier. Accessed: 31 Aug 2024. [Online]. Available: https://www.sciencedirect.com/science/article/pii/S0378775305004726
73. Prakash, A., Bahadur, D.: The role of ionic electrolytes on capacitive performance of ZnO-reduced graphene oxide nanohybrids with thermally tunable morphologies. ACS Appl. Mater. Interfaces **6**(3), 1394–1405 (2014). https://doi.org/10.1021/AM405031Y

74. Karuppasamy, K., et al.: Ionic liquid-based electrolytes for energy storage devices: a brief review on their limits and applications. Polymers **12**(4), 918 (2020). https://doi.org/10.3390/POLYM12040918
75. Chen, R., Yu, M., Sahu, R.P., Puri, I.K., Zhitomirsky, I.: The development of pseudocapacitor electrodes and devices with high active mass loading. Adv. Energy Mater. **10**(20) (2020). https://doi.org/10.1002/AENM.201903848
76. Girard, H.-L., Wang, H., d'Entremont, A., Pilon, L.: Enhancing faradaic charge storage contribution in hybrid pseudocapacitors. Elsevier. Accessed: 31 Aug 2024. [Online]. Available: https://www.sciencedirect.com/science/article/pii/S0013468615304825
77. Zhang, J., Cui, Y., Shan, G.: Metal oxide nanomaterials for pseudocapacitors (2019). Accessed: 31 Aug 2024. [Online]. Available: http://arxiv.org/abs/1905.01766
78. Han, S., Park, S., Yi, S.-H., Im, W.B.: Effect of potential and current on electrodeposited MnO_2 as a pseudocapacitor electrode: surface morphology/chemistry and stability. Elsevier. Accessed: 31 Aug 2024. [Online]. Available: https://www.sciencedirect.com/science/article/pii/S0925838820312019
79. Sun, X., et al.: Pseudocapacitance of amorphous TiO_2 thin films anchored to graphene and carbon nanotubes using atomic layer deposition. J. Phys. Chem. C **117**(44), 22497–22508 (2013). https://doi.org/10.1021/JP4066955
80. Bryan, A.M., Santino, L.M., Lu, Y., Acharya, S., D'Arcy, J.M.: Conducting polymers for pseudocapacitive energy storage. Chem. Mater. **28**(17), 5989–5998 (2016). https://doi.org/10.1021/ACS.CHEMMATER.6B01762
81. De Albuquerque, J.E., Mattoso, L.H.C., Balogh, D.T., Faria, R.M., Masters, J.G., MacDiarmid, A.G.: A simple method to estimate the oxidation state of polyanilines. Elsevier. Accessed: 31 Aug 2024. [Online]. Available: https://www.sciencedirect.com/science/article/pii/S0379677999002994
82. Shoaie, N., et al.: Electrochemical sensors and biosensors based on the use of polyaniline and its nanocomposites: a review on recent advances. Microchimi. Acta **186**(7) (2019). https://doi.org/10.1007/S00604-019-3588-1
83. Zhou, J., et al.: A conductive and highly deformable all-pseudocapacitive composite paper as supercapacitor electrode with improved areal and volumetric capacitance. Small **14**(51) (2018). https://doi.org/10.1002/SMLL.201803786
84. Wang, B., et al.: Graphene-based composites for electrochemical energy storage. Elsevier. Accessed: 31 Aug 2024. [Online]. Available: https://www.sciencedirect.com/science/article/pii/S2405829719309171
85. Yadav, S., Sharma, A.: Importance and challenges of hydrothermal technique for synthesis of transition metal oxides and composites as supercapacitor electrode materials. Elsevier. Accessed: 31 Aug 2024. [Online]. Available: https://www.sciencedirect.com/science/article/pii/S2352152X21009877
86. Kausar, A., Ahmad, I., Zhao, T., Aldaghri, O., Ibnaouf, K.H., Eisa, M.H.: Graphene nanocomposites as innovative materials for energy storage and conversion—design and headways. Int. J. Mol. Sci. **24**(14), 11593 (2023). https://doi.org/10.3390/IJMS241411593
87. Khomenko, V., Raymundo-Piñero, E., Béguin, F.: A new type of high energy asymmetric capacitor with nanoporous carbon electrodes in aqueous electrolyte. Elsevier. Accessed: 31 Aug 2024. [Online]. Available:https://www.sciencedirect.com/science/article/pii/S0378775310000182
88. Jung, H.-G., Venugopal, N., Scrosati, B., Sun, Y.-K.: A high energy and power density hybrid supercapacitor based on an advanced carbon-coated $Li_4Ti_5O_{12}$ electrode. Elsevier. Accessed: 31 Aug 2024. [Online]. Available: https://www.sciencedirect.com/science/article/pii/S0378775312013274
89. Suárez-Guevara, J., Ruiz, V., Gomez-Romero, P.: Hybrid energy storage: high voltage aqueous supercapacitors based on activated carbon–phosphotungstate hybrid materials. pubs.rsc.org. Accessed: 31 Aug 2024. [Online]. Available: https://pubs.rsc.org/en/content/articlehtml/2014/ta/c3ta14455k

90. Liu, Q., et al.: Recent progress and challenges of carbon materials for Zn-ion hybrid supercapacitors. Carbon Energy **2**(4), 521–539 (2020). https://doi.org/10.1002/cey2.69
91. Tallman, D.E., Petersen, S.L.: Composite electrodes for electroanalysis: principles and applications. Electroanalysis **2**(7), 499–510 (1990). https://doi.org/10.1002/ELAN.1140020702
92. Wu, Z., et al.: Graphene/Metal oxide composite electrode materials for energy storage. Elsevier. Accessed: 31 Aug 2024. [Online]. Available: https://www.sciencedirect.com/science/article/pii/S2211285511000176
93. Ko, J.M., Kim, M.: Electrochemical properties of MnO_2/activated carbon nanotube composite as an electrode material for supercapacitor. Elsevier. Accessed: 31 Aug 2024. [Online]. Available: https://www.sciencedirect.com/science/article/pii/S0254058408008614
94. Shown, I., Ganguly, A., Chen, L.-C., Chen, K.-H.: Conducting polymer-based flexible supercapacitor. Energy Sci. Eng. **3**(1), 2–26 (2015). https://doi.org/10.1002/ese3.50
95. Cui, M., Meng, X.: Overview of transition metal-based composite materials for supercapacitor electrodes (2020). pubs.rsc.org. https://doi.org/10.1039/d0na00573h
96. Li, Z., Xu, K., Pan, Y.: Recent development of supercapacitor electrode based on carbon materials. Nanotechnol. Rev. **8**(1), 35–49 (2019). https://doi.org/10.1515/NTREV-2019-0004/HTML
97. Jiang, J., Li, Y., Liu, J., Huang, X., Yuan, C., Lou, X.W.: Recent advances in metal oxide-based electrode architecture design for electrochemical energy storage. Adv. Mater. **24**(38), 5166–5180 (2012). https://doi.org/10.1002/ADMA.201202146
98. Chen, J., Lee, P.S., Chen, J., Lee, S.: Electrochemical supercapacitors: from mechanism understanding to multifunctional applications. Adv. Energy Mater. **11**(6) (2020). https://doi.org/10.1002/aenm.202003311
99. Wang, J., Li, F., Zhu, F., Schmidt, O.G.: Recent progress in micro-supercapacitor design, integration, and functionalization. Small Methods **3**(8) (2019). https://doi.org/10.1002/SMTD.201800367

Chapter 7
Microstructural Influence on Electrochemical Devices

7.1 Introduction

The performance and longevity of electrochemical devices, such as batteries, fuel cells, supercapacitors, and electrochemical sensors, are intricately linked to the microstructural properties of the materials from which they are constructed. Microstructure, encompassing the arrangement, size, shape, orientation, and distribution of grains, phases, pores, and defects within a material, profoundly influences a wide array of electrochemical behaviors [1]. These include ionic and electronic conductivity, electrochemical reaction kinetics, mechanical properties, and the overall efficiency and durability of devices. As the world increasingly relies on advanced energy storage and conversion technologies to power everything from portable electronics to electric vehicles and renewable energy systems, the need to optimize and engineer microstructures for superior electrochemical performance has never been more critical. Understanding the interplay between microstructure and electrochemical performance begins with recognizing that materials are not homogenous entities; they are composed of numerous microstructural features that can either enhance or hinder performance, depending on how they are configured. For instance, in lithium–ion batteries, the performance of electrode materials is heavily dependent on the microstructural characteristics of the active materials [2]. Nanoscale structuring of electrode materials can significantly increase the surface area available for electrochemical reactions, reduce the diffusion paths for ions, and improve the mechanical stability of the electrodes during charge–discharge cycles. This leads to higher capacities, faster charging rates, and prolonged cycle life, which are key performance metrics in battery technology [3].

Similarly, the microstructure of solid electrolytes in fuel cells plays a critical role in determining ionic conductivity, which is essential for efficient energy conversion. A well-designed microstructure can provide continuous pathways for ion transport while minimizing grain boundary resistance, leading to enhanced performance and efficiency [4]. In supercapacitors, the microstructural arrangement of porous

P. Phogat et al., *Electrochemical Devices*, Engineering Materials,
https://doi.org/10.1007/978-981-96-0527-9_7

carbon materials directly impacts the device's capacitance, energy density, and power density. By optimizing the pore size distribution and connectivity, it is possible to achieve rapid charge–discharge cycles and high energy storage capacity, which are crucial for applications requiring quick energy bursts [5]. Microstructural engineering is not just about optimizing performance; it also involves addressing the challenges that arise from the dynamic nature of microstructures during device operation. Over time, as electrochemical devices undergo numerous charge–discharge cycles, their microstructures can evolve due to processes such as phase transformations, grain growth, crack formation, and void generation. These changes can lead to a deterioration in performance, often manifesting as reduced capacity, increased internal resistance, or mechanical failure. For instance, in lithium–ion batteries, the repeated lithiation and delithiation of electrode materials can cause significant volume changes, leading to particle cracking, loss of electrical contact, and ultimately, capacity fade [6]. Understanding these degradation mechanisms and developing strategies to mitigate them is crucial for extending the life of electrochemical devices.

One of the most significant advancements in recent years is the use of machine learning (ML) and other data-driven approaches to study and predict the relationships between microstructure and electrochemical performance. Machine learning algorithms can analyze vast amounts of data from experimental studies, simulations, and real-world applications to identify patterns and correlations that may not be apparent through traditional analysis methods. By integrating ML into microstructural engineering, researchers can accelerate the design of new materials with optimized microstructures, predict how these structures will evolve under different operating conditions, and develop strategies to enhance device performance and durability [7]. This integration represents a new frontier in electrochemical device research, offering the potential to revolutionize how materials are designed and how their performance is predicted and optimized. As the demand for high-performance, reliable, and sustainable electrochemical devices continues to grow, the ability to precisely control and engineer microstructures will be a key driver of innovation. The development of advanced manufacturing techniques, such as additive manufacturing, atomic layer deposition, and other precision fabrication methods, has opened new possibilities for creating materials with tailor-made microstructures. These techniques allow for the fabrication of materials with controlled porosity, grain size, phase distribution, and other microstructural features, enabling the design of electrochemical devices that are finely tuned to specific applications and performance requirements [8]. This chapter will explore the critical role that microstructure plays in the performance of electrochemical devices, from the fundamental principles that govern the relationship between microstructure and electrochemical behavior to the latest advancements in microstructural engineering. We will examine how microstructural features can be engineered to enhance performance, address the challenges posed by microstructural evolution during device operation, and look ahead to emerging trends and future directions in this rapidly evolving field. By providing a comprehensive understanding of the impact of microstructure on electrochemical devices, this chapter aims to equip researchers, engineers, and practitioners with the knowledge needed to push the boundaries of current technologies and develop the next generation of electrochemical devices that are more efficient, durable, and sustainable.

The microstructure of materials is not merely a passive backdrop against which electrochemical processes occur; it is an active and dynamic factor that must be carefully engineered and controlled to achieve optimal performance in electrochemical devices. As we continue to push the limits of what these devices can achieve, the insights gained from microstructural studies will be essential for guiding the design and development of materials that meet the demanding requirements of modern energy technologies. This chapter sets the stage for understanding how microstructural properties can be harnessed to unlock new levels of performance and reliability in electrochemical devices, paving the way for a more sustainable and energy-efficient future.

7.2 Fundamentals of Microstructure in Electrochemical Systems

Microstructure is a fundamental concept in materials science, especially in the context of electrochemical systems where the microstructural properties of materials play a decisive role in determining their electrochemical behavior. The term "microstructure" refers to the arrangement and organization of different phases, grains, and defects within a material, typically observable at the microscopic or sub-microscopic scale. It represents the internal structure of materials at the microscopic scale, typically ranging from nanometers to micrometers [9].

7.2.1 Defining Microstructure: Grain Boundaries, Phases, and Defects

The microstructure of a material is composed of several key elements, including grain boundaries, phases, and defects as shown in Fig. 7.1. Grain boundaries are the interfaces between individual crystals, or grains, in a polycrystalline material. These boundaries can act as barriers or pathways for the movement of ions and electrons, depending on their structure and the presence of impurities or defects. The orientation, size, and distribution of grains are critical factors in determining the mechanical and electrochemical properties of the material. For instance, smaller grain sizes generally lead to higher grain boundary density, which can enhance certain properties, such as strength and ionic conductivity, but may also introduce challenges like increased grain boundary resistance. Phases refer to distinct regions within a material that have different physical or chemical properties. In many electrochemical systems, materials are composed of multiple phases, each contributing uniquely to the overall performance of the device [10]. For example, in lithium–ion batteries, the electrode materials often consist of mixed phases where one phase might provide high capacity while another offers better structural stability [11]. The distribution, size, and connectivity of these phases are crucial for optimizing the balance between performance and stability. Defects, which include vacancies, interstitials, dislocations, and more complex defect structures, are imperfections in the crystal lattice that can significantly impact a material's properties. In the context of electrochemical systems, defects can either be beneficial or detrimental, depending on their nature and concentration. For example, vacancies and interstitial defects can enhance ionic conductivity by providing additional pathways for ion movement. However, excessive defects can lead to degradation, such as increased resistance or mechanical failure [10]. Understanding and controlling the type and concentration of defects is therefore essential for optimizing the microstructure of materials used in electrochemical devices.

7.2.2 Key Microstructural Features in Energy Conversion and Storage Materials

In energy conversion and storage devices, certain microstructural features are particularly important due to their direct impact on device performance. One of the most critical features is the porosity of the material. Porosity refers to the presence of pores or voids within the material, which can significantly influence the surface area available for electrochemical reactions and the diffusion of ions and electrons [12]. For instance, in supercapacitors and batteries, a high surface area provided by a porous structure can enhance the capacity and power density of the device [13]. However, the size, shape, and distribution of pores must be carefully controlled to ensure that they do not compromise the mechanical integrity or increase the internal resistance of the material. Another key microstructural feature is the connectivity of conductive pathways within the material. In composite or hybrid materials, where different phases may offer distinct advantages (e.g., high conductivity, structural stability), the connectivity between these phases is crucial. For example, in a composite electrode material, the conductive phase must form a continuous network to facilitate efficient electron transport, while the active phase provides sites for the electrochemical reactions. The microstructural arrangement of these phases determines the overall conductivity and efficiency of the device [12].

Crystallographic orientation is also a significant microstructural feature, particularly in materials where anisotropic properties are desired [14]. In some electrochemical systems, such as certain types of solid electrolytes or electrode materials, the orientation of crystals can influence the directional conductivity of ions or electrons. By controlling the crystallographic orientation during material synthesis, it is possible to enhance the performance of the device in specific applications. Furthermore, the microstructural homogeneity or heterogeneity within a material can greatly affect its performance [15]. Homogeneous microstructures, where phases and grains

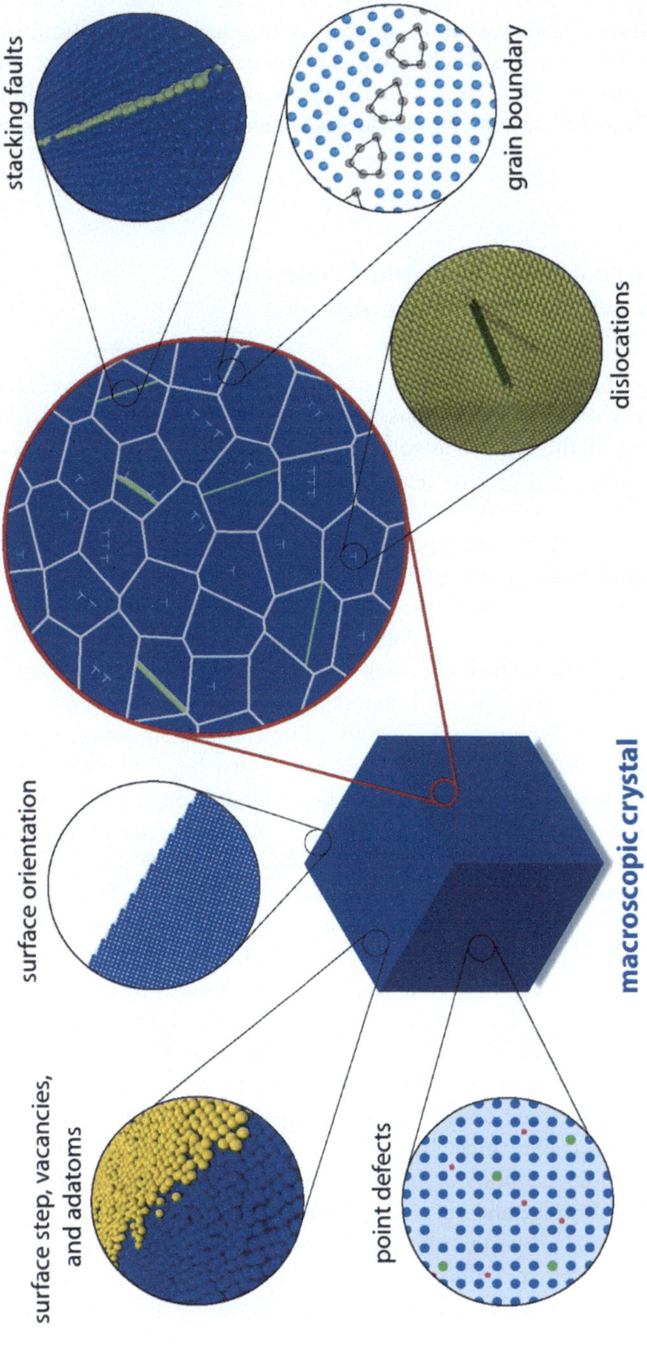

Fig. 7.1 Microstructural properties of materials reproduced from Ref. [10] Copyright © 2022 by the authors

are uniformly distributed, often provide consistent properties throughout the material, which is desirable in many applications. However, in some cases, a heterogeneous microstructure, with variations in grain size, phase composition, or defect distribution, can offer benefits, such as enhanced mechanical strength or improved tolerance to degradation [16]. The challenge lies in designing and controlling these microstructural features to achieve the desired balance between performance and durability.

7.2.3 Influence of Synthesis and Processing on Microstructural Properties

The synthesis and processing techniques used to create a material have a profound impact on its microstructure, and consequently, its electrochemical performance. Different synthesis methods, such as sol–gel processing, solid-state reactions, chemical vapor deposition, and electrochemical deposition, among others, can lead to varying microstructural outcomes. For instance, the choice of precursor materials, reaction conditions, and processing parameters can influence grain size, phase distribution, porosity, and defect concentration. One of the most critical aspects of material synthesis is the control of grain size. Techniques such as rapid quenching, high-energy ball milling, or the use of surfactants during synthesis can be employed to achieve nano-sized grains, which are often desirable for improving the performance of electrochemical devices. Nanostructured materials typically exhibit enhanced surface area, shorter diffusion paths for ions and electrons, and improved mechanical properties, all of which contribute to superior electrochemical performance [17]. Phase formation and distribution can also be controlled through careful selection of synthesis conditions. For example, in the synthesis of cathode materials for lithium–ion batteries, controlling the cooling rate and annealing temperature can determine the phase purity and the distribution of secondary phases [18]. Similarly, in solid oxide fuel cells, the sintering process plays a crucial role in achieving the desired phase composition and microstructure in the electrolyte and electrode materials [19].

Defect engineering is another area where synthesis and processing techniques have a significant impact. By adjusting parameters such as temperature, pressure, and atmosphere during synthesis, it is possible to introduce specific types of defects in controlled amounts. For instance, oxygen vacancies in perovskite oxides can be engineered to enhance ionic conductivity, which is beneficial for fuel cell applications [20]. However, care must be taken to avoid excessive defects that could lead to unwanted side reactions or mechanical instability. Processing techniques, such as annealing, doping, and surface modification, further allow for the refinement of microstructural properties after the initial synthesis. Annealing, for instance, can relieve internal stresses, reduce defects, and promote grain growth, leading to improved mechanical stability and performance. Doping, where foreign atoms are introduced into the crystal lattice, can modify the electrical, thermal, and electrochemical properties of the material by altering its microstructure. Surface modifications, such as coating or functionalization, can enhance the interfacial properties of the material, which is particularly important in applications like batteries and supercapacitors where surface reactions play a dominant role [21].

7.3 Microstructural Characterization Techniques

Understanding and optimizing microstructure require advanced characterization techniques that can provide detailed insights into the structural, chemical, and physical properties of materials at various length scales. This section covers the most relevant techniques for characterizing the microstructure of materials used in electrochemical devices.

7.3.1 Scanning Electron Microscopy (SEM)

Scanning Electron Microscopy (SEM) is a powerful imaging technique that provides high-resolution images of the surface morphology of samples by scanning them with a focused beam of electrons. Unlike conventional optical microscopy, which uses visible light, SEM utilizes electron beams to achieve much higher magnifications and resolutions, typically down to the nanometer scale as shown in Fig. 7.2. This capability makes SEM an indispensable tool for studying the microstructure of materials, particularly in the context of electrochemical systems. In SEM, a sample is first coated with a thin layer of conductive material, such as gold or carbon, to prevent charging effects that could distort the imaging process. The sample is then placed in a vacuum chamber where it is bombarded with a beam of electrons. The interaction between the electron beam and the sample generates secondary electrons, backscattered electrons, and characteristic X-rays. Secondary electrons, which are emitted from the surface of the sample, are collected to form an image that reveals the topography and surface features of the material. This allows for detailed visualization of surface morphology, including features such as surface roughness, porosity, and structural defects. SEM is particularly useful for examining the surface structure of electrodes in energy storage devices, such as batteries and supercapacitors. For example, SEM can reveal the morphology of electrode materials, including the distribution and connectivity of active particles, binder materials, and conductive additives. This information is crucial for understanding how the microstructure influences electrochemical performance, such as capacity, charge/discharge rates, and cycling stability. Additionally, SEM can be used to observe the effects of electrochemical cycling on electrode materials, such as the formation of cracks, delamination, and particle agglomeration, which are critical for assessing the longevity and reliability of the device. The resolution of SEM can be affected by several factors, including

Fig. 7.2 FE-SEM images of Au-NS modified CPE at different magnifications reproduced from Ref. [23] Copyright © 2023 by the authors

the accelerating voltage of the electron beam, the working distance, and the sample's conductive properties. Modern SEM systems equipped with field emission guns (FEGs) provide improved resolution and contrast, enabling the observation of finer details in complex microstructures. SEM can also be coupled with other techniques, such as Energy-Dispersive X-ray Spectroscopy (EDS), to obtain compositional information about the sample. This combination allows for a comprehensive analysis of both the morphology and composition of electrochemical materials, providing insights into how microstructural features correlate with electrochemical behavior [22].

7.3.2 Transmission Electron Microscopy (TEM)

Transmission Electron Microscopy (TEM) is a high-resolution imaging technique that allows for the observation of the internal structure of materials at the atomic scale. Unlike SEM, which primarily provides surface imaging, TEM involves transmitting a beam of electrons through an ultra-thin sample to generate images and diffraction patterns that reveal detailed information about the material's internal microstructure. TEM is particularly valuable for studying the fine structural details of materials used in electrochemical systems, such as nanoparticles, thin films, and composite materials. In TEM, the sample must be extremely thin, typically less than 100 nm thick, to allow the electron beam to pass through. This requirement necessitates precise sample preparation techniques, such as ultramicrotomy or focused ion beam (FIB) milling, to achieve the desired thickness while preserving the material's integrity. Once prepared, the sample is placed in a high-vacuum chamber and illuminated with a collimated beam of electrons. The transmitted electrons are then detected by a camera or a scintillator to produce images that reveal the material's internal structure [24].

TEM provides a wealth of information about the microstructure of materials, including crystal lattice arrangements, defect types, and phase boundaries. High-resolution TEM (HRTEM) can directly visualize atomic planes and resolve lattice fringes, allowing for the examination of crystallographic orientations, strain distributions, and phase compositions at the atomic level. This capability is particularly useful for characterizing nanostructured materials and understanding how their fine-scale features affect electrochemical performance. In the context of electrochemical systems, TEM can be used to investigate the microstructure of materials such as nanoparticles used in catalysts, battery electrodes, and supercapacitors. For example, TEM can reveal the size and shape of nanoparticles, the presence of defects or dislocations, and the distribution of different phases within a composite material. This information is crucial for optimizing the material's performance, as it helps in understanding how these microstructural features influence properties such as surface reactivity, ion diffusion, and electrical conductivity. TEM can also be used in conjunction with techniques such as Selected Area Electron Diffraction (SAED) and Energy-Dispersive X-ray Spectroscopy (EDS) to provide additional insights into the crystallographic and compositional characteristics of the material. SAED provides information about the crystal structure and phase composition, while EDS allows for elemental analysis, helping to correlate microstructural features with chemical composition. Overall, TEM is an essential tool for gaining a deep understanding of the internal microstructure of materials, offering insights that are critical for the design and optimization of electrochemical devices. By providing atomic-scale resolution and detailed information about crystallographic and compositional features, TEM enables researchers to explore the fundamental relationships between microstructure and electrochemical performance, leading to the development of more efficient and durable energy storage and conversion technologies [25].

7.3.3 X-Ray Diffraction (XRD)

X-ray Diffraction (XRD) is a fundamental technique used to analyze the crystal structure and microstructural properties of materials by measuring the scattering of X-rays as they interact with the periodic atomic planes in a crystalline sample. This method provides valuable insights into various aspects of microstructure, including phase composition, crystal structure, grain size, and strain, all of which are crucial for understanding and optimizing the performance of electrochemical systems. XRD operates on the principle of Bragg's Law, which relates the angles at which X-rays are diffracted by the crystal planes to the spacing between these planes. The diffraction pattern, or X-ray diffractogram, consists of a series of peaks, each corresponding to a specific set of lattice planes within the crystal. The position, intensity, and shape of these peaks provide information about the crystal structure, including lattice parameters, phase identification, and crystallite size. The peak positions are related to the interplanar spacing (d-spacing) of the crystal lattice, while the peak intensities provide insights into the arrangement and number of atoms within the unit cell. XRD is widely used in the study of electrochemical systems due to its ability to reveal detailed information about the crystalline structure and phase composition of materials used in energy storage and conversion devices [26]. Some key applications include:

Phase Identification and Quantification: XRD is essential for identifying the phases present in a material, which is crucial for understanding its electrochemical behavior. For example, in battery materials, XRD can be used to identify the different phases of the active electrode materials, such as lithium cobalt oxide or silicon. Accurate phase identification helps in optimizing the material's performance and stability by ensuring that the desired phases are present in the correct proportions.

Crystal Structure Analysis: XRD provides detailed information about the crystal structure, including lattice parameters and symmetry. This information is important for understanding how the crystal structure affects the material's electrochemical properties. For instance, in solid oxide fuel cells, the crystal structure of the electrolyte material influences its ionic conductivity, and XRD can help in tailoring the structure for improved performance.

Grain Size and Crystallite Size Determination: XRD can be used to estimate the grain size and crystallite size of polycrystalline materials using the Scherrer equation. Grain size affects the mechanical and electrochemical properties of materials, such as their conductivity and capacity. For example, in lithium–ion batteries, smaller grain sizes in electrode materials can enhance ion transport and increase capacity. XRD analysis helps in designing materials with optimal grain sizes for specific applications.

Strain and Stress Analysis: XRD can also be used to measure residual stress and strain in materials. Changes in the peak positions in the X-ray diffraction pattern can indicate the presence of internal stresses or strains, which can affect the performance and durability of electrochemical devices. For instance, during the cycling of lithium–ion batteries, mechanical stresses induced by volume changes can lead to degradation of the electrode materials. XRD helps in monitoring these changes and designing materials that can withstand such stresses.

Phase Transformations and Stability Studies: XRD is useful for studying phase transformations that occur during the operation of electrochemical devices. For example, during battery cycling, phase changes in electrode materials can affect performance. XRD allows for the monitoring of these phase transformations in situ, providing insights into the stability and degradation mechanisms of the materials.

Thin Film and Coating Analysis: In electrochemical devices that use thin films or coatings, XRD is used to analyze the thickness, texture, and phase composition of the films. For example, in fuel cells and batteries, thin film electrodes or solid electrolytes are commonly used, and XRD provides information on their structural integrity and phase composition, which are critical for device performance.

X-ray Diffraction (XRD) is a crucial tool for analyzing the microstructural properties of materials used in electrochemical systems. It provides detailed information about crystal structure, phase composition, grain size, and residual stress, all of which are essential for optimizing the performance and durability of electrochemical devices. By leveraging the capabilities of XRD, researchers can gain a deeper understanding of the relationship between microstructure and electrochemical behavior, leading to the development of more efficient and reliable energy storage and conversion technologies [26].

7.3.4 Small-Angle X-Ray Scattering (SAXS)

Small-Angle X-ray Scattering (SAXS) is an advanced technique used to investigate the nanoscale structural features of materials by measuring the scattering of X-rays at very small angles. SAXS provides crucial insights into the size, shape, and distribution of nanostructures within a material, making it particularly valuable for analyzing materials used in electrochemical systems where nanoscale features can significantly influence performance. SAXS operates on the principle that when X-rays are directed at a material, they are scattered by the electron density variations within the sample. In SAXS, the scattered X-rays are detected at very small angles relative to the incident beam. These small-angle scatterings arise from the presence of nanoscale inhomogeneities, such as nanoparticles, pores, or phase boundaries, which produce scattering patterns that provide information about their size and spatial distribution. SAXS is particularly useful for characterizing the microstructural properties of materials in electrochemical systems due to its sensitivity to nanoscale features [27]. Some key applications include:

(a) **Characterizing Nanoparticles**: SAXS is widely used to study nanoparticles, which are common in various electrochemical applications, such as catalysts, battery electrodes, and supercapacitors. SAXS provides information on the size, shape, and size distribution of nanoparticles. For example, in lithium–ion battery electrodes, SAXS can be used to measure the size and distribution of nanoparticles used in the electrode material, which influences the battery's capacity and charge/discharge rates.

(b) **Studying Porosity and Pore Size Distribution**: SAXS is effective for analyzing the porosity and pore size distribution in materials used in electrochemical devices. For instance, in supercapacitors and porous electrodes, SAXS can determine the size and distribution of pores, which affects the surface area available for electrochemical reactions and ion storage. By providing detailed information about the pore structure, SAXS helps in optimizing the material's performance.

(c) **Analyzing Phase Separation and Composite Materials**: In composite materials, SAXS can be used to investigate phase separation and the distribution of different components within the matrix. For example, in composite electrodes for fuel cells, SAXS can reveal the distribution of conductive additives and active phases, which affects the overall conductivity and reactivity of the electrode. SAXS helps in understanding how the microstructure of composites influences their electrochemical performance.

(d) **Monitoring Structural Changes During Operation**: SAXS can be employed to monitor structural changes in electrochemical materials during operation, such as during charge/discharge cycles in batteries or fuel cells. Changes in the SAXS pattern can indicate alterations in particle size, shape, or distribution, providing insights into phenomena such as particle aggregation, phase transformations, or degradation mechanisms. This real-time monitoring capability is valuable for assessing the stability and reliability of electrochemical devices.

(e) **Investigating Nanostructured Thin Films**: SAXS is also used to study nanostructured thin films and coatings, which are commonly used in electrochemical devices. For example, SAXS can characterize the nanostructure of thin film electrodes or solid electrolytes, providing information on film thickness, texture, and uniformity. This information is essential for ensuring the quality and performance of thin film-based electrochemical devices.

Interpreting SAXS data involves analyzing the scattering profile to extract information about the size, shape, and distribution of nanostructures. Key aspects of data interpretation include:

Particle Size and Shape Analysis: The intensity of the SAXS scattering profile as a function of the scattering vector q is related to the size and shape of the scattering particles. Models such as the Guinier model for small particles and the Porod model for larger structures are used to extract quantitative information about particle size and shape. The data can reveal whether the particles are spherical, rod-like, or have more complex shapes.

Size Distribution: SAXS data can be analyzed to obtain information about the size distribution of particles or pores within the sample. Techniques such as the regularization method or the inverse Fourier transform can be used to derive the size distribution from the scattering profile. This information helps in understanding the homogeneity or heterogeneity of the nanostructures.

Porosity and Pore Size Distribution: For porous materials, SAXS can provide information about the pore size distribution and porosity. The scattering pattern is analyzed to determine the size and distribution of pores, which is critical for optimizing materials used in energy storage and conversion devices.

Structural Changes: Changes in the SAXS profile during operation or under different conditions can provide insights into structural changes, such as particle growth, phase separation, or degradation. Monitoring these changes helps in understanding the stability and performance of electrochemical materials.

It is a valuable tool for analyzing the microstructural properties of materials used in electrochemical systems. By providing detailed information about nanoscale features, such as particle size, shape, distribution, and porosity, SAXS enables researchers to optimize the performance of energy storage and conversion devices. Its ability to monitor structural changes in real-time further enhances its utility in developing more efficient and reliable electrochemical technologies [27].

7.3.5 *Focused Ion Beam (FIB) Techniques for 3D Microstructural Analysis*

Focused Ion Beam Microscopy (FIB) is an advanced imaging and fabrication technique that uses a highly focused beam of ions to scan and analyze the surface and subsurface structures of materials. While similar to SEM in its ability to provide high-resolution imaging, FIB utilizes ions, typically gallium (Ga), instead of electrons. This distinction offers unique capabilities for studying and modifying the microstructure of materials used in electrochemical systems.

In FIB microscopy, a focused beam of ions is directed at the sample, causing sputtering of material from the surface. The ejected secondary electrons and ions are collected and analyzed to create high-resolution images of the sample's surface. Additionally, FIB can be used to precisely mill or etch material, allowing for the creation of cross-sections and detailed 3D reconstructions. Below are some of the applications of FIB.

Sample Preparation and Cross-Sectioning: One of the primary applications of FIB is in sample preparation. FIB can precisely cut and polish samples to create cross-sections, which are essential for analyzing the internal structure of electrochemical materials. For instance, in lithium–ion batteries, FIB can be used to slice through electrodes and analyze the distribution of active materials, conductive additives, and binder phases.

Imaging and Analysis: FIB provides high-resolution imaging of the sample's surface and near-surface regions. This capability is useful for examining the morphology of electrode materials, including surface roughness, particle size, and the interface between different phases. High-resolution images obtained from FIB can reveal details such as cracks, delamination, or the formation of solid-electrolyte interphases (SEI) in batteries.

Site-Specific Analysis: FIB allows for site-specific analysis by targeting particular regions of interest on the sample. This is particularly useful for investigating localized phenomena, such as the behavior of specific grains or interfaces in a composite electrode material. By focusing the ion beam on a specific area, researchers can gain insights into localized structural changes or degradation mechanisms.

3D Reconstruction: FIB is capable of generating 3D reconstructions of the sample by sequentially milling and imaging thin layers of the material. This technique, often referred to as FIB-SEM tomography, provides detailed 3D information about the

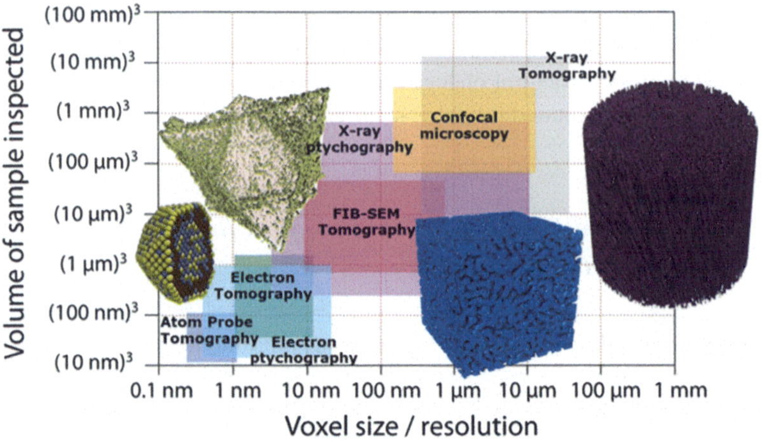

Fig. 7.3 Classification of various tomographic imaging methods. Reproduced from Ref. [29] Copyright © 2022 by the authors

internal structure of electrochemical materials, such as the distribution of nanoparticles, porosity, and the connectivity of phases. This comprehensive view is crucial for understanding the relationships between microstructure and electrochemical performance. Figure 7.3 shows various tomographic imaging methods.

While FIB offers high-resolution imaging and precise sample preparation capabilities, it also comes with some challenges. The ion milling process can induce artifacts such as surface damage or alteration of the material's microstructure. Additionally, the sample preparation process can be time-consuming and may require careful optimization to minimize these artifacts [28]. Figure 7.4a, b show SEM micrographs of large trenches milled with (a) showing a Ga^+ LMIS (ca. 10 h milling time) on the surface of a corundum sample, while (b) shows a femto-pulsed-laser (85 s milling time) on the surface of tungsten carbide. Figure 7.4c–f show SEM micrographs of a FIB-SEM imaging block with two lateral and one frontal trench prior to (c–e) and after (f) a slice and image routine in a FIB-SEM tomography experiment on a meso-macroporous $CoRu/Al_2O_3$ catalyst.

7.3.6 3D Tomography

Three-Dimensional (3D) Tomography is a powerful imaging technique that provides volumetric information about the internal structure of materials. Unlike traditional 2D imaging methods, 3D tomography reconstructs a series of 2D images obtained from different angles to create a detailed three-dimensional representation of the sample's internal features. 3D tomography involves capturing a series of 2D images, known as slices, at different angles around the sample. These slices are then reconstructed into a 3D model using computational algorithms. The result is a volumetric representation

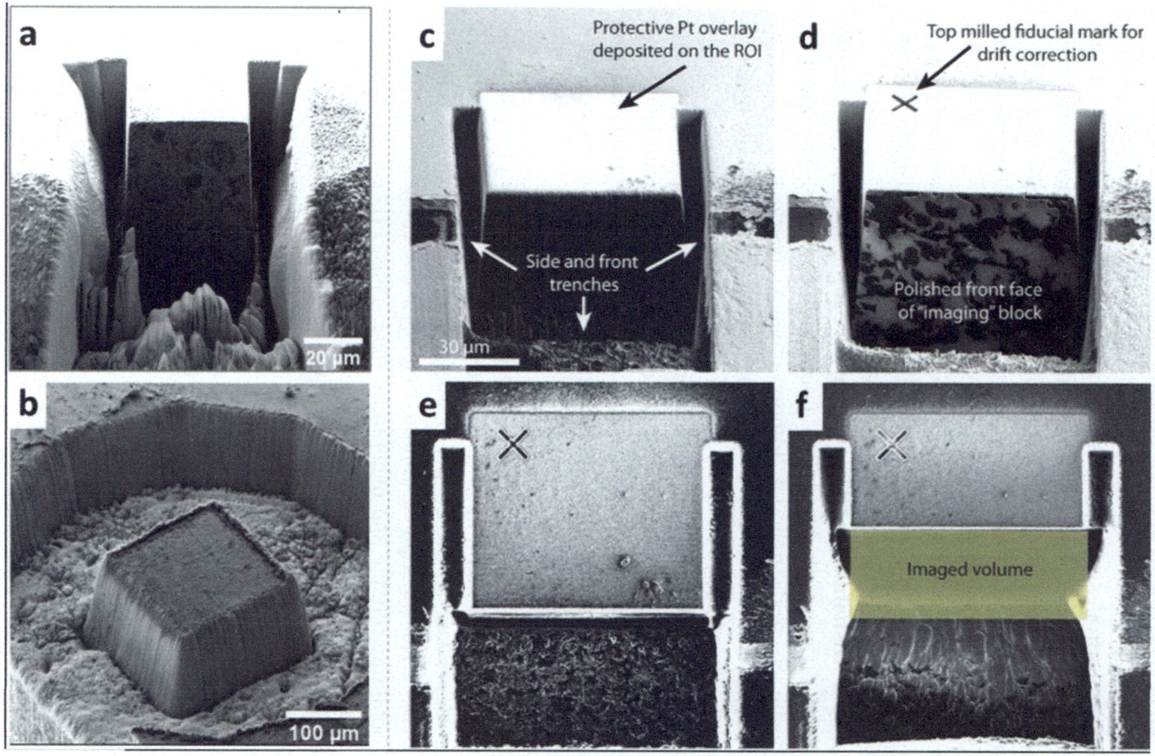

Fig. 7.4 **a**, **b** SEM micrographs of large trenches milled with **a** a Ga^+ LMIS (ca. 10 h milling time) on the surface of a corundum sample. Reproduced from Ref. [29] Copyright © 2022 The Authors. **b** A femto-pulsed-laser (85 s milling time) on the surface of tungsten carbide reproduced from Ref. [30] Copyright © 2020 The Authors. **c**–**f** SEM micrographs of a FIB-SEM imaging block with two lateral and one frontal trench prior to (**c**–**e**) and after (**f**) a slice and image routine in a FIB-SEM tomography experiment on a meso-macroporous CoRu/Al_2O_3 catalyst. The volume of sample eroded during the FIB-SEM experiment is marked on panel (**f**). Reproduced from Ref. [29] Copyright © 2022 The Authors

of the sample, which can be analyzed to study the internal microstructure and spatial distribution of different phases or components [31]. Below are the applications of 3D tomography in electrochemical systems.

(a) **Visualization of Internal Structures**: 3D tomography provides a detailed view of the internal structures of electrochemical materials, such as porous electrodes, battery electrodes, and catalyst supports. By visualizing the internal morphology, researchers can study the distribution of pores, active materials, and conductive pathways, which are critical for optimizing performance [32].
(b) **Analysis of Porosity and Connectivity**: In materials such as battery electrodes and supercapacitors, the porosity and connectivity of the active material are crucial for performance. 3D tomography allows for the measurement of pore size distribution, connectivity, and overall porosity, providing insights into how these factors influence electrochemical behavior [33].
(c) **Study of Phase Distribution**: 3D tomography can be used to analyze the distribution of different phases within composite materials. For example, in a composite electrode material, tomography can reveal how the active phase, conductive additives, and binder are distributed throughout the material, impacting its performance [34].
(d) **Monitoring Changes Over Time**: By combining 3D tomography with in situ techniques, researchers can monitor changes in the internal structure of electrochemical materials over time. This capability is useful for studying degradation mechanisms, phase transformations, and structural changes during the operation of electrochemical devices [31].

3D tomography requires the acquisition of a large number of 2D images, which can be time-consuming and data-intensive. Additionally, the resolution of the 3D reconstruction is dependent on the quality of the 2D images and the reconstruction algorithms used. Ensuring high-resolution imaging and accurate reconstruction is essential for obtaining reliable and meaningful data. Thus, FIB and 3D Tomography are powerful tools for analyzing the microstructural properties of materials used in electrochemical systems. FIB provides high-resolution imaging and precise sample preparation capabilities, while 3D tomography offers comprehensive volumetric information about internal structures [31]. Together, these techniques enable researchers to gain a deeper understanding of the relationships between microstructure and electrochemical performance, leading to the development of more efficient and reliable energy storage and conversion technologies.

7.4 Microstructural Influence on Electrochemical Behavior

The microstructure of a material has a profound impact on its electrochemical behavior, influencing properties such as charge transport, ion diffusion, electrode reaction kinetics, and overall device efficiency. The microstructural features of materials are pivotal in determining their charge transport and ion diffusion properties, which are fundamental to the performance of electrochemical devices. Understanding how these features influence the movement of electrons and ions can help in designing materials with optimized characteristics for energy storage and conversion applications. In polycrystalline materials, grain boundaries can significantly impact charge transport [35]. These boundaries are regions where the crystal lattice is misaligned, creating discontinuities in the material's structure. As a result, grain boundaries can act as barriers to the flow of charge carriers, such as electrons and ions, due to the localized increase in resistivity. This effect is particularly pronounced in materials where the grain boundaries are poorly conductive or contain impurities that further hinder charge transport. However, grain boundaries are not always detrimental to charge transport. In some cases, they can facilitate rapid ion migration. For instance, [36] the size and orientation of grains play a crucial role in determining how effectively these boundaries contribute to or impede charge transport. For materials where high conductivity is desired, such as in conductive electrodes or supercapacitors, optimizing grain size and boundary characteristics is essential for enhancing overall performance. Nanostructuring materials offers a promising approach to improving ion diffusion by minimizing the distance that ions must travel within the material. Nanostructured materials, such as nanoparticles, nanowires, and nanorods, provide shorter diffusion paths and a higher surface area compared to bulk materials. This increase in surface area not only exposes more active sites for electrochemical reactions but also facilitates faster ion migration.

In batteries, nanostructured electrodes enable more efficient lithiation and delithiation processes. The reduced diffusion distance for lithium ions enhances charge/discharge rates, leading to improved performance in terms of capacity and cycle life [37]. Similarly, in supercapacitors and fuel cells, the high surface area and porous nature of nanostructured materials contribute to better ion transport and storage capabilities [38]. The porosity and connectivity of these nanostructures are critical in ensuring that ions can move freely and access all available active sites, which is essential for maximizing energy storage and conversion efficiency. Porosity is another key microstructural feature that affects ionic conductivity. In porous electrodes, the presence of voids allows the electrolyte to penetrate the material more easily, providing multiple pathways for ion migration. This increased access to the electrolyte can enhance the material's ionic conductivity and, consequently, its performance in electrochemical devices. For example, in supercapacitors, porous carbon materials with high surface area and well-designed pore structures facilitate efficient ion movement and energy storage [39].

However, excessive porosity can introduce challenges. While it improves ionic conductivity, it can also reduce the mechanical strength of the material, leading to potential structural issues. Additionally, very high porosity may result in the formation of dead volumes within the material that do not contribute to charge storage or ion transport. Balancing the degree of porosity with the need for mechanical stability and functional efficiency is crucial. The design of porous materials requires careful consideration to ensure that the benefits of increased ionic conductivity are not offset by diminished structural integrity or reduced active surface area. The interplay between microstructure and electrochemical performance underscores the importance of integrating microstructural design with device functionality. Advances in material science and manufacturing techniques allow for the precise control of microstructural features, enabling the development of materials with optimized charge transport and ion diffusion properties. By tailoring grain size, nanostructure dimensions, and porosity, researchers can enhance the efficiency and effectiveness of electrochemical devices across various applications, from batteries and supercapacitors to fuel cells and beyond [40]. Thus, the microstructural properties of materials, including grain boundaries, nanostructures, and porosity, play a significant role in determining charge transport and ion diffusion characteristics. Understanding and manipulating these properties are essential for developing high-performance electrochemical devices with improved energy storage, conversion efficiency, and overall reliability.

7.5 Microstructural Defects and Their Impact on Electrochemical Kinetics

Microstructural defects in materials significantly influence their electrochemical kinetics, which is critical for the performance of electrochemical devices. These defects, including point defects, dislocations, and surface irregularities, can alter ion mobility, charge transport, and reaction kinetics in complex ways. Understanding their roles and effects is essential for optimizing material performance in energy storage and conversion systems.

7.5.1 Point Defects and Ion Mobility

Point defects, such as vacancies and interstitials, play a crucial role in modifying the ionic conductivity of materials. Vacancies are missing atoms in the crystal lattice, while interstitials are extra atoms situated in the spaces between the regular lattice positions. These defects create localized disruptions in the crystal structure, which can facilitate the movement of ions through the material. In solid electrolytes, for instance, a controlled concentration of point defects can enhance ion mobility by providing pathways for ion migration. This is particularly important in materials like lithium–ion conductors and polymer electrolytes, where high ionic conductivity is essential for effective device performance. However, an excessive concentration of point defects can adversely affect the material's stability and performance. High defect densities can lead to structural instability, making the material prone to degradation over time. This degradation may result in reduced ionic conductivity, mechanical failure, or diminished overall performance of the electrochemical device [41]. Therefore, achieving an optimal balance in defect concentration is crucial for maintaining both high ionic conductivity and material integrity.

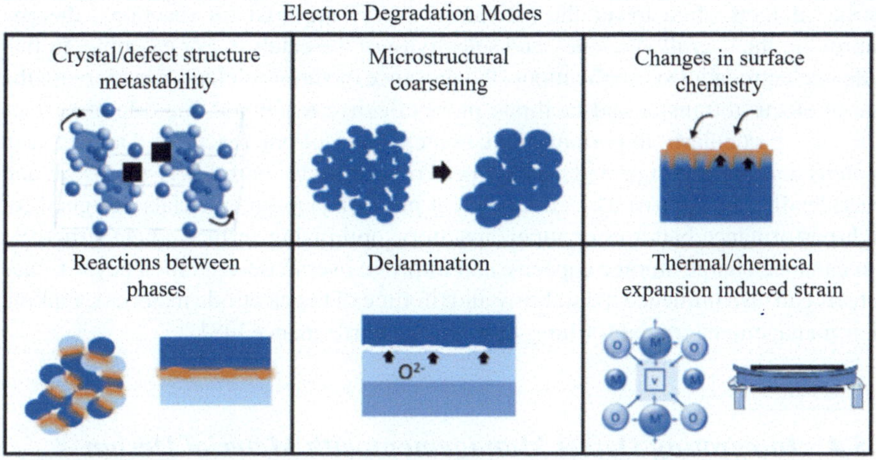

Fig. 7.5 Degradation modes in oxygen electrode. Reproduced from Ref. [45] Copyright © 2016 by the authors

7.5.2 Dislocations and Their Dual Role

Dislocations are linear defects within the crystal structure that significantly impact electrochemical kinetics. These defects can both hinder and facilitate electrochemical reactions, depending on their nature and the context of their application. On one hand, dislocations can act as traps for charge carriers, thereby reducing electrical conductivity. This trapping effect can impede the efficient movement of electrons or ions, leading to reduced performance in devices like batteries or supercapacitors. On the other hand, dislocations can also provide beneficial sites for nucleation and growth of reaction products. In some electrochemical processes, such as the formation of protective layers on electrodes, dislocations can serve as active sites for these processes. The presence of dislocations can promote the formation of desired phases or coatings that enhance the stability and performance of the electrode materials. For example, in lithium–ion batteries, dislocations may facilitate the formation of stable SEI (solid-electrolyte interphase) layers, which can protect the electrode material from further degradation [42]. Figure 7.5 depicts the degradation modes in oxygen electrode.

7.5.3 Surface Defects and Electrode Reactions

Surface defects, including steps, kinks, and vacancies, have a significant impact on the catalytic activity and efficiency of electrode materials. These surface irregularities create active sites that can enhance electrochemical reactions. In catalytic processes, surface defects often lower the activation energy required for reactions, thereby improving the overall efficiency and selectivity of the catalyst. For example, in fuel cells or electrocatalysis applications, the presence of surface defects can increase the rate of electron transfer and facilitate more effective reactions. The role of surface defects is particularly important in the context of electrode reactions. Defects such as steps and kinks can provide sites where reactants are more likely to adsorb and react, leading to improved electrochemical performance. In materials designed for high-performance batteries or supercapacitors, optimizing surface defect structures can enhance charge storage capacity and improve overall device efficiency. Surface defects can also influence the stability and lifetime of the electrode materials, making their management crucial for long-term device performance [43].

7.5.4 Integrating Defect Management with Material Design

The impact of microstructural defects as shown in Fig. 7.6 on electrochemical kinetics underscores the importance of integrating defect management with material design strategies. By carefully controlling the types and concentrations of defects, researchers can enhance the performance of electrochemical materials and devices. This involves not only optimizing defect densities but also understanding their specific roles in various electrochemical processes. Future advancements in material science should focus on developing methods to precisely control and exploit microstructural defects. This includes employing advanced fabrication techniques and characterization tools to tailor defect structures and improve their beneficial effects while mitigating potential drawbacks. By achieving a deeper understanding of how different defects influence electrochemical kinetics, researchers can design more efficient and reliable materials for energy storage and conversion applications. Thus, it can be inferred that microstructural defects, including point defects, dislocations, and surface irregularities, play a significant role in shaping the electrochemical kinetics of materials. Their effects on ion mobility, charge transport, and reaction rates are complex and multifaceted. By integrating defect management into material design, researchers can enhance the performance and durability of electrochemical devices, paving the way for more advanced and efficient energy storage and conversion technologies [44].

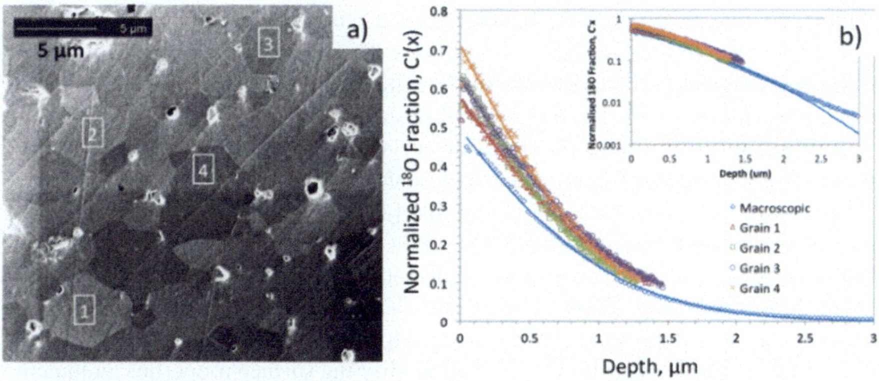

Fig. 7.6 **a** Microstructure, showing analyzed areas by ion-induced secondary electron image, **b** profiles of corresponding areas of image (**a**). Reproduced from Ref. [45] Copyright © 2016 by the authors

7.6 Surface Morphology and Its Effect on Electrode Reactions

Surface morphology plays a critical role in determining the electrochemical performance of electrodes. The texture, roughness, and overall topography of an electrode surface directly impact its reactivity, efficiency, and longevity in electrochemical devices. By understanding and optimizing surface morphology, researchers can significantly enhance the performance of energy storage and conversion systems. The surface roughness and texture of an electrode are key factors influencing its electrochemical performance. A rough surface with increased texture provides a larger surface area for electrochemical reactions, which can enhance the current density and overall efficiency of the device. This is particularly advantageous in applications where high rates of reaction are required, such as in high-capacity batteries or supercapacitors. The increased surface area allows for more active sites where electrochemical processes can occur, thereby improving the performance metrics of the device. However, excessive roughness can introduce complications. Uneven surfaces may lead to non-uniform current distribution during operation, which can cause localized degradation and affect the overall stability of the electrode. This is especially problematic in high-stress environments, such as high-rate cycling conditions in batteries. The mechanical stresses associated with cycling can exacerbate issues related to surface roughness, potentially leading to increased wear and reduced lifespan of the electrode. Therefore, while rough surfaces can enhance performance, careful control of surface texture is essential to avoid detrimental effects on device operation and longevity [46].

Surface engineering is a powerful tool for optimizing the reactivity of electrode materials. Techniques such as coating, etching, and deposition are employed to modify surface morphology and enhance electrochemical performance. For example, coating electrodes with nanostructured catalysts can significantly increase the number of active sites available for reactions. These nanostructured coatings often exhibit higher catalytic activity due to their increased surface area and improved interaction with reactants. Etching is another surface engineering technique used to create hierarchical structures on electrode surfaces. By selectively removing material, etching can produce micro- and nanoscale features that enhance the balance between surface area and mechanical stability [47]. This hierarchical structuring can improve the accessibility of active sites while maintaining the structural integrity of the electrode, leading to better overall performance and durability. In addition to coating and etching, deposition techniques such as chemical vapor deposition (CVD) and physical vapor deposition (PVD) are used to apply thin films or layers onto electrode surfaces. These deposited layers can modify the surface properties, enhancing reactivity and improving performance in specific electrochemical processes [48]. The choice of surface engineering technique and material depends on the desired properties and application of the electrode.

Grain boundaries in polycrystalline materials are another important aspect of surface morphology that affects electrochemical performance. These boundaries can act as barriers to charge transport, influencing the efficiency of electrochemical reactions. However, they also play a significant role in the reactivity of the material. The chemistry and structure of grain boundaries can create active sites for electrochemical processes, with certain configurations offering enhanced catalytic activity. The impact of grain boundaries on electrochemical performance is complex and depends on the specific material and application. In some cases, grain boundaries can facilitate the formation of protective layers or reaction intermediates, improving the stability and performance of the electrode. In other cases, they may hinder charge transport and reduce overall efficiency. Understanding the role of grain boundaries in electrochemical processes is essential for optimizing the performance of materials used in energy devices [49].

The influence of surface morphology on electrode reactions highlights the importance of integrating surface design with overall device engineering. By tailoring surface roughness, employing surface engineering techniques, and considering the effects of grain boundaries, researchers can develop electrodes with optimized performance characteristics. This integrated approach allows for the creation of advanced electrochemical devices that offer improved efficiency, durability, and reliability. Future research should focus on developing innovative surface modification techniques and exploring new materials that can enhance electrochemical performance through improved surface morphology [47]. Advances in fabrication technologies and characterization methods will enable more precise control over surface features, leading to further improvements in energy storage and conversion technologies. In summary, surface morphology, including roughness, texture, and grain boundary characteristics, plays a crucial role in determining the electrochemical performance of electrodes. By understanding and optimizing these features, researchers can enhance the efficiency and stability of electrochemical devices, paving the way for more advanced and effective energy technologies.

7.7 Grain Boundary Effects in Electrochemical Performance

Grain boundaries, the interfaces where different crystal grains meet within a polycrystalline material, are critical factors influencing electrochemical performance. These boundaries can significantly impact the efficiency and stability of electrochemical devices, such as batteries, supercapacitors, and fuel cells. Understanding the effects of grain boundaries on electrochemical processes is essential for optimizing material performance and designing advanced energy storage and conversion systems. Grain boundaries can act as barriers to charge transport due to the misalignment of crystal lattices and the presence of additional interfaces between grains. This misalignment creates regions of increased resistivity that can impede the flow of charge carriers, such as electrons and ions. In materials like electrode active materials for batteries or supercapacitors, the efficiency of charge transport is crucial for high performance. The presence of grain boundaries can thus lead to reduced electrical conductivity and overall efficiency of the device. However, the impact of grain boundaries on charge transport is not always detrimental. In some materials, particularly those engineered for specific applications, grain boundaries can facilitate certain types of ion or electron movement. For example, in some solid electrolytes, grain boundaries may provide pathways for ion conduction, enhancing the material's ionic conductivity. The effectiveness of these pathways depends on the nature of the grain boundaries and their interaction with the surrounding matrix [50].

Grain boundaries can also play a significant role in electrochemical reactions, often acting as active sites where these reactions can occur. The chemistry and structure of grain boundaries can create favorable conditions for certain electrochemical processes. For instance, in catalytic materials used in fuel cells or batteries, grain boundaries may serve as sites for the adsorption and reaction of reactants, improving the overall catalytic activity. The role of grain boundaries in electrochemical reactions can vary based on their composition and structure. In some cases, grain boundaries can enhance reaction kinetics by providing additional sites for the formation of reaction intermediates or products. This can be beneficial for processes that require high catalytic activity or specific reaction conditions. Conversely, the presence of grain boundaries may also lead to undesired reactions or the formation of non-conductive phases, which can detract from the overall performance of the material [51].

The stability and durability of materials in electrochemical devices are significantly influenced by grain boundaries. These boundaries can be sites of structural weakness or instability, particularly under high-stress conditions such as repeated charge–discharge cycles in batteries. Over time, grain boundaries can become sites of degradation, leading to reduced mechanical strength and compromised performance of the device. In some cases, grain boundaries may facilitate the formation of protective layers or phases that enhance material stability. For example, in lithium–ion batteries, the formation of a stable SEI (solid-electrolyte interphase) layer at grain boundaries can protect the electrode material from degradation and extend the battery's lifespan. However, the presence of unstable or poorly formed grain boundaries can lead to issues such as capacity fading, increased resistance, and overall reduced performance [52].

To harness the benefits of grain boundaries while mitigating their potential drawbacks, researchers focus on optimizing their characteristics through various material design strategies. This includes controlling grain size, boundary chemistry, and the overall microstructure of the material. Techniques such as annealing, doping, and precise fabrication methods can be employed to tailor grain boundaries to desired specifications, enhancing their positive effects and minimizing adverse impacts. Advances in material science and characterization techniques enable more detailed understanding and manipulation of grain boundaries. By employing tools like high-resolution electron microscopy and advanced diffraction methods, researchers can gain insights into the structure and behavior of grain boundaries, leading to improved design and performance of electrochemical materials.

7.8 Microstructural Effects in Specific Electrochemical Devices

In this section, we explore several case studies that highlight the impact of microstructural features on the performance of specific electrochemical devices. These case studies illustrate the practical implications of microstructural control and provide insights into the design strategies for optimizing device performance.

7.8.1 Lithium–Ion Batteries: Grain Structure and Cycling Stability

Lithium–ion batteries (LIBs) have become a cornerstone of modern energy storage technology, powering everything from portable electronics to electric vehicles. The performance and longevity of LIBs are profoundly influenced by the microstructural properties of their components, particularly the cathodes and anodes. Understanding and optimizing these microstructural features is essential for improving battery efficiency, capacity, and cycle life. The microstructure of both cathodes and anodes plays a pivotal role in determining the performance of lithium–ion batteries [53]. In cathode materials, such as lithium cobalt oxide (LCO) and lithium iron phosphate (LFP), the grain size and orientation significantly impact the rate of lithium intercalation and the overall capacity of the battery [54, 55]. Smaller grain sizes can enhance the rate of lithium–ion diffusion and increase the material's capacity, as they shorten the diffusion path for lithium ions. However, excessively fine grains may also lead to increased internal resistance and reduced mechanical stability. For anodes, materials such as graphite and silicon exhibit unique microstructural challenges. In graphite anodes, the layered structure facilitates lithium–ion insertion and extraction, but the performance is limited by the low capacity compared to other materials. Silicon anodes, on the other hand, offer much higher capacity but suffer from significant volume expansion and contraction during lithiation and delithiation cycles [56]. Reducing the grain size of silicon anodes can help alleviate these issues by accommodating volume changes more effectively and thereby improving cycling stability. Strategies such as forming silicon nanoparticles or using silicon-carbon composites can help mitigate the detrimental effects of volume expansion.

Grain boundaries in cathode materials, such as lithium nickel manganese cobalt oxide (NMC), can significantly influence battery degradation. These boundaries often act as sites for crack initiation and propagation during cycling. The mechanical stresses generated during charge and discharge cycles can exacerbate the formation of cracks along grain boundaries, leading to capacity fading and reduced battery life. The presence of cracks can also increase the likelihood of electrolyte degradation and internal short circuits, further compromising battery performance [57]. To address these issues, various strategies are employed to mitigate the impact of grain boundaries. One approach is grain boundary passivation, which involves modifying the grain boundaries to make them less prone to degradation. This can be achieved through coating the cathode material with a thin layer of protective material that prevents direct contact between the electrolyte and the grain boundaries [58]. Another strategy involves the use of advanced synthesis techniques to control grain size and distribution more precisely, thereby minimizing the formation of weak grain boundaries. Coatings and protective layers can also be applied to anodes to reduce the impact of grain boundaries and improve cycling stability. For instance, coating silicon anodes with a layer of carbon or other materials can buffer the mechanical stresses associated with volume changes, thereby enhancing the anode's lifespan and performance. The key to optimizing the microstructural properties of lithium–ion battery materials lies in balancing various factors such as grain size, orientation, and boundary characteristics [59]. This balance is crucial for achieving high capacity, excellent rate capability, and long cycle life. Advances in material science and fabrication techniques continue to provide new insights into how microstructural features can be tailored to enhance battery performance. Future research should focus on developing innovative materials and methods for controlling grain structure and boundary effects. This includes exploring new synthesis techniques, such as high-energy ball milling and advanced deposition methods, to create materials with optimized microstructures. Additionally, understanding the interaction between microstructural features and battery chemistry will be essential for designing more efficient and durable lithium–ion batteries. Figure 7.7 shows the SEM images of (**a**) gravel- and (**b**) rod-NMC secondary particles showing the interior grain arrangements, (**c**) SXRD patterns and (**d**) K-edge XANES of gravel- and rod-NMCs in the pristine state. Vertical dash lines indicate Ni white-line energy position. (**e**) Schematic of the experimental setup for in situ XANES-3DTXM measurements, and the representative 3D rendering of (**f**) interior morphology. (**g**) Ni white-line energy distribution from XANES-3DTXM of a secondary particle. The Ni white-line energies are color-coded, as blue and red stand for low and high oxidation states, respectively.

In summary, the microstructure of cathodes and anodes plays a crucial role in the performance and stability of lithium–ion batteries. By addressing the challenges associated with grain size, orientation, and grain boundaries, researchers can enhance the efficiency, capacity, and longevity of these essential energy storage devices. Through continued innovation and optimization, the next generation of lithium–ion batteries will offer even greater performance and reliability.

7.8.2 Fuel Cells: Catalyst Layer Microstructure and Performance Optimization

Fuel cells, particularly proton exchange membrane fuel cells (PEMFCs), are critical technologies for clean energy conversion, offering high efficiency and low emissions. The performance of these fuel cells is heavily influenced by the microstructure of the catalyst layer, which is integral to the electrochemical reaction process. Optimizing this microstructure is essential for enhancing power density, durability, and overall efficiency. The catalyst layer in PEMFCs typically consists of platinum nanoparticles supported on a carbon substrate, combined with an ionomer phase that facilitates proton transport [61]. The distribution and dispersion of platinum nanoparticles within this layer are fundamental to the catalyst's performance. Uniformly distributed nanoparticles ensure that more active sites are available for the electrochemical reactions, leading to increased catalytic activity and power output. However, the stability of these nanoparticles is also a concern, as agglomeration can reduce the effective surface area and, consequently, the catalytic efficiency. The porosity of the carbon support material plays a crucial role in the microstructure of the catalyst layer [62]. Adequate porosity ensures sufficient access to the catalyst sites and facilitates effective gas diffusion from the reactant gases to the catalyst particles. This improves the overall reaction kinetics. Conversely, excessive porosity may lead to reduced mechanical strength and potential instability of the catalyst layer. The ionomer phase, typically composed of a polymer like Nafion, acts as a proton conductor and binder within the catalyst layer [39]. Its distribution and connectivity are vital for maintaining proton conductivity and ensuring proper interaction between the catalyst particles and the proton exchange membrane. An optimal ionomer content and distribution can enhance both the catalytic activity and the stability of the catalyst layer, contributing to the long-term performance of the fuel cell [61].

Effective water management is crucial for the performance of PEMFCs, as both excess water and insufficient hydration can adversely affect the cell's efficiency and longevity [63]. Water is produced during the electrochemical reaction in the fuel cell, and its management within the electrode structure significantly impacts the overall performance. The microstructure of the fuel cell electrodes, including pore size distribution and the hydrophobic/hydrophilic balance, plays a critical role in managing

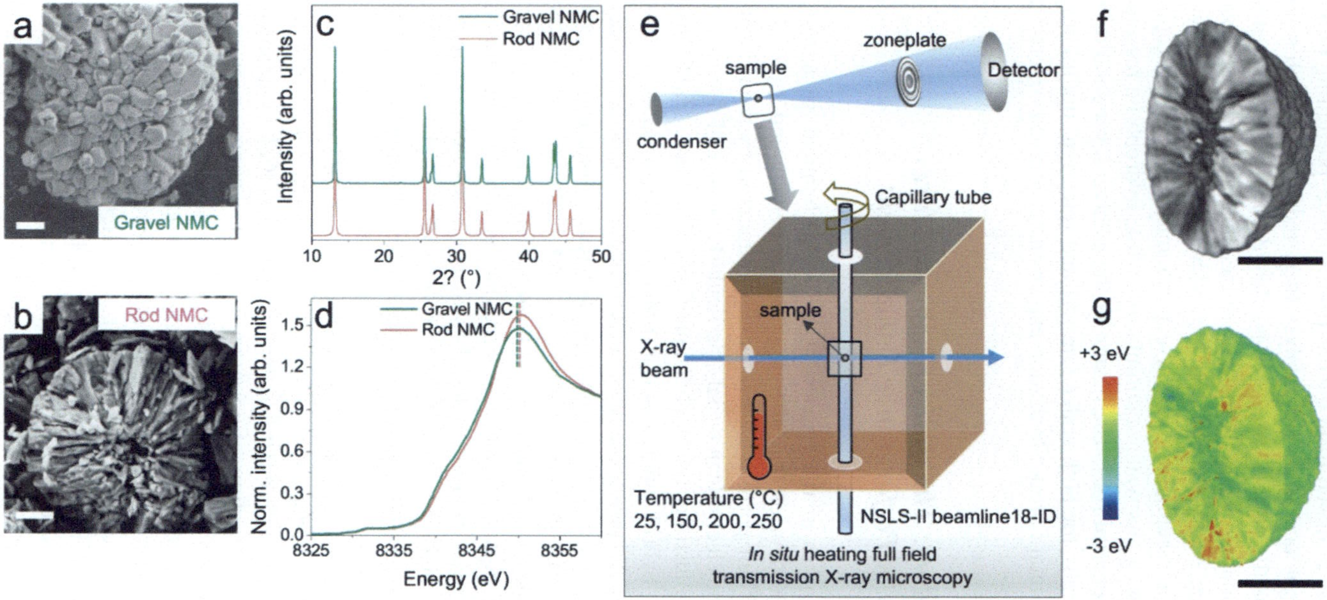

Fig. 7.7 SEM images of **a** gravel- and **b** rod-NMC secondary particles showing the interior grain arrangements, with a scale bar of 1 μm. **c** SXRD patterns and **d** K-edge XANES of gravel- and rod-NMCs in the pristine state. Vertical dash lines indicate Ni white-line energy position. **e** Schematic of the experimental setup for in situ XANES-3DTXM measurements, and the representative 3D rendering of **f** interior morphology. **g** Ni white-line energy distribution from XANES-3DTXM of a secondary particle, with a scale bar of 5 μm. The Ni white-line energies are color-coded, as blue and red stand for low and high oxidation states, respectively reproduced from Ref. [60] Copyright © 2022 by the authors

water [64]. Properly designed pore structures ensure that water produced during the reaction can be efficiently removed from the electrode surface, preventing flooding. Flooding can obstruct gas access to the catalyst sites, reducing the cell's power output and efficiency. On the other hand, excessive dehydration of the catalyst layer can diminish proton conductivity and increase resistance, leading to reduced performance. Achieving a balance between hydrophobic and hydrophilic properties within the electrode structure is essential for effective water management. Hydrophobic regions can help prevent excess water accumulation, while hydrophilic regions aid in water transport and removal [65]. The design of the catalyst layer should therefore include considerations for both water removal and retention, optimizing the electrode's ability to maintain proper hydration and ensure continuous operation.

7.8.3 Supercapacitors: The Role of Nanostructures and Porosity in Performance Enhancement

Supercapacitors, also known as ultracapacitors or electrochemical capacitors, are advanced energy storage devices that offer high power density and rapid charge/discharge capabilities. Their performance is heavily influenced by the microstructural characteristics of the electrode materials, particularly the role of nanostructures and porosity. Figure 7.8 illustrates the mechanism of charge storage of (a) EDLC, (b) pseudocapacitor and (c) hybrid supercapacitor. Enhancing these features is crucial for optimizing the energy and power density of supercapacitors. Nanostructured electrodes are central to improving the performance of supercapacitors. These materials, which include activated carbon, carbon nanotubes, and graphene, are selected for their high surface area and exceptional electrical conductivity. The high surface area is essential for achieving high capacitance, as it provides more active sites for charge storage. For example, activated carbon materials with nanostructured pores can offer specific surface areas in the range of 1000–2000 m^2/g, which significantly enhances their energy storage capacity [66]. The microstructure of these nanostructured materials, including the size, shape, and distribution of nanostructures, directly impacts the supercapacitor's performance. Carbon nanotubes, with their cylindrical nanostructure, provide excellent conductivity and high surface area but require careful alignment and density control to maximize their effectiveness [67]. Graphene, with its two-dimensional planar structure, offers superior electrical conductivity and high surface area, contributing to enhanced power and energy density [68]. Figure 7.9 shows graphene based nanocomposites for supercapacitor applications. The specific arrangement of these nanostructures affects both the electrochemical performance and the mechanical stability of the supercapacitor electrodes.

The porosity of electrode materials plays a crucial role in supercapacitor performance by affecting ion accessibility and transport. A well-designed porous structure allows for rapid ion movement to and from the active sites on the electrode surface, which is essential for reducing internal resistance and enhancing charge/discharge

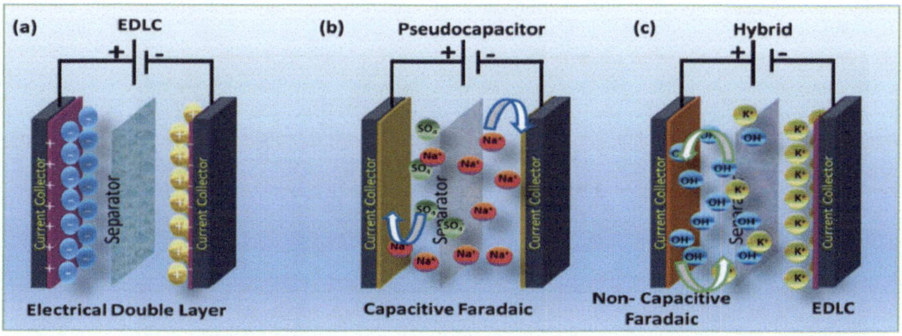

Fig. 7.8 Mechanism of charge storage of **a** EDLC, **b** pseudocapacitor and **c** hybrid supercapacitor reproduced with permission from Ref. [39] Copyright © Royal society of chemistry

rates. Porous materials with appropriate pore sizes can facilitate effective ion transport, leading to improved power density and efficiency [12]. However, balancing porosity is critical, as excessive porosity can compromise the mechanical stability and volumetric capacitance of the electrode material. For instance, while high porosity increases the surface area available for charge storage, it can also lead to structural weakness and reduced volumetric density. Thus, optimizing the porosity involves achieving a balance that maximizes the surface area while maintaining sufficient mechanical strength to ensure long-term stability and performance.

The distribution of pore sizes within the electrode material is a key factor in determining supercapacitor performance. Micropores, with diameters less than 2 nm, contribute to high capacitance by increasing the surface area available for electrochemical reactions [69]. However, they may also restrict ion transport due to their limited accessibility, which can hinder the rate of charge and discharge. In contrast, mesopores (2–50 nm) and macropores (greater than 50 nm) facilitate ion diffusion and reduce internal resistance, making them particularly important for high-power applications [69]. Mesopores provide pathways for faster ion movement, while macropores help in reducing the overall resistance by allowing easier access to the electrode surface. An optimal combination of these pore sizes, known as hierarchical porosity, can significantly enhance both energy density and charge/discharge rates. Hierarchical porosity integrates micropores for high capacitance with mesopores and macropores for improved ion transport and power delivery.

7.8.4 Electrochemical Sensors: Sensitivity and Selectivity Through Microstructural Design

Electrochemical sensors are critical tools in various applications, including environmental monitoring, medical diagnostics, and industrial process control. Their performance, particularly in terms of sensitivity and selectivity, is profoundly influenced

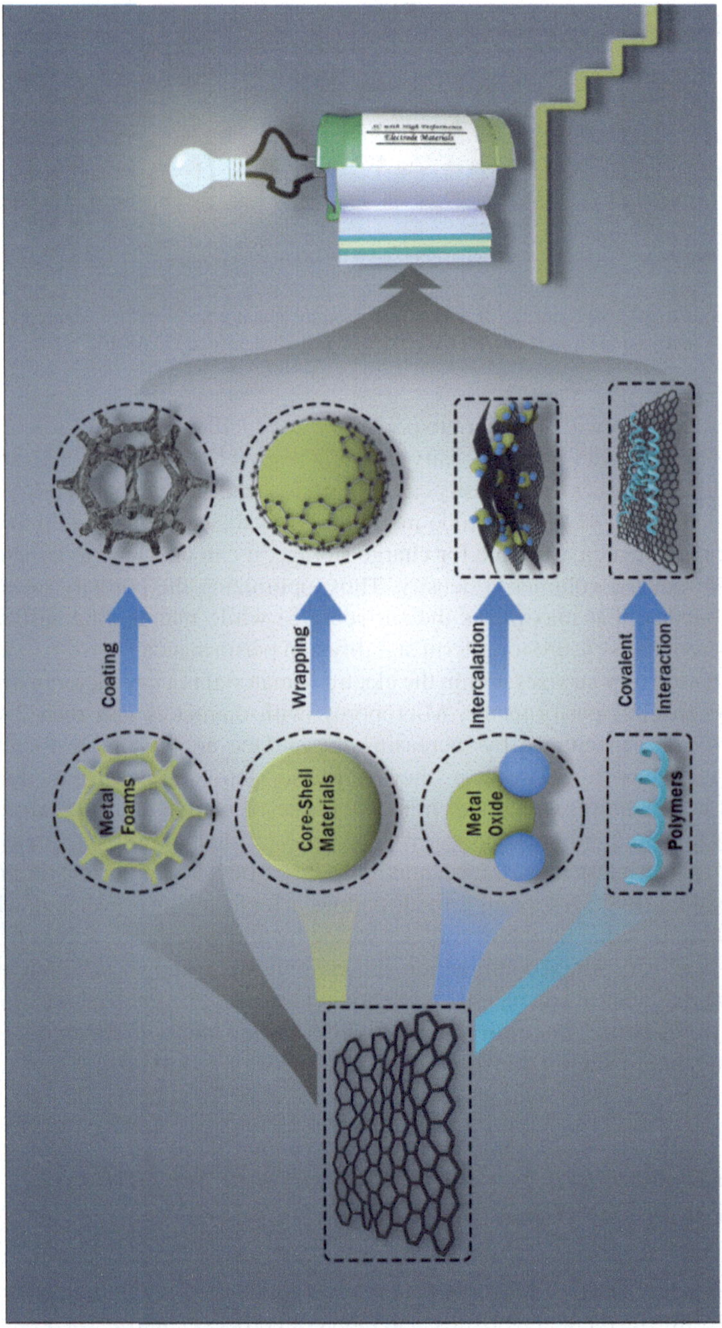

Fig. 7.9 Graphene based nanocomposites for supercapacitor applications reproduced from Ref. [70] Copyright © 2020 by the authors

by the microstructural characteristics of the sensing materials. By optimizing these microstructural features, it is possible to enhance sensor performance significantly. The sensitivity of electrochemical sensors largely depends on the microstructure of the sensing material. Materials with a high surface area and reactivity, such as nanostructured metal oxides, conductive polymers, and carbon-based materials, are often employed to improve sensor performance [71]. The microstructural design of these materials influences the density of active sites available for analyte adsorption and reaction, which directly impacts the sensor's ability to detect low concentrations of target substances. Nanostructuring techniques, such as creating nanoparticle-based composites or developing porous nanostructures, can substantially enhance sensitivity. For instance, the surface roughness and porosity of nanostructured materials can be tailored to increase the number of active sites. This, in turn, improves the sensor's response to analytes at lower concentrations. By engineering the microstructure to optimize surface area and create favorable conditions for analyte interaction, sensors can achieve higher sensitivity and more accurate detection.

Selectivity in electrochemical sensors is crucial for distinguishing between different analytes in complex mixtures. Surface functionalization plays a key role in enhancing selectivity by attaching specific functional groups or molecular receptors to the sensor's surface [72]. The microstructure of the sensing material affects the distribution and interaction of these functional groups with target molecules. For example, in biosensors, the microstructure of the electrode material influences the immobilization of enzymes or antibodies, which are essential for specific binding to biological molecules. Uniform functionalization and optimal orientation of these receptors are vital for achieving high selectivity. By controlling the microstructure to ensure precise placement and interaction of functional groups, sensors can be tailored to detect specific analytes with high selectivity. The stability of electrochemical sensors under operating conditions is another critical factor influenced by their microstructure. Sensors often operate in harsh environments, such as high temperatures or corrosive media, which can lead to degradation of the sensing material [73]. Therefore, it is essential to design sensors with stable microstructures that resist such adverse conditions. Nanostructured materials, such as metal oxides used in gas sensors, must maintain their microstructural integrity to ensure consistent performance over time. Techniques such as doping, surface coating, and the use of protective layers can enhance the durability and stability of these materials. For instance, doping with certain elements can improve resistance to environmental factors, while surface coatings can provide additional protection against corrosion and wear [74]. In conclusion, the microstructural design of electrochemical sensors plays a pivotal role in determining their sensitivity and selectivity. By optimizing nanostructures, surface functionalization, and microstructural stability, it is possible to develop sensors with enhanced performance and durability. Advances in microstructural engineering continue to drive improvements in sensor technology, making them more effective for a wide range of applications.

7.8.5 Solid Oxide Fuel Cells (SOFCs): Microstructure and Ionic Conductivity

Solid oxide fuel cells (SOFCs) represent a key technology in the field of clean energy conversion, leveraging high-temperature electrochemical reactions to generate electricity. The performance of SOFCs is significantly influenced by the microstructural characteristics of their components, including the electrolyte, anode, and cathode. This section delves into how microstructural properties affect the efficiency and functionality of SOFCs. The electrolyte in an SOFC must exhibit high ionic conductivity while maintaining low electronic conductivity to ensure effective ion transport between the anode and cathode. The microstructure of the electrolyte material plays a crucial role in determining its ionic conductivity. Dense, well-sintered electrolytes with minimal grain boundaries and defects are preferred, as grain boundaries can impede ion migration and contribute to increased resistance [75]. To achieve optimal ionic conductivity, the choice of electrolyte material and processing conditions is vital. Sintering temperature and time significantly impact the grain size and density of the electrolyte. High sintering temperatures generally reduce the number of grain boundaries and improve densification, thereby enhancing ionic conductivity [76]. Additionally, materials such as yttria-stabilized zirconia (YSZ) are commonly used for their excellent ionic transport properties, which are influenced by their microstructural characteristics. The anode in an SOFC is responsible for the oxidation of the fuel, and its microstructure is critical in determining the efficiency of this process [77]. A well-optimized anode microstructure maximizes the surface area available for electrochemical reactions while facilitating efficient fuel diffusion through porous structures. Porosity in the anode allows for effective gas diffusion and reaction with the electrolyte. However, the size, distribution, and connectivity of the pores must be carefully controlled. Large pores can reduce the mechanical stability of the anode and increase the risk of delamination or collapse under thermal cycling [78]. Conversely, an optimal pore structure ensures sufficient connectivity between the electrolyte and the fuel source, reducing mass transport losses and improving overall performance. The mechanical stability of the anode is also an essential consideration, particularly in systems subjected to repeated thermal cycling. The anode must withstand thermal expansion and contraction without significant degradation. Therefore, a balance must be struck between pore size, distribution, and mechanical strength to ensure long-term durability and reliability.

The cathode in an SOFC is crucial for catalyzing the oxygen reduction reaction (ORR), which is central to the cell's operation. The microstructure of the cathode material—comprising its porosity, grain size, and phase distribution—significantly affects the efficiency of the ORR. A porous cathode structure with interconnected pathways is essential for efficient oxygen diffusion and ion transport. The porosity must be optimized to balance between providing sufficient active sites for the ORR and maintaining a structure that supports good mechanical integrity. Techniques such as co-sintering and infiltration can be employed to tailor the microstructure of the cathode, enhancing its catalytic activity and durability. Co-sintering methods allow for the integration of different materials into the cathode structure, improving its performance by enhancing both ionic and electronic conductivity. Infiltration techniques can be used to introduce additional catalytic materials into the cathode, further boosting the ORR kinetics [79].

7.9 Microstructural Engineering for Enhanced Electrochemical Performance

The field of electrochemical devices, ranging from batteries and supercapacitors to fuel cells and sensors, has seen significant advancements through the application of microstructural engineering. By tailoring the microstructure of materials, researchers can optimize properties such as conductivity, durability, and energy density, leading to enhanced device performance. This section delves into the key strategies of microstructural engineering that have proven to be instrumental in pushing the boundaries of electrochemical performance. Nanostructuring involves the manipulation of materials at the nanoscale to achieve specific structural features that can significantly enhance their properties and performance in various applications, including electrochemical devices. This approach leverages the unique properties of nanomaterials, which often differ markedly from their bulk counterparts, to optimize performance in energy storage and conversion systems.

Nanostructuring encompasses a range of techniques designed to create materials with structures at the nanometer scale. These techniques include chemical vapor deposition (CVD), sol–gel processes, electrospinning, and various forms of nanolithography. Each method offers different advantages for tailoring the size, shape, and arrangement of nanostructures. For instance, CVD allows for the deposition of thin films with controlled thickness and uniformity, which is critical for applications in thin film batteries and supercapacitors. The sol–gel process facilitates the creation of nanoparticle suspensions and porous structures, useful for electrodes in fuel cells and batteries. Electrospinning produces nanofibers with high surface area-to-volume ratios, enhancing the performance of electrodes and catalysis. Nanolithography enables the fabrication of nanoscale patterns and structures with high precision, beneficial for microelectrochemical systems and sensors. Nanostructuring significantly influences various aspects of electrochemical performance, including surface area, charge transfer kinetics, and material stability. These effects stem from the unique properties of nanomaterials, which often include high surface-to-volume ratios, quantum effects, and enhanced reactivity.

(a) **Increased Surface Area**: Nanostructured materials, such as nanoparticles, nanowires, and nanorods, offer a dramatically increased surface area compared to bulk materials. This increased surface area provides more active sites for electrochemical reactions, improving the performance of electrodes in batteries, supercapacitors, and fuel cells. For example, in lithium–ion batteries, nanostructured anodes can accommodate higher capacities and deliver faster charge/

discharge rates due to the increased surface area available for lithium–ion intercalation [80].

(b) **Enhanced Charge Transfer Kinetics**: The reduced dimensions of nanostructured materials facilitate shorter electron and ion diffusion paths, which enhances charge transfer kinetics. This improved charge transfer is critical for high-power applications, such as supercapacitors and fuel cells, where rapid charge and discharge processes are required. Nanostructuring can also help in reducing the resistance of electrochemical interfaces, leading to more efficient energy conversion and storage [81].

(c) **Improved Material Stability**: Nanostructuring can also contribute to the stability of electrochemical materials. For instance, nanostructured materials often exhibit enhanced mechanical stability due to their unique structural features. In lithium–ion batteries, nanoscale electrodes can better accommodate the volume changes that occur during charge and discharge cycles, reducing the risk of material degradation and extending the battery's lifespan [82].

(d) **Facilitation of Multiple Reaction Pathways**: Nanostructured materials can provide multiple reaction pathways and accessible active sites, which is advantageous for catalysis in fuel cells and batteries. For example, in catalytic applications, such as oxygen reduction reactions (ORR) in fuel cells, nanostructured catalysts can enhance catalytic activity by exposing more active sites and improving reaction kinetics [83].

(e) **Tailoring Material Properties**: Nanostructuring allows for the precise tuning of material properties to meet specific application requirements. By controlling the size, shape, and distribution of nanostructures, researchers can tailor properties such as electronic conductivity, ionic conductivity, and mechanical strength. This customization is essential for optimizing the performance of electrochemical devices across different operating conditions and applications [84].

While nanostructuring offers significant advantages, it also presents challenges. The synthesis of nanostructured materials often requires precise control over processing conditions to achieve desired properties. Additionally, scaling up the production of nanostructured materials while maintaining uniformity and quality can be challenging. There are also concerns related to the stability and environmental impact of nanomaterials, which must be addressed to ensure their safe and sustainable use. The nanostructuring represents a powerful approach for enhancing the performance of electrochemical devices by leveraging the unique properties of materials at the nanoscale. By increasing surface area, improving charge transfer kinetics, and tailoring material properties, nanostructured materials can significantly boost the efficiency and effectiveness of energy storage and conversion technologies. However, careful consideration of synthesis methods, scalability, and environmental impact is essential to fully realize the potential of nanostructuring in electrochemical applications.

7.9.1 Composite and Hybrid Material Strategies

Composite and hybrid materials are pivotal in advancing the performance of electrochemical devices by combining distinct materials to harness their complementary properties. These strategies enable the creation of materials that exhibit superior characteristics compared to their individual components, enhancing the efficiency, stability, and functionality of energy storage and conversion systems. Composite materials involve the combination of two or more distinct phases, typically a matrix and a reinforcement, to achieve properties that surpass those of the individual constituents [85]. In the context of electrochemical applications, the matrix often consists of polymers, metals, or ceramics, while the reinforcement could be nanoparticles, fibers, or other additives. For instance, in lithium–ion batteries, composites of active materials such as lithium iron phosphate with conductive additives like carbon black are used to enhance electronic conductivity and improve charge/discharge rates [55]. This combination not only boosts the electrical performance but also accommodates the volumetric changes that occur during cycling, thereby extending the battery's lifespan. Similarly, in supercapacitors, composite electrodes that integrate activated carbon with conducting polymers or metal oxides exhibit high surface areas and improved electrical conductivity. This leads to enhanced energy and power densities, which are crucial for applications requiring rapid charge and discharge cycles.

In fuel cells, the incorporation of metals like platinum with carbon supports in composite catalysts improves both catalytic activity and durability, ensuring efficient fuel oxidation and reducing catalyst degradation. Hybrid materials, on the other hand, blend organic and inorganic components to exploit their synergistic effects [86]. These materials can harness the high conductivity of inorganic substances along with the flexibility and processability of organic elements. For example, organic–inorganic hybrids that combine conducting polymers with metal oxides or carbon nanomaterials create systems with enhanced electronic and ionic conductivities. These hybrids are particularly beneficial in batteries and supercapacitors where flexibility and high performance are essential. Metal–organic frameworks (MOFs), another type of hybrid material, feature highly tunable structures and large surface areas. Their high porosity and ability to host various guest species make them suitable for applications such as gas storage, sensing, and catalysis [87]. In batteries, hybrid materials that integrate high-capacity active substances with conductive polymers or carbon-based materials improve overall performance. A notable example is the combination of silicon with carbon nanotubes, which enhances both conductivity and cycle stability. For supercapacitors, hybrids that merge pseudocapacitive metal oxides with high-surface-area carbon materials offer high specific capacitance and power density, leveraging the advantages of both components to achieve improved energy storage capabilities [67].

In catalytic applications, hybrid catalysts that combine noble metals with support materials such as graphene or carbon nanotubes demonstrate enhanced catalytic activity and stability, making them effective in various electrochemical reactions including fuel cell processes and water splitting [88]. Despite the significant advantages, the development of composite and hybrid materials presents several challenges. Ensuring compatibility between different components is critical to achieving the desired properties and performance. The synthesis of these materials must be carefully controlled to prevent issues such as phase separation or poor adhesion between components. Moreover, scaling up the production while maintaining uniformity and quality can be complex. Additionally, the stability and environmental impact of these materials need to be addressed to ensure their safe and sustainable use [89]. Overall, composite and hybrid materials represent a powerful approach to enhancing the performance of electrochemical devices. By combining distinct materials to leverage their complementary properties, these strategies enable the development of advanced energy storage and conversion systems with improved efficiency and functionality.

7.9.2 *Tailoring Microstructure for High Energy Density and Stability*

Tailoring the microstructure of materials is pivotal for optimizing the performance of electrochemical devices, particularly when aiming to achieve high energy density and long-term stability. The design and manipulation of microstructural features play a crucial role in influencing key properties such as charge storage capacity, conductivity, and mechanical integrity, which collectively impact the efficiency and reliability of energy storage and conversion systems. To achieve high energy density in electrochemical devices, several strategies revolve around enhancing the material's ability to store and release energy. One prominent method is nanostructuring, which involves reducing the material dimensions to the nanoscale. Nanostructured materials, due to their significantly increased surface area-to-volume ratio, offer more active sites for electrochemical reactions [90]. This characteristic is particularly beneficial in lithium–ion batteries, where nanostructured anodes and cathodes can deliver higher capacities and faster charge/discharge rates. The nanoscale dimensions facilitate more efficient ion diffusion and better accommodation of volume changes during cycling, thus improving the overall energy density of the device.

Another effective approach is the control of porosity within the material. By introducing controlled porosity, it is possible to enhance the surface area available for electrochemical processes. This strategy is commonly utilized in supercapacitors, where electrodes made from materials with high surface areas, such as activated carbon or metal oxides, provide increased charge storage capacity. Designing the pore size and distribution to align with the specific needs of the device ensures optimal performance and energy density. High surface area materials inherently contribute to increased energy density. Materials such as porous carbon, metal–organic frameworks (MOFs), and certain metal oxides exhibit large surface areas, which facilitate greater interaction with the electrolyte and thus improve energy storage or conversion capabilities. The incorporation of such materials into electrochemical devices can significantly enhance their performance [91]. Ensuring stability over the long term is as crucial as achieving high energy density. Figure 7.10 shows the influence of the Energy Density for Selective Laser Melting on the Microstructure and Mechanical Properties of Stainless Steel. The stability of electrochemical devices is largely influenced by their microstructural design. Mechanical stability is particularly important; materials designed to accommodate volumetric changes without cracking or delaminating tend to exhibit better long-term performance. For example, in lithium–ion batteries, electrodes with nanostructured or composite designs can handle the expansion and contraction associated with lithium–ion intercalation and deintercalation more effectively, thereby extending the battery's cycle life [92]. The interfaces within electrochemical devices, such as those between different phases or between the electrode and electrolyte, are critical to stability. Tailoring the microstructure to optimize these interfaces can minimize resistance and prevent performance degradation. For instance, incorporating coatings or buffer layers can improve the stability of the electrode–electrolyte interface and reduce unwanted side reactions. Microstructural features can also be engineered to mitigate common degradation mechanisms. For example, in high-capacity electrode materials, specific designs can reduce the formation of by-products or secondary phases that impair performance [93]. Additionally, addressing issues like the formation of solid-electrolyte interphases (SEIs) can help maintain high energy density and performance over extended periods. Thermal stability is another essential factor. Tailoring the microstructure to enhance thermal stability can prevent issues such as thermal runaway or degradation under high operating temperatures. Advanced processing techniques can be employed to improve the thermal conductivity and stability of the materials used in electrochemical devices. Despite the advantages, tailoring microstructure involves several challenges. Achieving the desired microstructural features requires precise control over material synthesis and processing conditions [94]. Furthermore, scaling up these techniques and managing the cost of advanced materials or fabrication methods can be complex. The engineered micro-structures as shown in Fig. 7.11 will provide a maintained performance across long-term cycling and varying operating conditions is also a significant consideration. Figure 7.11 shows FEESEM images of (**a**) Ti–48Al–2Cr–2Nb–1.5V–0.1Gd, (**b**) Ti–48Al–2Cr–2Nb, (**c**) copper powder. In summary, the strategic tailoring of microstructure is essential for optimizing both the energy density and stability of electrochemical materials. Through techniques such as nanostructuring, porosity control, and enhancing mechanical and thermal stability, researchers can develop advanced materials that improve the performance and longevity of energy storage and conversion devices. Addressing the associated challenges ensures that these tailored materials can be effectively applied in practical scenarios, advancing the field of electrochemical technology.

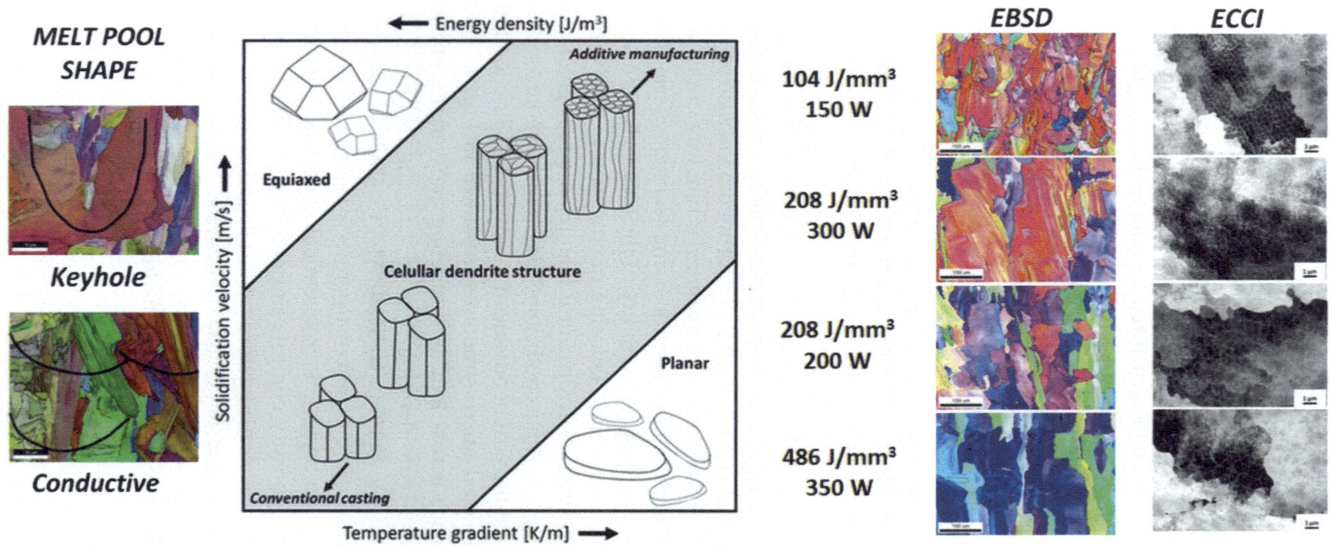

Fig. 7.10 Influence of the energy density for selective laser melting on the microstructure and mechanical properties of stainless steel. Reproduced from Ref. [95] Copyright © 2020 by the authors

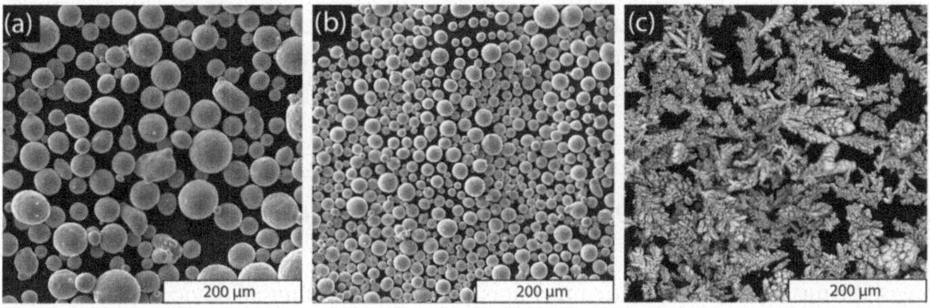

Fig. 7.11 FEESEM images of **a** Ti–48Al–2Cr–2Nb–1.5V–0.1Gd, **b** Ti–48Al–2Cr–2Nb, **c** copper powder. Reproduced from Ref. [96] Copyright © 2020 by the authors

7.10 Linking Microstructure to Electrochemical Degradation Mechanisms

Understanding the link between microstructure and electrochemical degradation mechanisms is essential for enhancing the durability and performance of electrochemical devices. Degradation often arises from changes in the microstructure during operation, which can lead to a decline in the device's efficiency and lifespan. This section explores the evolution of microstructure during cycling and aging, the associated failure mechanisms, and strategies to mitigate degradation through microstructural design.

7.10.1 Microstructural Evolution During Cycling and Aging

Electrochemical devices, such as batteries and supercapacitors, undergo repeated charge–discharge cycles during their operation. These cycles induce significant changes in the microstructure of the active materials. For instance, in lithium–ion batteries, continuous insertion and extraction of lithium ions can cause mechanical stress, leading to the formation of cracks, particle pulverization, and phase transformations. These changes compromise the structural integrity of the electrode, resulting in a loss of capacity over time. Aging, a result of prolonged use, also impacts the microstructure of electrochemical materials. In addition to cycling-induced changes, aging can lead to phenomena such as dendrite formation in lithium–metal batteries or the growth of SEI layers. These microstructural evolutions not only affect the electrochemical performance but also pose safety risks, such as short-circuiting due to dendrite penetration. The extent and nature of microstructural evolution during cycling and aging depend on the material's composition, structure, and morphology. For example, in SOFCs, the grain growth of ceramic electrolytes over time can reduce the number of grain boundaries, leading to a decrease in ionic conductivity [97]. Similarly, in supercapacitors, the agglomeration of nanostructured carbon materials can reduce the effective surface area available for charge storage, thereby lowering the capacitance [98]. Understanding these material-specific behaviors is crucial for designing more resilient electrochemical devices.

7.10.2 Failure Mechanisms Related to Structural Changes

One of the most common failure mechanisms in electrochemical devices is the formation and propagation of cracks within the active materials. These cracks often originate from the mechanical stress caused by the volume changes during ion insertion and extraction. For example, in silicon-based anodes for lithium–ion batteries, the significant volume expansion during lithiation can cause severe cracking, leading to the isolation of active material and a subsequent loss of capacity [99]. Crack propagation is exacerbated by repeated cycling, eventually leading to the mechanical failure of the electrode [100]. Phase transformations are another critical failure mechanism linked to microstructural changes [101]. During electrochemical cycling, materials may undergo transformations from one phase to another, accompanied by changes in crystal structure, volume, and mechanical properties. In some cases, these transformations are irreversible, leading to the degradation of electrochemical performance. For instance, in high-capacity cathode materials like Li-rich layered oxides, the transition from a layered structure to a spinel or rock-salt phase during cycling can result in reduced capacity and increased resistance. The interfaces between different components in electrochemical devices, such as the electrode–electrolyte interface, are particularly susceptible to degradation. Microstructural changes at these interfaces, such as the formation of SEI layers in lithium–ion batteries or the delamination of electrode layers in SOFCs, can significantly impact device performance. Interfacial degradation often leads to increased resistance, reduced ion transport, and ultimately, the failure of the electrochemical device. In nanostructured materials, particle aggregation is a common issue that arises from changes in microstructure during operation. Aggregation reduces the effective surface area available for electrochemical reactions, leading to a decline in performance. For example, in supercapacitors, the aggregation of carbon nanotubes or graphene sheets can decrease the surface area and porosity, thereby lowering the capacitance and energy density. This phenomenon is particularly pronounced in high-power applications where the material is subjected to rapid cycling [97].

7.10.3 Strategies to Mitigate Degradation Through Microstructural Design

To mitigate degradation, one effective strategy is to design nanostructured materials that can accommodate the mechanical stress associated with cycling. For example, in silicon anodes for lithium–ion batteries, nanostructuring the silicon into nanowires or hollow nanoparticles can help alleviate the stress caused by volume expansion [56]. This approach allows the material to undergo large changes in volume without fracturing, thereby enhancing its cycle life. Additionally, using conductive coatings or embedding silicon nanoparticles in a flexible matrix can further improve structural integrity [102]. Another approach to mitigating degradation is to use phase-stabilized materials that resist undesirable phase transformations during cycling. This can be achieved by doping, alloying, or designing materials with inherent phase stability. For instance, in cathode materials, doping with elements like aluminum or magnesium can stabilize the crystal structure and prevent the phase transformations that lead to capacity fade [103]. Similarly, in solid-state electrolytes, designing materials with a stable cubic phase can enhance ionic conductivity and prevent grain growth during operation. Improving the stability of interfaces through microstructural design is crucial for preventing interfacial degradation. One strategy is to use coatings that protect the electrode surface and maintain a stable interface with the electrolyte. For example, in lithium–ion batteries, applying a thin layer of a stable oxide or phosphate material on the cathode surface can prevent the formation of a resistive SEI layer [104]. Another approach is to engineer the microstructure of the electrode to enhance adhesion and prevent delamination. In SOFCs, for instance, using graded or composite electrode materials can improve mechanical stability and reduce the risk of delamination [105].

Controlling the particle size and morphology of active materials is another effective strategy for mitigating degradation. Smaller particles with optimized shapes can reduce the stress associated with volume changes and minimize the formation of cracks. For example, using nanosized particles in lithium–ion battery electrodes can reduce the diffusion distance for lithium ions, thereby lowering the mechanical strain during cycling [106]. Additionally, designing materials with porous or hierarchical structures can help accommodate volume changes and maintain high surface area, enhancing both performance and stability.

Implementing advanced manufacturing techniques, such as atomic layer deposition (ALD) or molecular layer deposition (MLD), allows for precise control over the microstructure of materials. These techniques can be used to create thin, conformal coatings on electrodes, which protect against degradation while maintaining electrochemical performance [107]. Additionally, techniques like 3D printing or electrospinning can be employed to fabricate materials with complex architectures that enhance stress accommodation, reduce particle aggregation, and improve interfacial stability. By leveraging these advanced manufacturing methods, researchers can design electrochemical devices with tailored microstructures that resist degradation over extended periods of operation.

7.11 Challenges and Future Directions

The interplay between microstructure and electrochemical performance in devices is a dynamic and evolving field, but it comes with its set of challenges. As the demand for high-performance electrochemical devices grows, so does the need to overcome the limitations in microstructural characterization and modeling, to integrate advanced predictive tools like machine learning, and to stay ahead of emerging trends in research.

One of the primary challenges in this field is the accurate characterization and modeling of microstructures at different scales. While advances in imaging techniques like TEM, XRD, and AFM have significantly improved our ability to visualize and analyze microstructures, there are still significant gaps. High-resolution characterization techniques often provide a wealth of data, but this data can be difficult to interpret in the context of electrochemical performance. Additionally, these techniques are often time-consuming, expensive, and may not fully capture dynamic processes occurring during device operation, such as the evolution of microstructures under real-time cycling conditions. Modeling these microstructural changes also presents considerable difficulties. Current models often rely on simplifying assumptions that may not fully represent the complex interactions between microstructural features and electrochemical processes. For example, in modeling the degradation mechanisms in lithium–ion batteries, assumptions about uniform stress distribution or ion diffusion might not hold true across different scales of microstructure. Furthermore, there is a need for multi-scale models that can bridge the gap between atomic-level phenomena and macroscopic device behavior, which remains a significant challenge [108].

The integration of machine learning (ML) into microstructure-electrochemistry research presents a promising pathway to address some of these challenges. ML algorithms can handle large datasets generated from experimental and simulation studies, enabling the identification of complex, non-linear relationships between microstructural features and electrochemical performance. This capability is particularly valuable for predictive modeling, where traditional approaches may fall short. For instance, ML can be used to predict the impact of microstructural changes on battery life or the efficiency of fuel cells by learning from vast datasets of cycling data, microstructural images, and performance metrics. Moreover, ML can assist in optimizing microstructural design by rapidly screening through potential material compositions and configurations, predicting their performance outcomes without the need for exhaustive experimentation. However, the application of ML in this domain is not without challenges. One of the main hurdles is the need for large, high-quality datasets that are often required to train ML models effectively. The heterogeneity in data sources, coupled with the complexity of microstructural information, can make it difficult to develop robust models. Additionally, there is a need for interpretability in ML predictions to ensure that the models provide insights that are actionable and grounded in physical principles rather than just empirical correlations [109].

Looking ahead, several emerging trends are likely to shape the future of microstructure-electrochemistry research. One significant trend is the development of in situ and operando characterization techniques. These techniques allow for the real-time observation of microstructural changes during device operation, providing invaluable insights into the dynamic processes that govern electrochemical performance. Advances in synchrotron-based X-ray techniques, environmental TEM, and other real-time imaging methods are expected to play a critical role in this area [110]. Another emerging trend is the shift towards sustainable and green electrochemical technologies, which brings new challenges and opportunities in microstructural design. For example, the push towards solid-state batteries and the use of bio-derived materials in energy storage devices require a rethinking of microstructural strategies to maintain performance while ensuring environmental sustainability [111]. This trend is also driving research into more sustainable manufacturing processes that can produce optimized microstructures with lower environmental impact. Additionally, there is growing interest in the use of advanced manufacturing techniques, such as additive manufacturing and atomic layer deposition, to create materials with precisely controlled microstructures. These techniques offer the potential to engineer microstructures with unprecedented precision, enabling the design of materials that can meet the specific demands of next-generation electrochemical devices [14]. In summary, while the field of microstructure-electrochemistry research is poised for significant advancements, it also faces considerable challenges. Overcoming the limitations in characterization and modeling, leveraging machine learning for predictive insights, and staying attuned to emerging trends will be crucial for driving the next wave of innovations in this area. By addressing these challenges and embracing future directions, researchers can unlock new possibilities for designing more efficient, durable, and sustainable electrochemical devices.

7.12 Conclusion

The role of microstructural properties in electrochemical studies and devices is undeniably profound. Throughout this chapter, we have explored how the microstructure of materials—from the atomic scale to the macroscopic scale—directly influences the performance, durability, and overall efficiency of electrochemical devices. The intricate relationship between microstructure and electrochemical behavior underscores the need for precise control and engineering of microstructural features to optimize device functionality.

Microstructural properties, including grain size, phase distribution, porosity, and the nature of interfaces, critically affect key electrochemical parameters such as ionic and electronic conductivity, reaction kinetics, and mechanical stability. For instance, nanostructuring can enhance surface area and improve charge transport, while composite and hybrid material strategies can synergize the strengths of different phases to create more robust and efficient devices. Tailoring microstructure for specific applications, such as achieving high energy density or enhancing stability under cycling, is essential for meeting the growing demands of advanced energy storage and conversion technologies.

The degradation mechanisms that arise from microstructural changes during operation highlight the importance of understanding and controlling these properties. Cracking, phase transformations, and interfacial degradation are just a few of the challenges that can be mitigated through thoughtful microstructural design. Moreover, the integration of advanced manufacturing techniques and predictive tools, such as machine learning, opens new avenues for optimizing microstructural properties and anticipating their impact on device performance.

As the field of electrochemical devices continues to evolve, the ability to engineer microstructures with precision will be pivotal in driving the next generation of high-performance, durable, and sustainable technologies. By harnessing the full potential of microstructural engineering, we can push the boundaries of what is possible in energy storage, conversion, and beyond, ultimately contributing to a more efficient and resilient energy future.

References

1. Greco, G., et al.: Influence of the electrode nano/microstructure on the electrochemical properties of graphite in aluminum batteries. J. Mater. Chem. A Mater. **6**(45), 22673–22680 (2018). https://doi.org/10.1039/C8TA08319C
2. Wang, K.X., Li, X.H., Chen, J.S.: Surface and interface engineering of electrode materials for lithium-ion batteries. Adv. Mater. **27**(3), 527–545 (2015). https://doi.org/10.1002/ADMA.201402962
3. Chen, G., Yan, L., Luo, H., Guo, S.: Nanoscale engineering of heterostructured anode materials for boosting lithium-ion storage. Adv. Mater. **28**(35), 7580–7602 (2016). https://doi.org/10.1002/ADMA.201600164
4. Figueiredo, F.M.L., Marques, F.M.B.: Electrolytes for solid oxide fuel cells. Wiley Interdiscip. Rev. Energy Environ. **2**(1), 52–72 (2013). https://doi.org/10.1002/WENE.23
5. Li, J., et al.: Fabrication of high performance structural N-doped hierarchical porous carbon for supercapacitors. Carbon N Y **164**, 42–50 (2020). https://doi.org/10.1016/J.CARBON.2020.03.044
6. Pender, J.P., et al.: Electrode degradation in lithium-ion batteries. ACS Nano **14**(2), 1243–1295 (2020). https://doi.org/10.1021/ACSNANO.9B04365/ASSET/IMAGES/LARGE/NN9B04365_0028.JPEG
7. Mistry, A., Franco, A.A., Cooper, S.J., Roberts, S.A., Viswanathan, V.: How machine learning will revolutionize electrochemical sciences. ACS Energy Lett. **6**(4), 1422–1431 (2021). https://doi.org/10.1021/ACSENERGYLETT.1C00194/ASSET/IMAGES/LARGE/NZ1C00194_0005.JPEG
8. Chizari, S., et al.: Current challenges and potential directions towards precision microscale additive manufacturing—Part III: Energy induced deposition and hybrid electrochemical processes. Precis. Eng. **68**, 174–186 (2021). https://doi.org/10.1016/J.PRECISIONENG.2020.12.013
9. Clemens, H., Mayer, S., Scheu, C.: Microstructure and properties of engineering materials. In: Neutrons and Synchrotron Radiation in Engineering Materials Science: From Fundamentals to Applications, 2nd edn, pp. 1–20 (2017). https://doi.org/10.1002/9783527684489.CH1
10. Freitas, R., Cao, Y.: Machine-learning potentials for crystal defects. MRS Commun. **12**(5), 510–520 (2022). https://doi.org/10.1557/S43579-022-00221-5/FIGURES/4

11. Zhou, L., et al.: Recent developments on and prospects for electrode materials with hierarchical structures for lithium-ion batteries. Adv. Energy Mater. **8**(6), 1701415 (2018). https://doi.org/10.1002/AENM.201701415
12. Vu, A., Qian, Y., Stein, A.: Porous electrode materials for lithium-ion batteries—how to prepare them and what makes them special. Adv. Energy Mater. **2**(9), 1056–1085 (2012). https://doi.org/10.1002/AENM.201200320
13. Niu, J., et al.: Biomass-derived mesopore-dominant porous carbons with large specific surface area and high defect density as high performance electrode materials for Li-ion batteries and supercapacitors. Nano Energy **36**, 322–330 (2017). https://doi.org/10.1016/J.NANOEN.2017.04.042
14. Kok, Y., et al.: Anisotropy and heterogeneity of microstructure and mechanical properties in metal additive manufacturing: a critical review. Mater. Des. **139**, 565–586 (2018). https://doi.org/10.1016/J.MATDES.2017.11.021
15. Zheng, J., Archer, L.A.: Crystallographically textured electrodes for rechargeable batteries: symmetry, fabrication, and characterization. Chem. Rev. **122**(18), 14440–14470 (2022). https://doi.org/10.1021/ACS.CHEMREV.2C00022/ASSET/IMAGES/MEDIUM/CR2C00022_0026.GIF
16. Sathiyamoorthi, P., Kim, H.S.: High-entropy alloys with heterogeneous microstructure: processing and mechanical properties. Prog. Mater. Sci. **123**, 100709 (2022). https://doi.org/10.1016/J.PMATSCI.2020.100709
17. Shreya, Phogat, P., Jha, R., Singh, S.: Enhanced electrochemical performance and charge-transfer dynamics of 2D MoS_2/WO_3 nanocomposites for futuristic energy applications. ACS Appl. Nano Mater. (2024). https://doi.org/10.1021/ACSANM.3C06017
18. Dong, H., Koenig, G.M.: A review on synthesis and engineering of crystal precursors produced via coprecipitation for multicomponent lithium-ion battery cathode materials. CrystEngComm **22**(9), 1514–1530 (2020). https://doi.org/10.1039/C9CE00679F
19. Hanif, M.B., Rauf, S., Motola, M., Babar, Z.U.D., Li, C.J., Li, C.X.: Recent progress of perovskite-based electrolyte materials for solid oxide fuel cells and performance optimizing strategies for energy storage applications. Mater. Res. Bull. **146**, 111612 (2022). https://doi.org/10.1016/J.MATERRESBULL.2021.111612
20. Ji, Q., Bi, L., Zhang, J., Cao, H., Zhao, X.S.: The role of oxygen vacancies of ABO_3 perovskite oxides in the oxygen reduction reaction. Energy Environ. Sci. **13**(5), 1408–1428 (2020). https://doi.org/10.1039/D0EE00092B
21. Share, K., Westover, A., Li, M., Pint, C.L.: Surface engineering of nanomaterials for improved energy storage—a review. Chem. Eng. Sci. **154**, 3–19 (2016). https://doi.org/10.1016/J.CES.2016.05.034
22. Mohammed, A., Abdullah, A.: Scanning electron microscopy (SEM): a review
23. Zaimbashi, R., Tajik, S., Beitollahi, H., Torkzadeh-Mahani, M.: Fabrication of a novel and ultrasensitive label-free electrochemical aptasensor based on gold nanostructure for detection of homocysteine. Biosensors **13**(2), 244 (2023). https://doi.org/10.3390/BIOS13020244
24. Lei, C.: Transmission electron microscopy. In: Nanotechnology Research Methods for Foods and Bioproducts, pp. 127–144 (2012). https://doi.org/10.1002/9781118229347.CH7
25. Grimley, E.D., LeBeau, J.M.: Transmission electron microscopy (STEM and TEM). In: Ferroelectricity in Doped Hafnium Oxide: Materials, Properties and Devices, pp. 317–340 (2019). https://doi.org/10.1016/B978-0-08-102430-0.00015-2
26. Bunaciu, A.A., Gabriela Udriştioiu, E., Aboul-Enein, H.Y.: X-Ray diffraction: instrumentation and applications. Crit. Rev. Anal. Chem. **45**(4), 289–299 (2015). https://doi.org/10.1080/10408347.2014.949616
27. Hura, G.L., et al.: Robust, high-throughput solution structural analyses by small angle X-ray scattering (SAXS). Nat. Methods **6**(8), 606–612 (2009). https://doi.org/10.1038/nmeth.1353
28. Adams, F., Barbante, C.: Electron-based imaging techniques. Compr. Anal. Chem. **69**, 269–313 (2015). https://doi.org/10.1016/B978-0-444-63439-9.00007-4
29. Rodenas, T., Prieto, G.: FIB-SEM tomography in catalysis and electrochemistry. Catal. Today **405–406**, 2–13 (2022). https://doi.org/10.1016/J.CATTOD.2022.09.013

30. Tordoff, B., et al.: The LaserFIB: new application opportunities combining a high-performance FIB-SEM with femtosecond laser processing in an integrated second chamber. Appl. Microsc. **50**(1), 1–11 (2020). https://doi.org/10.1186/S42649-020-00044-5/FIGURES/9
31. Ylä-Anttila, P., Vihinen, H., Jokitalo, E., Eskelinen, E.L.: 3D tomography reveals connections between the phagophore and endoplasmic reticulum. Autophagy **5**(8), 1180–1185 (2009). https://doi.org/10.4161/AUTO.5.8.10274
32. Kunishima, N., Takeda, Y., Hirose, R., Kalasová, D., Šalplachta, J., Omote, K.: Visualization of internal 3D structure of small live seed on germination by laboratory-based X-ray microscopy with phase contrast computed tomography. Plant Methods **16**(1), 1–10 (2020). https://doi.org/10.1186/S13007-020-0557-Y/FIGURES/5
33. Pires, L.F., Roque, W.L., Rosa, J.A., Mooney, S.J.: 3D analysis of the soil porous architecture under long term contrasting management systems by X-ray computed tomography. Soil Tillage Res. **191**, 197–206 (2019). https://doi.org/10.1016/J.STILL.2019.02.018
34. Zhao, W., Chen, Y., Shen, L., Yi, A.Y.: Investigation of the refractive index distribution in precision compression glass molding by use of 3D tomography. Meas. Sci. Technol. **20**(5), 055109 (2009). https://doi.org/10.1088/0957-0233/20/5/055109
35. Greuter, F., Blatter, G.: Electrical properties of grain boundaries in polycrystalline compound semiconductors. Semicond. Sci. Technol. **5**(2), 111 (1990). https://doi.org/10.1088/0268-1242/5/2/001
36. He, X., Sun, H., Ding, X., Zhao, K.: Grain boundaries and their impact on Li kinetics in layered-oxide cathodes for Li-ion batteries. J. Phys. Chem. C **125**(19), 10284–10294 (2021). https://doi.org/10.1021/ACS.JPCC.1C02400/SUPPL_FILE/JP1C02400_SI_001.PDF
37. Kim, M.G., Cho, J.: Reversible and high-capacity nanostructured electrode materials for Li-ion batteries. Adv. Funct. Mater. **19**(10), 1497–1514 (2009). https://doi.org/10.1002/ADFM.200801095
38. Hou, J., Shao, Y., Ellis, M.W., Moore, R.B., Yi, B.: Graphene-based electrochemical energy conversion and storage: fuel cells, supercapacitors and lithium ion batteries. Phys. Chem. Chem. Phys. **13**(34), 15384–15402 (2011). https://doi.org/10.1039/C1CP21915D
39. Swain, N., Saravanakumar, B., Kundu, M., Schmidt-Mende, L., Ramadoss, A.: Recent trends in template assisted 3D porous materials for electrochemical supercapacitors. J. Mater. Chem. A Mater. **9**(45), 25286–25324 (2021). https://doi.org/10.1039/D1TA06122D
40. Bandyopadhyay, A., Mitra, I., Avila, J.D., Upadhyayula, M., Bose, S.: Porous metal implants: processing, properties, and challenges. Int. J. Extreme Manufact. **5**(3), 032014 (2023). https://doi.org/10.1088/2631-7990/ACDD35
41. Yang, Y., et al.: Elastic properties, defect thermodynamics, electrochemical window, phase stability, and Li^+ Mobility of Li_3PS_4: insights from first-principles calculations. ACS Appl. Mater. Interfaces **8**(38), 25229–25242 (2016). https://doi.org/10.1021/ACSAMI.6B06754/SUPPL_FILE/AM6B06754_SI_001.PDF
42. Zhou, L., et al.: Dislocation effect boosting the electrochemical properties of Prussian blue analogues for 2.6 V high-voltage aqueous zinc-based batteries. ACS Appl. Mater. Interfaces (2024). https://doi.org/10.1021/ACSAMI.4C07693/SUPPL_FILE/AM4C07693_SI_001.PDF
43. Szroeder, P., Sagalianov, I.Y., Radchenko, T.M., Tatarenko, V.A., Prylutskyy, Y.I., Strupiński, W.: Effect of uniaxial stress on the electrochemical properties of graphene with point defects. Appl. Surf. Sci. **442**, 185–188 (2018). https://doi.org/10.1016/J.APSUSC.2018.02.150
44. Hall, P.J., et al.: Energy storage in electrochemical capacitors: designing functional materials to improve performance. Energy Environ. Sci. **3**(9), 1238–1251 (2010). https://doi.org/10.1039/C0EE00004C
45. Perry, N.H., Ishihara, T.: Roles of bulk and surface chemistry in the oxygen exchange kinetics and related properties of mixed conducting perovskite oxide electrodes. Materials **9**(10), 858 (2016). https://doi.org/10.3390/MA9100858
46. Jannesari, H., Emami, M.D., Ziegler, C.: Effect of electrolyte transport properties and variations in the morphological parameters on the variation of side reaction rate across the anode electrode and the aging of lithium ion batteries. J. Power. Sources **196**(22), 9654–9664 (2011). https://doi.org/10.1016/J.JPOWSOUR.2011.07.026

47. Cho, H.M., Chen, M.V., MacRae, A.C., Meng, Y.S.: Effect of surface modification on nanostructured $LiNi_{0.5}Mn_{1.5}O_4$ spinel materials. ACS Appl. Mater. Interfaces **7**(30), 16231–16239 (2015). https://doi.org/10.1021/ACSAMI.5B01392/SUPPL_FILE/AM5B01392_SI_001.PDF
48. Fu, L.J., et al.: Surface modifications of electrode materials for lithium ion batteries. Solid State Sci. **8**(2), 113–128 (2006). https://doi.org/10.1016/J.SOLIDSTATESCIENCES.2005.10.019
49. Cheng, L., et al.: Effect of surface microstructure on electrochemical performance of garnet solid electrolytes. ACS Appl. Mater. Interfaces **7**(3), 2073–2081 (2015). https://doi.org/10.1021/AM508111R/SUPPL_FILE/AM508111R_SI_001.PDF
50. Cui, Q., Yi, D., Wang, H., Zhang, J., Xu, J., Wang, B.: Effects of grain size and secondary phase on corrosion behavior and electrochemical performance of Mg-3Al-5Pb-1Ga-Y sacrificial anode. J. Rare Earths **37**(12), 1341–1350 (2019). https://doi.org/10.1016/J.JRE.2018.11.012
51. Choi, M., Koo, J.Y., Ahn, M., Lee, W.: Effects of grain boundaries at the electrolyte/cathode interfaces on oxygen reduction reaction kinetics of solid oxide fuel cells. Bull. Korean Chem. Soc. **38**(4), 423–428 (2017). https://doi.org/10.1002/BKCS.11102
52. Wang, Y., Li, M., Ren, H.: Interfacial structure and energy determine the heterogeneity in the electrochemical metal dissolution activity at grain boundary. Chem. Mater. **35**(11), 4243–4249 (2023). https://doi.org/10.1021/ACS.CHEMMATER.3C00220/SUPPL_FILE/CM3C00220_SI_002.AVI
53. Yan, P., et al.: Tailoring grain boundary structures and chemistry of Ni-rich layered cathodes for enhanced cycle stability of lithium-ion batteries. Nat. Energy **3**(7), 600–605 (2018). https://doi.org/10.1038/s41560-018-0191-3
54. Wang, L., Chen, B., Ma, J., Cui, G., Chen, L.: Reviving lithium cobalt oxide-based lithium secondary batteries-toward a higher energy density. Chem. Soc. Rev. **47**(17), 6505–6602 (2018). https://doi.org/10.1039/C8CS00322J
55. Jugović, D., Uskoković, D.: A review of recent developments in the synthesis procedures of lithium iron phosphate powders. J. Power. Sources **190**(2), 538–544 (2009). https://doi.org/10.1016/J.JPOWSOUR.2009.01.074
56. Rahman, M.A., Song, G., Bhatt, A.I., Wong, Y.C., Wen, C.: Nanostructured silicon anodes for high-performance lithium-ion batteries. Adv. Funct. Mater. **26**(5), 647–678 (2016). https://doi.org/10.1002/ADFM.201502959
57. Phillip, N.D., Westover, A.S., Daniel, C., Veith, G.M.: Structural degradation of high voltage lithium nickel manganese cobalt oxide (NMC) cathodes in solid-state batteries and implications for next generation energy storage. ACS Appl. Energy Mater. **3**(2), 1768–1774 (2020). https://doi.org/10.1021/ACSAEM.9B02230/SUPPL_FILE/AE9B02230_SI_001.PDF
58. Niu, T., et al.: Stable high-performance perovskite solar cells via grain boundary passivation. Adv. Mater. **30**(16), 1706576 (2018). https://doi.org/10.1002/ADMA.201706576
59. Almar, L., Joos, J., Weber, A., Ivers-Tiffée, E.: Microstructural feature analysis of commercial Li-ion battery cathodes by focused ion beam tomography. J. Power. Sources **427**, 1–14 (2019). https://doi.org/10.1016/J.JPOWSOUR.2019.04.019
60. Hou, D., et al.: Effect of the grain arrangements on the thermal stability of polycrystalline nickel-rich lithium-based battery cathodes. Nat. Commun. **13**(1), 1–11 (2022). https://doi.org/10.1038/s41467-022-30935-y
61. Shahgaldi, S., Hamelin, J.: Improved carbon nanostructures as a novel catalyst support in the cathode side of PEMFC: a critical review. Carbon N Y **94**, 705–728 (2015). https://doi.org/10.1016/J.CARBON.2015.07.055
62. Shreya, Phogat, P., Jha, R., Singh, S.: Carbon nanospheres-induced enhanced capacitive dynamics in $C/WS_2/WO_3$ nanocomposites for high-performance electrochemical capacitors. Mater. Sci. Eng. B **304**, 117390 (2024). https://doi.org/10.1016/J.MSEB.2024.117390
63. Ji, M., Wei, Z.: A review of water management in polymer electrolyte membrane fuel cells. Energies **2**(4), 1057–1106 (2009). https://doi.org/10.3390/EN20401057
64. Okonkwo, P.C., Otor, C.: A review of gas diffusion layer properties and water management in proton exchange membrane fuel cell system. Int. J. Energy Res. **45**(3), 3780–3800 (2021). https://doi.org/10.1002/ER.6227

65. Dai, W., et al.: A review on water balance in the membrane electrode assembly of proton exchange membrane fuel cells. Int. J. Hydrogen Energy **34**(23), 9461–9478 (2009). https://doi.org/10.1016/J.IJHYDENE.2009.09.017
66. Dujearic-Stephane, K., et al.: The effect of modifications of activated carbon materials on the capacitive performance: surface, microstructure, and wettability. J. Compos. Sci. **5**(3), 66 (2021). https://doi.org/10.3390/JCS5030066
67. Yao, F., Pham, D.T., Lee, Y.H.: Carbon-based materials for lithium-ion batteries, electrochemical capacitors, and their hybrid devices. ChemSusChem **8**(14), 2284–2311 (2015). https://doi.org/10.1002/CSSC.201403490
68. Tang, H.L., et al.: Multilayer graphene-WSe_2 heterostructures for WSe_2 transistors. ACS Nano **11**(12), 12817–12823 (2017). https://doi.org/10.1021/ACSNANO.7B07755/SUPPL_FILE/NN7B07755_SI_001.PDF
69. Wang, H., et al.: Micro-meso porous structured carbon nanofibers with ultra-high surface area and large supercapacitor electrode capacitance. J. Power. Sources **482**, 228986 (2021). https://doi.org/10.1016/J.JPOWSOUR.2020.228986
70. Tiwari, S.K., Thakur, S.K., De Adhikari, A., Zhu, Y., Wang, N.: Current research of graphene-based nanocomposites and their application for supercapacitors. Nanomater. **10**(10), 2046 (2020). https://doi.org/10.3390/NANO10102046
71. Bhati, V.S., Hojamberdiev, M., Kumar, M.: Enhanced sensing performance of ZnO nanostructures-based gas sensors: a review. Energy Rep. **6**, 46–62 (2020). https://doi.org/10.1016/J.EGYR.2019.08.070
72. Dong, C., Zhao, R., Yao, L., Ran, Y., Zhang, X., Wang, Y.: A review on WO_3 based gas sensors: morphology control and enhanced sensing properties. J. Alloys Compd. **820**, 153194 (2020). https://doi.org/10.1016/J.JALLCOM.2019.153194
73. Labidi, A., et al.: Impedance spectroscopy on WO_3 gas sensor. Sens. Actuators B Chem. **106**(2), 713–718 (2005). https://doi.org/10.1016/J.SNB.2004.09.022
74. Singh, S., Dogra, N., Sharma, S.: A sensitive H_2S sensor using MoS_2/WO_3 composite. Mater. Today Proc. **28**, 8–10 (2020). https://doi.org/10.1016/J.MATPR.2019.12.104
75. Patra, S., Narayanasamy, J., Panneerselvam, T., Murugan, R.: Review—microstructural modification in lithium garnet solid-state electrolytes: emerging trends. J. Electrochem. Soc. **169**(3), 030548 (2022). https://doi.org/10.1149/1945-7111/AC5C99
76. Botros, M., Djenadic, R., Clemens, O., Möller, M., Hahn, H.: Field assisted sintering of fine-grained $Li_{7-3x}La_3Zr_2Al_xO_{12}$ solid electrolyte and the influence of the microstructure on the electrochemical performance. J. Power. Sources **309**, 108–115 (2016). https://doi.org/10.1016/J.JPOWSOUR.2016.01.086
77. Ren, X., Pan, W.: Mechanical properties of high-temperature-degraded yttria-stabilized zirconia. Acta Mater. **69**, 397–406 (2014). https://doi.org/10.1016/J.ACTAMAT.2014.01.017
78. Tran, C., Yang, X.Q., Qu, D.: Investigation of the gas-diffusion-electrode used as lithium/air cathode in non-aqueous electrolyte and the importance of carbon material porosity. J. Power. Sources **195**(7), 2057–2063 (2010). https://doi.org/10.1016/J.JPOWSOUR.2009.10.012
79. Zhang, W., Hu, Y.H.: Recent progress in design and fabrication of SOFC cathodes for efficient catalytic oxygen reduction. Catal. Today **409**, 71–86 (2023). https://doi.org/10.1016/J.CATTOD.2022.05.008
80. Phogat, P., Shreya, Jha, R., Singh, S.: Supercapacitive studies of hybrid materials based on cadmium deuteroxide chloride (CdDOCl) with activated carbon. J. Mater. Sci. **59**(26), 11757–11780 (2024). https://doi.org/10.1007/S10853-024-09874-0/METRICS
81. Shreya., Phogat, P., Jha, R., Singh, S.: Emerging advances and future prospects of two dimensional nanomaterials based solar cells. J. Alloys Compd. 175063 (2024). https://doi.org/10.1016/J.JALLCOM.2024.175063
82. Phogat, P., Shreya, Jha, R., Singh, S.: Synthesis of novel ZnO nanoparticles with optimized band gap of 1.4 eV for high-sensitivity photo electrochemical detection. Mater. Today Sustain. **27**, 100823 (2024). https://doi.org/10.1016/J.MTSUST.2024.100823
83. Xia, Z., An, L., Chen, P., Xia, D.: Non-Pt nanostructured catalysts for oxygen reduction reaction: synthesis, catalytic activity and its key factors. Adv. Energy Mater. **6**(17), 1600458 (2016). https://doi.org/10.1002/AENM.201600458

84. Shreya, Phogat, P., Jha, R., Singh, S.: Electrochemical study of cerium and iron doped MoO3 nanoparticles showing potential for supercapacitor application. Next Mater. **5**, 100260 (2024). https://doi.org/10.1016/J.NXMATE.2024.100260
85. Shreya, Phogat, P., Jha, R., Singh, S.: Electrochemical analysis of hydrothermally synthesized 2D/1D WS_2/WO_3 nanocomposites for solar cell application. J. Phys. Chem. Solids **192**, 112110 (2024). https://doi.org/10.1016/J.JPCS.2024.112110
86. Qiao, Z., Wang, C., Zeng, Y., Spendelow, J.S., Wu, G.: Advanced nanocarbons for enhanced performance and durability of platinum catalysts in proton exchange membrane fuel cells. Small **17**(48), 2006805 (2021). https://doi.org/10.1002/SMLL.202006805
87. Zhou, H.C.J., Kitagawa, S.: Metal–organic frameworks (MOFs). Chem. Soc. Rev. **43**(16), 5415–5418 (2014). https://doi.org/10.1039/C4CS90059F
88. Cheng, Y., Fan, Y., Pei, Y., Qiao, M.: Graphene-supported metal/metal oxide nanohybrids: synthesis and applications in heterogeneous catalysis. Catal. Sci. Technol. **5**(8), 3903–3916 (2015). https://doi.org/10.1039/C5CY00630A
89. Melchionna, M., Marchesan, S., Prato, M., Fornasiero, P.: Carbon nanotubes and catalysis: the many facets of a successful marriage. Catal. Sci. Technol. **5**(8), 3859–3875 (2015). https://doi.org/10.1039/C5CY00651A
90. Phogat, P., Shreya, Jha, R., Singh, S.: Synthesis and characterization of C@CdS core-shell structures for high-performance capacitors. Next Mater. **5**, 100246 (2024). https://doi.org/10.1016/J.NXMATE.2024.100246
91. Bae, J.H., Han, J.H., Chung, T.D.: Electrochemistry at nanoporous interfaces: new opportunity for electrocatalysis. Phys. Chem. Chem. Phys. **14**(2), 448–463 (2011). https://doi.org/10.1039/C1CP22927C
92. Chen, R., Luo, R., Huang, Y., Wu, F., Li, L.: Advanced high energy density secondary batteries with multi-electron reaction materials. Adv. Sci. **3**(10), 1600051 (2016). https://doi.org/10.1002/ADVS.201600051
93. Hy, S., Liu, H., Zhang, M., Qian, D., Hwang, B.J., Meng, Y.S.: Performance and design considerations for lithium excess layered oxide positive electrode materials for lithium ion batteries. Energy Environ. Sci. **9**(6), 1931–1954 (2016). https://doi.org/10.1039/C5EE03573B
94. Velumani, D., Bansal, A.: Thermal behavior of lithium- and sodium-ion batteries: a review on heat generation, battery degradation, thermal runway—perspective and future directions. Energy Fuels **36**(23), 14000–14029 (2022). https://doi.org/10.1021/ACS.ENERGYFUELS.2C02889/ASSET/IMAGES/MEDIUM/EF2C02889_0020.GIF
95. Donik, Č., Kraner, J., Paulin, I., Godec, M.: Influence of the energy density for selective laser melting on the microstructure and mechanical properties of stainless steel. Metals **10**(7), 919 (2020). https://doi.org/10.3390/MET10070919
96. Polozov, I., Sokolova, V., Gracheva, A., Popovich, A.: Tailoring the microstructure of laser-additive-manufactured titanium aluminide alloys via in situ alloying and parameter variation. Metals **13**(8), 1429 (2023). https://doi.org/10.3390/MET13081429
97. Zhu, Y., et al.: Electrode/electrolyte interphases in high-temperature batteries: a review. Energy Environ. Sci. **16**(7), 2825–2855 (2023). https://doi.org/10.1039/D3EE00439B
98. Cheng, F., Yang, X., Zhang, S., Lu, W.: Boosting the supercapacitor performances of activated carbon with carbon nanomaterials. J. Power Sources **450**, 227678 (2020). https://doi.org/10.1016/J.JPOWSOUR.2019.227678
99. Zhang, C., et al.: Challenges and recent progress on silicon-based anode materials for next-generation lithium-ion batteries. Small Struct. **2**(6), 2100009 (2021). https://doi.org/10.1002/SSTR.202100009
100. Xu, R., Zhao, K.: Corrosive fracture of electrodes in Li-ion batteries. J. Mech. Phys. Solids **121**, 258–280 (2018). https://doi.org/10.1016/J.JMPS.2018.07.021
101. Liens, A., et al.: Phase transformation induces plasticity with negligible damage in ceria-stabilized zirconia-based ceramics. Acta Mater. **183**, 261–273 (2020). https://doi.org/10.1016/J.ACTAMAT.2019.10.046
102. Dai, X., et al.: Silicon nanoparticles encapsulated in multifunctional crosslinked nano-silica/carbon hybrid matrix as a high-performance anode for Li-ion batteries. Chem. Eng. J. **418**, 129468 (2021). https://doi.org/10.1016/J.CEJ.2021.129468

103. Shao, Z., Cao, X., Luo, H., Jin, P.: Recent progress in the phase-transition mechanism and modulation of vanadium dioxide materials. NPG Asia Mater. **10**(7), 581–605 (2018). https://doi.org/10.1038/s41427-018-0061-2
104. Han, J.G., Lee, S.J., Lee, J., Kim, J.S., Lee, K.T., Choi, N.S.: Tunable and robust phosphite-derived surface film to protect lithium-rich cathodes in lithium-ion batteries. ACS Appl. Mater. Interfaces **7**(15), 8319–8329 (2015). https://doi.org/10.1021/ACSAMI.5B01770/SUPPL_FILE/AM5B01770_SI_001.PDF
105. Khanna, V.K.: Adhesion–delamination phenomena at the surfaces and interfaces in micro-electronics and MEMS structures and packaged devices. J. Phys. D Appl. Phys. **44**(3), 034004 (2010). https://doi.org/10.1088/0022-3727/44/3/034004
106. Poizot, P., Laruelle, S., Grugeon, S., Dupont, L., Tarascon, J.M.: Nano-sized transition-metal oxides as negative-electrode materials for lithium-ion batteries. Nature **407**(6803), 496–499 (2000). https://doi.org/10.1038/35035045
107. Zhao, Y., et al.: Atomic/Molecular layer deposition for energy storage and conversion. Chem. Soc. Rev. **50**(6), 3889–3956 (2021). https://doi.org/10.1039/D0CS00156B
108. Bostanabad, R., et al.: Computational microstructure characterization and reconstruction: review of the state-of-the-art techniques. Prog. Mater. Sci. **95**, 1–41 (2018). https://doi.org/10.1016/J.PMATSCI.2018.01.005
109. Taylor, C., et al.: Predicting corrosion fatigue behavior using a Bayesian network that integrates microstructure, electrochemistry and fracture mechanics, 15 Apr 2018. OnePetro. Accessed: 12 Sept 2024. [Online]. Available: https://dx.doi.org/
110. Liu, D., et al.: Review of recent development of in situ/operando characterization techniques for lithium battery research. Adv. Mater. **31**(28), 1806620 (2019). https://doi.org/10.1002/ADMA.201806620
111. Ferrari, S., et al.: Solid-state post Li metal ion batteries: a sustainable forthcoming reality? Adv. Energy Mater. **11**(43), 2100785 (2021). https://doi.org/10.1002/AENM.202100785

Chapter 8
Future Perspectives and Challenges

"Future Perspectives and Challenges" explores the evolving landscape of electrochemical devices, highlighting emerging trends, addressing persistent challenges, and identifying opportunities for innovation and collaboration. The chapter begins by discussing emerging trends in electrochemical devices, including advancements in materials, device architectures, and integration with emerging technologies. It then delves into the critical challenges facing the field, focusing on issues of stability, scalability, and cost that impact the widespread adoption of electrochemical technologies. Strategies for addressing these challenges are examined, drawing on insights from materials science, engineering, and policy considerations. Furthermore, the chapter emphasizes the importance of collaboration and interdisciplinary research in overcoming these challenges and unlocking new opportunities for innovation. By providing a forward-looking perspective, this chapter aims to inspire researchers, engineers, and policymakers to navigate the complexities of the field and drive transformative advancements in electrochemistry.

8.1 Introduction

In recent decades, electrochemical devices have evolved remarkably, propelling forward a myriad of technologies that are increasingly integral to modern life. These advancements span a wide array of applications, from energy storage systems [1–3] and fuel cells [4, 5] to sensors [6–8] and medical devices [9, 10]. This chapter delves into the forefront of electrochemical device technology, exploring both the promising future and the challenges that lie ahead. Electrochemical devices harness the interplay of electrical and chemical processes to achieve a range of functions, and their advancements have been transformative [11]. Central to this evolution are three primary areas: materials science, device design, and performance enhancement. Materials Science, one of the most significant strides in electrochemical devices has

P. Phogat et al., *Electrochemical Devices*, Engineering Materials,
https://doi.org/10.1007/978-981-96-0527-9_8

been in the development of new materials. Traditional materials like lead–acid and nickel–cadmium batteries have given way to advanced materials such as lithium–ion [12], solid-state electrolytes, and nanostructured electrodes. These materials offer improved energy density, longer life cycles, and greater safety. For instance, lithium–ion batteries, which power everything from smartphones to electric vehicles, have benefited from innovations in anode and cathode materials, leading to higher capacities and faster charging times. Additionally, research into solid-state batteries promises to enhance safety by eliminating flammable liquid electrolytes and improving overall performance.

Device Design, innovations in design have significantly impacted the functionality and integration of electrochemical devices. The rise of flexible and wearable electronics exemplifies this trend, with devices becoming more adaptable to various form factors and applications [13]. Advances in microfabrication techniques have enabled the creation of miniature electrochemical sensors that can be seamlessly integrated into everyday objects. Moreover, the integration of electrochemical devices with other technologies, such as the Internet of Things (IoT), has enabled smart, data-driven applications [14]. These developments highlight a shift toward multifunctional devices that can communicate and interact with their environment in real-time. Performance Enhancement, enhancing the performance of electrochemical devices involves not just improving individual components but also optimizing entire systems. Innovations in battery management systems (BMS) have led to better monitoring and control of battery health, extending lifespan and efficiency [15]. For fuel cells, advancements in catalysts and membrane technologies have improved their power output and operational stability. Similarly, electrochemical sensors have seen improvements in sensitivity and selectivity, making them more effective for detecting a wider range of substances at lower concentrations.

While the current state of electrochemical devices is impressive, the future promises even greater potential. However, realizing this potential requires a thorough understanding of both the opportunities and the challenges that lie ahead. Exploring future perspectives in electrochemical devices is crucial for several reasons. First, ongoing research aims to address the limitations of current technologies. For example, while lithium–ion batteries are widespread, they face issues related to energy density and resource constraints. Future advancements in materials and design could lead to the development of next-generation batteries with even higher capacities and faster charging times. Similarly, emerging technologies such as hydrogen fuel cells and advanced supercapacitors hold the promise of new applications and improvements in energy storage and conversion. By examining these future trends, researchers and industry professionals can better prepare for and influence the direction of technological progress. Despite the promising future, there are significant challenges that must be addressed to ensure the continued advancement and successful commercialization of electrochemical devices. Material sustainability is a major concern, as many of the materials used in current devices are rare or environmentally hazardous. Developing alternative materials and improving recycling processes are essential for reducing the environmental impact of electrochemical technologies. Additionally, scaling up from laboratory research to large-scale production presents technical and economic hurdles. Cost reduction strategies and advancements in manufacturing techniques are critical for making these technologies commercially viable. Furthermore, regulatory and safety considerations play a pivotal role in the widespread adoption of new electrochemical devices. Ensuring compliance with industry standards and addressing safety concerns are fundamental for gaining consumer trust and achieving market success.

This chapter aims to provide a comprehensive exploration of the future perspectives and challenges associated with next-generation electrochemical devices. The objectives are threefold. We will examine the latest advancements in materials science, device design, and performance metrics. By reviewing these trends, the chapter will provide a clear understanding of the current state of the field and highlight the innovative technologies that are shaping the future of electrochemical devices. The chapter will address the primary challenges that need to be overcome for the successful development and commercialization of future electrochemical devices. This includes exploring issues related to material sustainability, cost reduction, manufacturing scalability, and regulatory compliance. Finally, the chapter will outline potential research directions and interdisciplinary approaches that could drive future advancements. By identifying areas for further investigation and collaboration, the chapter aims to inspire ongoing research and innovation in the field.

In summary, this chapter seeks to provide a forward-looking perspective on the field of electrochemical devices, emphasizing both the exciting possibilities and the critical challenges that must be addressed. By integrating insights from current advancements and future projections, the chapter aims to contribute to a deeper understanding of the dynamic landscape of electrochemical technologies and their role in shaping the future.

8.2 Emerging Trends in Electrochemical Technologies

8.2.1 Advances in Material Science

Advancements in materials science have significantly transformed electrochemical devices, driving innovations that enhance performance and expand applications. Two key areas of development—novel electrode materials and the impact of nanotechnology—have played pivotal roles in this evolution. Novel Electrode Materials have emerged as a cornerstone of progress in electrochemical devices. Traditionally, electrodes were limited to materials such as graphite, lead, and nickel. However, recent research has introduced a range of novel materials that offer superior performance characteristics. Lithium–ion batteries, for example, have benefited from the development of advanced anode and cathode materials. Silicon-based anodes have emerged as a promising alternative to traditional graphite, offering a substantial increase in capacity due to silicon's higher theoretical capacity for lithium–ion storage [16]. Similarly, the use of high-capacity cathode materials like lithium iron phosphate

($LiFePO_4$) and lithium cobalt oxide ($LiCoO_2$) has improved the overall energy density and stability of batteries. In the realm of supercapacitors, materials such as graphene and carbon nanotubes have been explored for their exceptional electrical conductivity and large surface areas, which enhance charge storage and delivery [17]. Additionally, in fuel cells, research into novel catalysts and membranes has led to improved efficiency and reduced costs. For instance, platinum-based catalysts have been replaced or supplemented with alternative materials like nickel and cobalt, reducing reliance on expensive precious metals while maintaining high performance.

Nanotechnology has further revolutionized materials science in electrochemical devices by enabling unprecedented levels of precision and functionality. The application of nanotechnology has led to the development of nanostructured materials that offer enhanced properties compared to their bulk counterparts. Nanomaterials, such as nanoparticles, nanotubes, and nanowires, possess unique characteristics, including high surface area-to-volume ratios and enhanced electrical and thermal conductivity [18]. These properties are particularly beneficial in electrochemical applications. For example, in lithium–ion batteries, nanostructured electrodes can improve charge and discharge rates by providing more efficient pathways for electron and ion transport. Nanotechnology has also contributed to the development of advanced coatings and composites that improve the stability and performance of electrodes. In supercapacitors, nanomaterials such as graphene and carbon nanotubes contribute to faster charge–discharge cycles and higher energy densities [19]. Moreover, nanotechnology enables the fabrication of high-precision sensors and devices with improved sensitivity and selectivity, which is crucial for applications in environmental monitoring, healthcare, and industrial processes. Overall, the integration of novel electrode materials and nanotechnology into electrochemical devices represents a significant leap forward in materials science. These advancements not only enhance the performance and efficiency of existing technologies but also open up new possibilities for the development of next-generation devices. As research continues to push the boundaries of materials science, we can expect further innovations that will drive the evolution of electrochemical technologies and their applications across various industries.

8.2.2 Innovative Design Concepts

Innovative design concepts in electrochemical devices have ushered in a new era of functionality and versatility, particularly through advancements in flexible and wearable technologies and the miniaturization and integration of devices. These design innovations are transforming how electrochemical devices are used and integrated into everyday life, opening up new possibilities for applications and user interactions.

Flexible and wearable electrochemical devices represent a significant leap forward in technology, offering enhanced adaptability and comfort. Unlike traditional rigid devices, flexible electronics can conform to various shapes and surfaces, enabling them to be seamlessly integrated into clothing, accessories, or even directly onto the skin. This flexibility is achieved through the use of advanced materials such as conductive polymers, organic semiconductors, and flexible substrates like polyimide or silicone. These materials allow for the creation of thin, lightweight devices that can bend, stretch, and flex without compromising functionality. One prominent example of flexible electrochemical devices is the development of wearable sensors that monitor physiological parameters such as heart rate, glucose levels, or body temperature. These sensors can be embedded in fabrics or attached to the skin, providing continuous health monitoring with minimal discomfort. For instance, smart bandages and patches can track wound healing or drug delivery in real-time, offering valuable insights for medical professionals [20]. Moreover, flexible batteries and energy storage systems are being designed to power these wearable devices, ensuring that they remain functional without adding bulk or rigidity. The integration of flexible electrochemical devices with wireless communication technologies further enhances their utility. Wearable health monitors, for example, can transmit data to smartphones or healthcare providers via Bluetooth or other wireless protocols, enabling real-time tracking and remote diagnosis. This connectivity not only improves the convenience of using such devices but also facilitates the collection of large-scale health data for research and personalized medicine [21].

Miniaturization and integration are key design principles driving the advancement of electrochemical devices, enabling them to be more compact and multifunctional. The trend towards miniaturization involves reducing the size of devices and their components while maintaining or even improving their performance. This approach has been facilitated by advancements in microfabrication techniques and materials science, which allow for the precise and efficient production of tiny electrochemical cells, sensors, and other components. Miniaturized electrochemical devices have numerous applications, particularly in the fields of diagnostics and environmental monitoring. For example, micro-sensors can be used for detecting trace amounts of pollutants or toxins in air or water, providing critical data for environmental protection and public health. Similarly, compact biosensors can be employed for point-of-care diagnostics, offering quick and accurate results for diseases such as diabetes or infections. The integration of multiple functions into a single device is another significant innovation. By combining electrochemical sensing, energy storage, and data processing capabilities, designers can create multifunctional devices that offer enhanced utility and convenience as depicted in Fig. 8.1. For instance, integrated sensors and energy harvesters can be used in smart textiles to monitor and power wearable electronics simultaneously. This integration not only reduces the need for multiple separate devices but also streamlines the user experience by consolidating various functionalities into a single, cohesive system [22].

Overall, the innovative design concepts of flexible and wearable devices, along with miniaturization and integration, are pushing the boundaries of what electrochemical devices can achieve. These advancements are leading to more adaptable, efficient, and multifunctional technologies that are increasingly embedded in our daily lives, paving the way for a future where electrochemical devices are seamlessly integrated into a wide range of applications and environments.

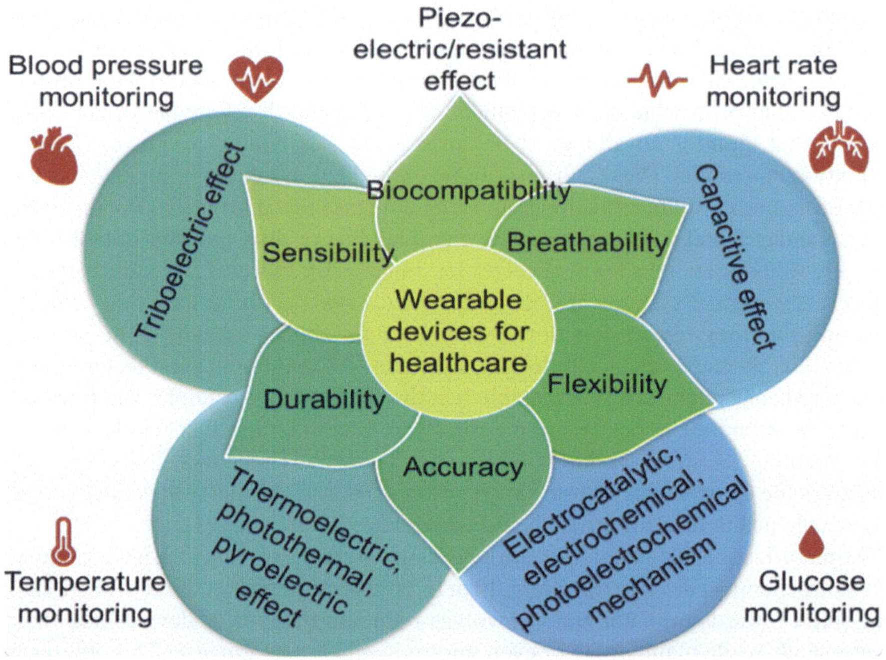

Fig. 8.1 Wearable monitoring devices for healthcare purposes. Reprinted with from [23]. Copyright © 1999–2024 John Wiley & Sons, Inc. or related companies. All rights reserved, including rights for text and data mining and training of artificial intelligence technologies or similar technologies

8.2.3 Enhanced Performance Metrics

The quest for enhanced performance metrics in electrochemical devices has been a driving force behind recent innovations, with energy density improvements and considerations of efficiency and sustainability being at the forefront. Energy Density Improvements are crucial for the advancement of electrochemical devices, especially for applications where space and weight constraints are critical, such as in portable electronics and electric vehicles. Energy density, which measures the amount of energy stored per unit volume or weight, has seen significant strides with the development of advanced materials and technologies. Lithium–ion batteries, for instance, have become a cornerstone of modern energy storage due to their high energy density compared to traditional battery technologies [24]. However, the pursuit of even higher energy densities has led researchers to explore novel materials and designs. Solid-state batteries, for example, utilize solid electrolytes instead of liquid ones, allowing for higher energy densities and improved safety [25]. Additionally, the integration of nanomaterials and advanced electrode architectures has been shown to enhance the energy storage capacity by increasing the surface area available for electrochemical reactions. These innovations not only extend the range of electric vehicles but also enable longer-lasting and more efficient portable devices, meeting the growing demand for higher performance in a compact form factor. Efficiency and Sustainability Considerations are integral to the future development of electrochemical devices. Efficiency, in this context, refers to the effectiveness of a device in converting stored energy into usable power and minimizing losses during this process. Advances in electrode materials, such as high-conductivity compounds and optimized catalytic materials, have led to improved efficiency in various electrochemical systems, from batteries to fuel cells. For instance, the development of high-performance catalysts in fuel cells has reduced energy losses and enhanced overall system efficiency, making them a more viable option for clean energy applications. Sustainability is another critical factor driving innovation in electrochemical devices [26]. As the demand for these technologies grows, so does the need to address environmental and resource-related challenges. The extraction and processing of materials used in current electrochemical devices, such as lithium, cobalt, and rare earth elements, pose significant environmental and ethical concerns. Researchers are focusing on developing alternative materials that are more abundant and less environmentally damaging. For example, the use of abundant materials like sodium and magnesium in battery technologies is being explored as a way to mitigate resource constraints and reduce environmental impact. Additionally, recycling and reuse of materials are becoming increasingly important, with efforts being made to improve the efficiency of recycling processes and to develop closed-loop systems that minimize waste. In conclusion, the enhanced performance metrics of electrochemical devices, particularly in terms of energy density improvements and efficiency, are pivotal for advancing technology and meeting the growing demands of modern applications. At the same time, addressing sustainability considerations is essential for ensuring that these advancements are environmentally responsible and economically feasible. By focusing on these aspects, the field of electrochemical devices can continue to evolve, offering higher performance while also promoting a more sustainable and efficient future [27].

8.3 Integration of Electrochemical Devices with Modern Technologies

The integration of electrochemical devices with modern technologies has been a transformative force, reshaping how these devices are utilized and enhancing their functionality. Three key areas where this integration is particularly impactful are the Internet of Things (IoT), artificial intelligence (AI), and wearable technology. Each of these domains leverages the unique capabilities of electrochemical devices to drive innovation and create more sophisticated and efficient systems.

The Internet of Things (IoT) represents a paradigm shift in how devices interact with each other and the environment, creating a network of interconnected systems that communicate and exchange data as illustrated in Fig. 8.2. Electrochemical devices are central to this transformation, particularly through the development of smart sensors. These sensors, which are increasingly used in various applications from environmental monitoring to industrial processes, rely on electrochemical principles to detect and quantify specific substances with high precision. One of the most significant advancements in this area is the ability to integrate smart sensors with IoT platforms, enabling real-time data collection and analysis. Electrochemical sensors can measure parameters such as pH, ion concentration, or gas levels, and this data can be transmitted to cloud-based systems for further processing. By integrating these sensors with IoT infrastructure, users can access continuous, real-time information about their environments, which is crucial for applications ranging from smart agriculture to urban air quality monitoring. Data integration through IoT platforms allows for the aggregation of information from multiple sensors, providing a comprehensive overview of various environmental or operational parameters. This integration facilitates advanced analytics and decision-making processes, leading to more informed and timely responses. For instance, in industrial settings, IoT-enabled electrochemical sensors can monitor the condition of equipment, track chemical reactions, and detect anomalies, thereby optimizing operations and improving safety [28].

Artificial intelligence (AI) is another transformative technology that, when combined with electrochemical devices, significantly enhances their capabilities. Machine learning, a subset of AI, plays a critical role in predictive maintenance, an area where electrochemical devices have seen notable advancements. Predictive maintenance involves using data-driven insights to predict equipment failures before they occur, allowing for proactive maintenance and reducing downtime. Machine learning algorithms can analyze data from electrochemical sensors to identify patterns and anomalies that may indicate potential issues. For example, in battery management systems, machine learning models can predict battery degradation and performance issues by analyzing historical data and real-time measurements from electrochemical sensors. These models can recognize subtle changes in performance that may precede a failure, enabling timely interventions. By incorporating AI-driven predictive maintenance, organizations can extend the lifespan of electrochemical devices, enhance reliability, and minimize operational disruptions. AI also helps in optimizing the performance of electrochemical devices by adjusting operational parameters based on real-time data, thus ensuring that devices operate at their peak efficiency. This integration of AI with electrochemical technologies is paving the way for more intelligent and autonomous systems that can self-monitor and self-optimize [30, 31].

Wearable technology is another area where electrochemical devices are making a significant impact, particularly in health monitoring applications. Wearable electrochemical sensors are increasingly used in health and fitness devices to monitor various physiological parameters, providing users with valuable insights into their health and well-being. Electrochemical sensors in wearable devices can measure biomarkers such as glucose, lactate, and electrolyte levels, offering real-time feedback on an individual's health status. For instance, continuous glucose monitoring systems use electrochemical sensors to provide diabetics with real-time glucose levels, allowing for better management of their condition. Similarly, wearable sensors can track hydration levels and electrolyte balance, which is crucial for athletes and individuals with

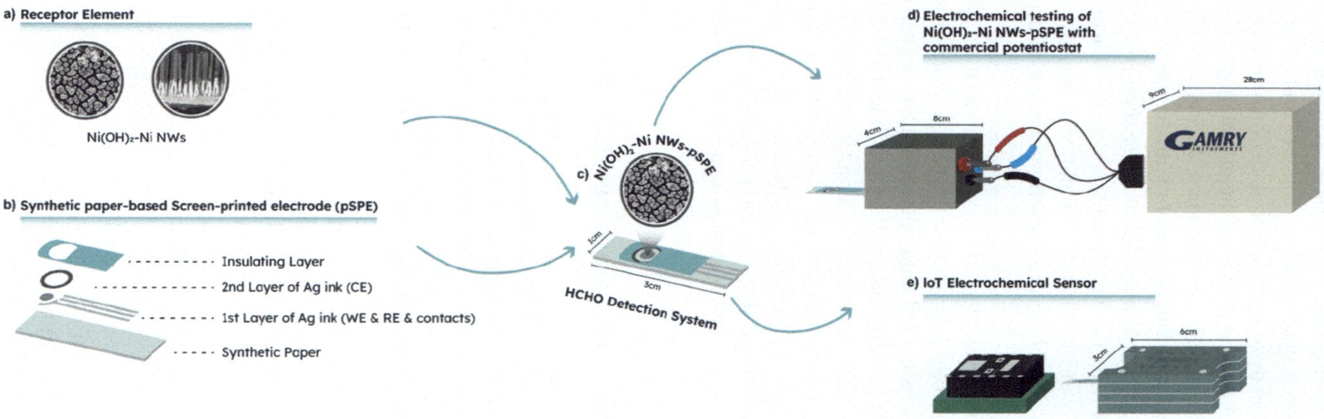

Fig. 8.2 Schematic fabrication process of IoT electrochemical sensor. Reprinted with permission from [29]. © 2023 by the authors. Licensee MDPI, Basel, Switzerland

specific health conditions. The integration of these sensors with wearable technology enables seamless health monitoring and data collection. Wearable devices equipped with electrochemical sensors can sync with smartphones and other digital platforms, allowing users to access and analyze their health data conveniently. This integration also supports the development of personalized health management plans based on real-time data, enhancing overall health outcomes. Moreover, the convergence of wearable technology and electrochemical sensors with AI and IoT further amplifies their benefits. For instance, AI algorithms can analyze data from wearable devices to provide personalized health recommendations and early warnings for potential health issues. IoT connectivity ensures that health data can be shared with healthcare providers or fitness coaches, enabling more comprehensive and coordinated care [32, 33].

The integration of electrochemical devices with modern technologies such as IoT, AI, and wearable technology has opened up new avenues for innovation and application. Through smart sensors and data integration, AI-driven predictive maintenance, and advanced health monitoring applications, electrochemical devices are becoming more sophisticated and capable. This convergence not only enhances the functionality and efficiency of these devices but also paves the way for more intelligent, connected, and personalized systems. As technology continues to advance, the synergy between electrochemical devices and modern technologies promises to drive further innovation and create new opportunities across various fields.

8.4 Challenges in Scaling and Commercialization

The journey from innovative electrochemical devices in the laboratory to successful commercial products involves navigating a complex landscape of challenges. Key among these are material sourcing and sustainability, cost reduction strategies, and manufacturing techniques. Each of these elements plays a critical role in determining whether new technologies can be scaled up effectively and introduced to the market at a feasible price.

Material Sourcing and Sustainability stands as one of the foremost challenges in the commercialization of electrochemical devices. As technology advances, the demand for high-performance materials such as lithium, cobalt, and rare earth elements has surged. These materials are crucial for achieving the high energy densities and efficiency required for modern batteries and fuel cells. However, the availability of these resources is limited, and their extraction often entails significant environmental impact. Mining operations for these materials can lead to habitat destruction, water pollution, and soil degradation. Furthermore, the geopolitical concentration of these resources in specific regions poses risks related to supply chain stability. For instance, the majority of cobalt is sourced from the Democratic Republic of Congo, a region fraught with political and ethical challenges. To address these issues, there is a pressing need for sustainable material sourcing practices, including the development of alternative materials and improved recycling technologies. Research into less environmentally damaging materials and the advancement of closed-loop recycling systems could mitigate the impact of material scarcity and enhance the overall sustainability of electrochemical devices [34].

Cost Reduction Strategies are another critical aspect of scaling and commercialization. The transition from lab-scale prototypes to mass-produced devices often encounters financial hurdles. Initial research and development costs can be prohibitively high, and scaling up production introduces additional expenses. Economies of scale play a crucial role in reducing costs; as production volumes increase, the per-unit cost of materials and manufacturing generally decreases. However, achieving these economies of scale requires significant investment in production infrastructure and technology. Efficient production techniques and process optimization are essential to make commercial-scale production viable. For instance, innovations in automated manufacturing processes, such as roll-to-roll processing for flexible electronics, can lower labor costs and increase throughput. Additionally, advances in process engineering that enhance yield and reduce waste contribute to cost savings. Strategic partnerships with suppliers and manufacturers can also play a role in negotiating better terms and reducing overall costs. Ultimately, a combination of technological advancements and strategic financial management is necessary to bring down costs and make new electrochemical technologies commercially competitive [35].

Manufacturing Techniques present their own set of challenges when scaling up from laboratory to industrial production. Laboratory-scale devices are often crafted with precision using bespoke methods that are not easily transferable to mass production. One major hurdle is maintaining consistency and quality across large production runs. Variability in material properties, process conditions, and quality control can lead to significant differences in device performance. To address these issues, manufacturers must develop robust quality assurance protocols and standardized production processes. Additionally, scaling up often requires new manufacturing techniques and equipment that can handle larger volumes while maintaining the desired level of precision. For example, the transition from small-batch synthesis of battery materials to large-scale production involves adapting chemical processes to industrial reactors and optimizing material handling procedures. Moreover, the integration of automation and advanced manufacturing technologies, such as precision robotics and digital twins, can enhance production efficiency and consistency. Overcoming these manufacturing challenges involves a careful balance of technology development, process optimization, and rigorous quality control to ensure that scaled-up products meet performance and reliability standards [36].

The path from laboratory innovation to market-ready electrochemical devices is fraught with challenges related to material sourcing, cost reduction, and manufacturing techniques. Addressing these challenges requires a multifaceted approach that includes sustainable material practices, strategic cost management, and advanced manufacturing solutions. By overcoming these hurdles, researchers and industry professionals can pave the way for the successful commercialization of next-generation electrochemical technologies, ultimately contributing to advancements in energy storage, conversion, and beyond [37].

8.5 Future Research Directions

In the rapidly evolving field of electrochemical devices, interdisciplinary collaboration has become an essential driving force behind innovation and progress. The complexity of developing next-generation devices necessitates the integration of diverse expertise, particularly from material scientists, engineers, and chemists. These collaborations are crucial in addressing the multifaceted challenges posed by modern electrochemical systems, where advancements in one area often depend on breakthroughs in another.

Material Scientists: Material scientists play a critical role in the development of electrochemical devices by exploring and synthesizing new materials that can enhance performance. Their work often focuses on developing materials with specific properties, such as high conductivity, stability, and energy density, which are essential for improving the efficiency and longevity of devices like batteries and fuel cells. For instance, the discovery and optimization of lithium–ion battery materials have been heavily reliant on the insights provided by material scientists [38]. These professionals investigate the atomic and molecular structure of materials, determining how they interact with other components in the device and how these interactions can be optimized for better performance.

Engineers: Engineers are responsible for the design, integration, and scaling up of electrochemical devices. Their work involves translating the materials developed by scientists into functional devices that can be manufactured at scale. This includes designing the architecture of the device, ensuring that it meets the required specifications, and developing methods for mass production. Engineers must also address the practical aspects of device deployment, such as durability, safety, and cost-effectiveness. For example, in the development of lithium–ion batteries for electric vehicles [39], engineers must ensure that the batteries can be produced on a large scale while maintaining high performance, safety, and reliability. They also work on integrating these batteries into the vehicle's overall system, considering factors such as thermal management and energy efficiency [40].

Chemists: Chemists contribute to electrochemical device development by exploring the chemical processes that occur within these systems. Their work is crucial for understanding the mechanisms of energy conversion and storage, as well as for developing new electrolytes and catalysts that can enhance device performance. Chemists also play a key role in addressing challenges related to the stability and safety of electrochemical devices. For example, in the development of fuel cells, chemists work on optimizing the chemical reactions that occur at the electrodes, aiming to improve the efficiency of energy conversion while reducing the production of harmful by-products. The collaboration between these disciplines is vital for advancing electrochemical technologies. By working together, material scientists, engineers, and chemists can address the complex challenges that arise in the development of next-generation devices. This interdisciplinary approach fosters innovation by combining different perspectives and expertise, leading to the creation of more efficient, durable, and sustainable electrochemical systems [41].

Long-Term Stability and Durability

As electrochemical devices become increasingly integral to various industries, the importance of long-term stability and durability cannot be overstated. These factors are critical for ensuring that devices can perform reliably over extended periods, minimizing the need for frequent maintenance or replacement. Achieving long-term stability and durability involves addressing several key challenges, including material degradation, device architecture, and operational conditions [42].

Material Degradation: One of the primary challenges in enhancing the stability and durability of electrochemical devices is material degradation. Over time, the materials used in these devices can undergo physical and chemical changes that reduce their performance. For example, in lithium–ion batteries, the repeated charging and discharging cycles can lead to the formation of solid-electrolyte interphase (SEI) layers on the anode, which can impede ion transport and reduce the battery's capacity [43]. Similarly, in fuel cells, the catalysts used to accelerate chemical reactions can degrade, leading to a decrease in efficiency. Addressing these issues requires a deep understanding of the mechanisms of degradation and the development of materials that are more resistant to these processes. For instance, researchers are exploring the use of advanced coatings and nanostructured materials that can enhance the durability of electrodes and other critical components.

Device Architecture: The architecture of electrochemical devices also plays a crucial role in their long-term stability and durability. The design of the device must ensure that the materials and components are arranged in a way that minimizes stress and degradation over time. This includes optimizing the thickness and composition of the electrodes, improving the distribution of electrolytes, and ensuring effective thermal management. For example, in solid-state batteries, the use of solid electrolytes instead of liquid ones can improve stability by reducing the risk of leakage and thermal runaway [44]. Engineers are also exploring novel designs, such as 3D architectures, which can enhance the performance and lifespan of electrochemical devices by improving the uniformity of current distribution and reducing the risk of localized degradation.

Operational Conditions: The operational conditions under which electrochemical devices are used can significantly impact their stability and durability. Factors such as temperature, pressure, and load cycles can affect the performance of the materials and components, leading to accelerated degradation. To enhance long-term stability, it is essential to optimize the operational conditions and develop devices that can withstand harsh environments. For example, in the case of batteries used in electric vehicles, thermal management systems are crucial for maintaining the temperature within an optimal range, preventing overheating and extending the batterys lifespan. Similarly, fuel cells used in industrial applications must be designed to operate reliably under varying pressure and load conditions. Enhancing the long-term stability and durability of electrochemical devices requires a comprehensive approach that addresses material degradation, device architecture, and operational conditions. By improving the resilience of these devices, we can reduce the need for frequent maintenance, lower the total cost of ownership, and increase the reliability of critical systems across various industries [45].

Innovative Applications

The advancement of electrochemical devices opens up a wide range of innovative applications across multiple fields and industries. These applications are not only transforming existing technologies but are also paving the way for new industries and economic opportunities. As electrochemical devices become more efficient, durable, and versatile, their potential applications are expanding into areas that were previously considered unfeasible [46].

Energy Storage and Management: One of the most significant areas of innovation for electrochemical devices is in energy storage and management. The development of high-capacity batteries and supercapacitors is driving the growth of renewable energy systems by enabling efficient storage of energy generated from sources such as solar and wind. This is critical for addressing the intermittency of renewable energy and ensuring a stable supply of power. Advanced batteries are also being integrated into smart grids, where they can provide grid balancing services, store excess energy, and ensure reliable power supply during peak demand periods. Additionally, the rise of electric vehicles (EVs) is heavily dependent on the advancement of battery technology. The development of high-performance batteries with longer ranges, faster charging times, and lower costs is essential for the widespread adoption of EVs, which are key to reducing greenhouse gas emissions and combating climate change [47].

Healthcare and Medical Devices: Electrochemical devices are also finding innovative applications in the healthcare and medical fields. For instance, the development of bioelectrochemical sensors is revolutionizing medical diagnostics by enabling real-time monitoring of biomarkers in bodily fluids. These sensors are being used in wearable devices that can continuously monitor a patient's health, providing valuable data for early diagnosis and personalized treatment. Electrochemical devices are also being used in implantable medical devices, such as pacemakers and neurostimulators, where they provide reliable and long-lasting power sources. Furthermore, advancements in electrochemical devices are enabling the development of lab-on-a-chip technologies, which can perform complex biochemical analyses at a fraction of the time and cost of traditional laboratory methods. These innovations are contributing to the growth of personalized medicine, where treatments are tailored to the individual needs of patients based on real-time data [48].

Environmental Monitoring and Remediation: The application of electrochemical devices in environmental monitoring and remediation is another area of significant innovation. Electrochemical sensors are being developed to detect pollutants and contaminants in air, water, and soil with high sensitivity and selectivity. These sensors can provide real-time data on environmental conditions, enabling more effective monitoring and management of pollution. For example, portable electrochemical sensors can be used to monitor air quality in urban areas, providing data that can be used to implement pollution control measures. Additionally, electrochemical devices are being developed for use in environmental remediation, where they can help to degrade or remove harmful substances from the environment. For instance, electrochemical processes can be used to treat wastewater by breaking down organic pollutants and removing heavy metals. These applications are critical for addressing environmental challenges and ensuring a sustainable future [49].

Wearable and Flexible Electronics: The development of wearable and flexible electronics is another area where electrochemical devices are making a significant impact. The advancement of flexible batteries, supercapacitors, and sensors is enabling the creation of wearable devices that can monitor various physiological parameters, such as heart rate, body temperature, and blood oxygen levels. These devices are becoming increasingly popular in the fitness and healthcare industries, where they provide users with real-time feedback on their health and performance. Additionally, flexible electrochemical devices are being integrated into smart textiles, which can monitor the wearer's environment and respond to changes in temperature, humidity, and other factors. These innovations are opening up new possibilities for wearable technology, from smart clothing that can regulate body temperature to medical devices that can monitor chronic conditions and alert users to potential health issues [50].

Industrial and Military Applications: Electrochemical devices are also finding innovative applications in industrial and military settings. In the industrial sector, advanced batteries and fuel cells are being used to power equipment and machinery in remote locations, where reliable access to the electrical grid is not available. These devices are also being used in backup power systems to ensure the continuity of critical operations during power outages. In the military sector, electrochemical devices are being developed for use in portable power systems that can support soldiers in the field, providing reliable energy for communications equipment, sensors, and other critical technologies. Additionally, the development of high-energy-density batteries is enabling the creation of more efficient and powerful electric vehicles, drones, and other military assets.

Agriculture and Food Security: Innovative applications of electrochemical devices are also emerging in agriculture, where they are being used to enhance food security and improve sustainability. Electrochemical sensors are being developed to monitor soil health, detecting nutrient levels, moisture content, and other parameters that are critical for crop growth. These sensors can provide farmers with real-time data, enabling more precise and efficient use of water, fertilizers, and pesticides. Additionally, electrochemical devices are being used in controlled-environment agriculture, such as vertical farming and hydroponics, where they can monitor and control conditions to optimize crop yields. These innovations are contributing to the growth of sustainable agriculture practices that can help meet the increasing global demand for food.

The interdisciplinary approach, long-term stability, and durability are crucial factors for the future of electrochemical devices. By leveraging the expertise of material scientists, engineers, and chemists, the development of these devices can be accelerated, addressing the complex challenges associated with their performance, scalability, and environmental impact. Enhancing the long-term stability and durability of electrochemical devices ensures their reliability and effectiveness in real-world applications, while innovative applications open up new possibilities across various industries, driving economic growth and technological progress. As we continue to explore the potential of electrochemical devices, it is essential to embrace interdisciplinary collaboration, focus on improving stability and durability, and explore new applications that can transform industries and improve quality of life.

8.6 Conclusion

The future of electrochemical devices is brimming with promise, yet it is equally fraught with challenges that demand innovative solutions and interdisciplinary collaboration. As we have explored throughout this chapter, the advancements in materials science, engineering design, and chemical processes have brought us to the cusp of a new era in electrochemical technology. These innovations are paving the way for more efficient, durable, and versatile devices that can meet the growing demands of energy storage, environmental sustainability, healthcare, and numerous other fields. However, to fully realize the potential of these next-generation electrochemical devices, it is imperative to address the challenges that remain. Long-term stability and durability are critical for ensuring that these devices can perform reliably over extended periods, reducing maintenance costs and environmental impact. Achieving this requires a concerted effort to understand and mitigate material degradation, optimize device architectures, and refine operational conditions. Interdisciplinary approaches will be key to overcoming these challenges. By fostering collaborations between material scientists, engineers, and chemists, we can accelerate the development of new materials, innovative designs, and advanced manufacturing techniques. These collaborations not only enhance the performance of electrochemical devices but also drive the discovery of novel applications that can transform industries and improve quality of life. The future perspectives of electrochemical devices are vast and varied, extending into fields such as renewable energy, healthcare, environmental monitoring, and beyond. As we move forward, it is essential to continue exploring these emerging applications while addressing the critical issues of sustainability, cost, and safety. By doing so, we can ensure that electrochemical devices not only meet the needs of today but also contribute to a more sustainable and technologically advanced future. In summary, the road ahead for electrochemical devices is one of immense opportunity and responsibility. The advancements and challenges discussed in this chapter highlight the need for continued research, innovation, and collaboration. By embracing these principles, we can unlock the full potential of electrochemical technologies, paving the way for a future where these devices play a central role in powering our world, protecting our environment, and enhancing our lives.

References

1. Sumdani, M.G., Islam, M.R., Yahaya, A.N.A., Safie, S.I.: Recent advancements in synthesis, properties, and applications of conductive polymers for electrochemical energy storage devices: a review. Polym. Eng. Sci. **62**(2), 269–303 (2022)
2. Islam, S., Mia, M.M., Shah, S.S., Naher, S., Shaikh, M.N., Aziz, M.A., Ahammad, A.J.S.: Recent advancements in electrochemical deposition of metal-based electrode materials for electrochemical supercapacitors. Chem. Rec. **22**(7), e202200013 (2022)
3. Majumdar, D., Mandal, M., Bhattacharya, S.K.: Journey from supercapacitors to supercapatteries: recent advancements in electrochemical energy storage systems. Emergent. Mater. **3**(3), 347–367 (2020)
4. Saikia, K., Kakati, B.K., Boro, B., Verma, A.: Current advances and applications of fuel cell technologies. In: Recent Advancements in Biofuels and Bioenergy Utilization, pp. 303–337 (2018)
5. Tan, S., Lee, W., Wong, C., Chua, Y.J.: Exploring the potential and advancements of hydrogen fuel cells: outcomes and applications. Fusion Multi. Res. Int. J. **4**(1), 407–419 (2023)
6. Ramya, M., Senthil Kumar, P., Rangasamy, G., Uma Shankar, V., Rajesh, G., Nirmala, K., Saravanan, A., Krishnapandi, A.: A recent advancement on the applications of nanomaterials in electrochemical sensors and biosensors. Chemosphere **308**, 136416 (2022)
7. Baranwal, J., Barse, B., Gatto, G., Broncova, G., Kumar, A.: Electrochemical sensors and their applications: a review. Chemosensors **10**(9), 363 (2022)
8. Sumitha, M.S., Xavier, T.S.: Recent advances in electrochemical biosensors—a brief review. Hybrid Adv. **2**, 100023 (2023)
9. Manickam, P., Kanagavel, V., Sonawane, A., Thipperudraswamy, S.P., Bhansali, S.: Electrochemical systems for healthcare applications. In: Bioelectrochemical Interface Engineering, pp. 385–409 (2019)
10. Singh, A.B., Khandelwal, C., Dangayach, G.S.: Revolutionizing healthcare materials: innovations in processing, advancements, and challenges for enhanced medical device integration and performance (2024). https://doi.org/10.1177/25165984241256234
11. Zhang, Y., Xie, M., Adamaki, V., Khanbareh, H., Bowen, C.R.: Control of electro-chemical processes using energy harvesting materials and devices. Chem. Soc. Rev. **46**(24), 7757–7786 (2017)
12. Mahmud, S., Rahman, M., Kamruzzaman, M., Ali, M.O., Emon, M.S.A., Khatun, H., Ali, M.R.: Recent advances in lithium-ion battery materials for improved electrochemical performance: a review. Results Eng. **15**, 100472 (2022)
13. Kim, J., Kumar, R., Bandodkar, A.J., Wang, J.: Advanced materials for printed wearable electrochemical devices: a review. Adv. Electron. Mater. **3**(1), 1600260 (2017)
14. Faham, S., Salimi, A., Ghavami, R.: Electrochemical-based remote biomarker monitoring: toward internet of wearable things in telemedicine. Talanta **253**, 123892 (2023)
15. Waseem, M., Ahmad, M., Parveen, A., Suhaib, M.: Battery technologies and functionality of battery management system for EVs: current status, key challenges, and future prospectives. J. Power. Sources **580**, 233349 (2023)

16. Zhang, C., Wang, F., Han, J., Bai, S., Tan, J., Liu, J., Li, F.: Challenges and recent progress on silicon-based anode materials for next-generation lithium-ion batteries. Small Struct. **2**(6), 2100009 (2021)
17. Wang, F., Wu, X., Yuan, X., Liu, Z., Zhang, Y., Fu, L., Zhu, Y., Zhou, Q., Wu, Y., Huang, W.: Latest advances in supercapacitors: from new electrode materials to novel device designs. Chem. Soc. Rev. **46**(22), 6816–6854 (2017)
18. Saleh, H.M., Hassan, A.I.: Synthesis and characterization of nanomaterials for application in cost-effective electrochemical devices. Sustainability **15**(14), 10891 (2023)
19. Kamila, S., Jena, B.K., Basu, S.: Advances in electrochemical energy storage device: supercapacitor. In: Photoconductivity and Photoconductive Materials: Fundamentals, Energy Storage, pp. 119–147 (2021)
20. Teymourian, H., Parrilla, M., Sempionatto, J.R., Montiel, N.F., Barfidokht, A., Van Echelpoel, R., De Wael, K., Wang, J.: Wearable electrochemical sensors for the monitoring and screening of drugs. ACS Sens. **5**(9), 2679–2700 (2020)
21. Mirlou, F., Beker, L.: Wearable electrochemical sensors for healthcare monitoring: a review of current developments and future prospects. IEEE Trans. Mol. Biol. Multiscale Commun. **9**(3), 364–373 (2023)
22. Dei, M., Muñoz, X.: Electrochemical miniaturized devices. In: Micro- and Nanotechnology Enabled Applications for Portable Miniaturized Analytical Systems, pp. 109–140 (2022)
23. Dai, B., Gao, C., Xie, Y.: Flexible wearable devices for intelligent health monitoring. View **3**, 20220027 (2022). https://doi.org/10.1002/VIW.20220027
24. Machín, A., Morant, C., Márquez, F.: Advancements and challenges in solid-state battery technology: an in-depth review of solid electrolytes and anode innovations. Batteries **10**(1), 29 (2024)
25. Parameswaran, A.K., Azadmanjiri, J., Palaniyandy, N., Pal, B., Palaniswami, S., Dekanovsky, L., Wu, B., Sofer, Z.: Recent progress of nanotechnology in the research framework of all-solid-state batteries. Nano Energy **105**, 107994 (2023)
26. Sandoval-González, A., Bustos, E., Martínez-Sánchez, C.: Basic and advanced considerations of energy storage devices. In: Engineering Materials, pp. 63–80 (2022)
27. Verma, D., Singh, K.R., Yadav, A.K., Nayak, V., Singh, J., Solanki, P.R., Singh, R.P.: Internet of things (IoT) in nano-integrated wearable biosensor devices for healthcare applications. Biosens. Bioelectron. X **11**, 100153 (2022)
28. Voitechovič, E., Pauliukaite, R.: Electrochemical multisensor systems and arrays in the era of artificial intelligence. Curr. Opin. Electrochem. **42**, 101411 (2023)
29. Kaushik, A.K., Dhau, J.S., Gohel, H., Mishra, Y.K., Kateb, B., Kim, N.Y., Goswami, D.Y.: Electrochemical SARS-CoV-2 sensing at point-of-care and artificial intelligence for intelligent COVID-19 management. ACS Appl. Bio Mater. **3**(11), 7306–7325 (2020)
30. Wu, J., Liu, H., Chen, W., Ma, B., Ju, H.: Device integration of electrochemical biosensors. Nat. Rev. Bioeng. **1**(5), 346–360 (2023)
31. Deroco, P.B., Wachholz Junior, D., Kubota, L.T.: Paper-based wearable electrochemical sensors: a new generation of analytical devices. Electroanalysis **35**(1), e202200177 (2023)
32. Yoo, H.D., Markevich, E., Salitra, G., Sharon, D., Aurbach, D.: On the challenge of developing advanced technologies for electrochemical energy storage and conversion. Mater. Today **17**(3), 110–121 (2014)
33. Lv, W., Li, Z., Deng, Y., Yang, Q.H., Kang, F.: Graphene-based materials for electrochemical energy storage devices: opportunities and challenges. Energy Storage Mater. **2**, 107–138 (2016)
34. Ayerbe, E., Berecibar, M., Clark, S., Franco, A.A., Ruhland, J.: Digitalization of battery manufacturing: current status, challenges, and opportunities. Adv. Energy Mater. **12**(17), 2102696 (2022)
35. Jadhav, D.A., Park, S.G., Pandit, S., Yang, E., Ali Abdelkareem, M., Jang, J.K., Chae, K.J.: Scalability of microbial electrochemical technologies: applications and challenges. Bioresour. Technol. **345**, 126498 (2022)
36. Khan, F.M.N.U., Rasul, M.G., Sayem, A.S.M., Mandal, N.K.: Design and optimization of lithium-ion battery as an efficient energy storage device for electric vehicles: a comprehensive review. J. Energy Storage **71**, 108033 (2023)

37. Ahasan Habib, A.K.M., Motakabber, S.M.A., Ibrahimy, M.I.: A comparative study of electrochemical battery for electric vehicles applications. In: 2019 IEEE International Conference on Power, Electrical, and Electronics and Industrial Applications (PEEIACON), pp. 43–47 (2019)
38. Dawahdeh, A.I., Al-Nimr, M.A.: A novel energy harvesting and battery thermal management in hybrid vehicles using a thermally regenerative electrochemical device. Energy **270**, 126865 (2023)
39. Badwal, S.P.S., Giddey, S.S., Munnings, C., Bhatt, A.I., Hollenkamp, A.F.: Emerging electrochemical energy conversion and storage technologies. Front. Chem. **2**(SEP), 97517 (2014)
40. Sharma, R., Kumar, H., Kumar, G., Sharma, S., Aneja, R., Sharma, A.K., Kumar, R., Kumar, P.: Progress and challenges in electrochemical energy storage devices: fabrication, electrode material, and economic aspects. Chem. Eng. J. **468**, 143706 (2023)
41. Wang, P., Yan, D., Wang, C., Ding, H., Dong, H., Wang, J., Wu, S., Cui, X., Li, C., Zhao, D., Li, S.: Study of the formation and evolution of solid electrolyte interface via in-situ electrochemical impedance spectroscopy. Appl. Surf. Sci. **596**, 153572 (2022)
42. Pandey, G.P., Brown, J.E., Li, J.: Architectural design for flexible solid-state batteries. In: ACS Symposium Series, vol. 1414, pp. 289–309 (2022)
43. Vichard, L., Steiner, N.Y., Zerhouni, N., Hissel, D.: Hybrid fuel cell system degradation modeling methods: a comprehensive review. J. Power. Sources **506**, 230071 (2021)
44. Xie, Y., Zhang, W., Gu, S., Yan, Y., Ma, Z.F.: Process engineering in electrochemical energy devices innovation. Chin. J. Chem. Eng. **24**(1), 39–47 (2016)
45. Chemali, E., Preindl, M., Malysz, P., Emadi, A.: Electrochemical and electrostatic energy storage and management systems for electric drive vehicles: state-of-the-art review and future trends. IEEE J. Emerg. Sel. Top. Power Electron. **4**(3), 1117–1134 (2016)
46. Hernández-Rodríguez, J.F., Rojas, D., Escarpa, A.: Electrochemical sensing directions for next-generation healthcare: trends, challenges, and frontiers. Anal. Chem. **93**(1), 167–183 (2021)
47. Raditya, A.N., O'Hare, D.: Review—electrochemical sensor biofouling in environmental sensor networks: characterisation, remediation and lessons from biomedical devices. J. Electrochem. Soc. **167**(12), 127503 (2020)
48. Wang, C., Xia, K., Wang, H., Liang, X., Yin, Z., Zhang, Y.: Advanced carbon for flexible and wearable electronics. Adv. Mater. **31**(9), 1801072 (2019)
49. Edwards, E., Brantley, C., Ruffin, P.B.: Overview of nanotechnology in military and aerospace applications. In: Nanotechnology Commercialization, pp. 133–176 (2017)
50. Nehru, R., Chen, C.W., Dong, C.D.: A review of smart electrochemical devices for pesticide detection in agricultural food and runoff contaminants. Sci. Total Environ. **935**, 173360 (2024)

SPRINGER NATURE

GPSR Compliance

The European Union's (EU) General Product Safety Regulation (GPSR) is a set of rules that requires consumer products to be safe and our obligations to ensure this.

If you have any concerns about our products, you can contact us on ProductSafety@springernature.com

In case Publisher is established outside the EU, the EU authorized representative is:

Springer Nature Customer Service Center GmbH
Europaplatz 3
69115 Heidelberg, Germany

The manufacturer's authorised representative in the EU is Springer
Nature Customer Service Centre GmbH, Europaplatz 3, 69115 Heidelberg,
Germany. If you have any concerns regarding our products, please
contact ProductSafety@springernature.com

Printed and bound by CPI Group (UK) Ltd, Croydon, CR0 4YY

01/10/2026

02225932-0006